T0179597

Analysis and Design of Machine Elements

Analysis and Design of Machine Elements

Wei Jiang
Dalian University of Technology, China

This edition first published 2019
© 2019 John Wiley & Sons Singapore Pte. Ltd

The right of Wei Jiang to be identified as the author of this work has been asserted in accordance with law.

Registered Offices
John Wiley & Sons Singapore Pte. Ltd, 1 Fusionopolis Walk, #07-01 Solaris South Tower, Singapore 138628
John Wiley & Sons, Inc., 111 River Street, Hoboken, NJ 07030, USA

Editorial Office
The Atrium, Southern Gate, Chichester, West Sussex, PO19 8SQ, UK

For details of our global editorial offices, customer services, and more information about Wiley products visit us at www.wiley.com.

Wiley also publishes its books in a variety of electronic formats and by print-on-demand. Some content that appears in standard print versions of this book may not be available in other formats.

Library of Congress Cataloging-in-Publication data applied for

ISBN: 9781119276074

Cover Design: Wiley
Cover Image: © Marilyn Nieves/Getty Images

Set in 10/12pt WarnockPro by SPi Global, Chennai, India

Printed in Singapore by C.O.S. Printers Pte Ltd

10 9 8 7 6 5 4 3 2 1

Contents

Preface

This book *Analysis and Design of Machine Elements* has been written for courses on mechanical design, a compulsory course for students majoring in mechanical engineering. Since safety has become an increasing concern nowadays, the book presents the subject in an up-to-date manner with a strong emphasis on failure analysis and prevention-based machine element design. It aims to provide students with basic concepts, principles and philosophy in analysing, selecting and designing safe, efficient and workable machine elements, and to expose them to the detailed design methods, skills and tools necessary to convert concepts into practical devices.

The aim is achieved by introducing design methodology and fundamentals including strength analysis, failure theory, as well as material selection and heat treatments in Part I, followed by 12 self-contained chapters on the application of fundamentals to the design of specific machine elements in Part II, covering threaded fasteners and joints, keys, splines, belts, chains, gears, worm gearings, shafts, contact rolling bearings and sliding bearings, couplings and clutches, springs and so on. For clarity and pedagogy, each of the self-contained chapters is organized in an almost consistent style, that is:

The *Introduction* provides a brief description of applications, characteristics and structures of the specific machine elements.

Working Condition Analysis includes kinematic, dynamic, force, stress and failure analyses of machine elements considered in each chapter, stressing physical interpretation of mathematical derivations and promoting understanding of the working mechanism of the discussed elements.

Load Carrying Capacities discusses capacity analyses within the limitation of design criteria determined by failure modes for the element under consideration.

Design Methods introduces step-by-step design procedures for practical design of machine elements, including guidelines on materials and heat treatment selection, design variable selection and determination and so on. Design variables are determined by analytical calculation according to the design criteria, combined with compatibility considerations with surrounding elements by structural design. When standard components are commercially available, steps necessary for their specification and selection are provided. The design methods and procedures are demonstrated by design cases in each chapter.

The *Problems* section provides students with different kinds of exercises, including review questions, objective questions, calculation questions, structural design problems, practice design problems and computer-aided design problems. These problems are designed to help students reinforce what they have learnt from different aspects, and to increase their analytical and design capabilities. Computer-aided design problems are provided in particular to encourage students to use computers in design. A solutions manual is available that contains answers to some of the problems.

Such organization emphasizes the importance of safety in machine element design, helps students who may have had little exposure to machines, structures or industrial practice to build a clear structure of the course and to link analysis with design.

Design requires extensive use of knowledge in mechanical engineering science, such as mechanical drawing, materials properties, statics and dynamics, mechanics of materials, synthesis and analysis of mechanisms and machines, manufacturing processes, geometric dimensioning and tolerancing, fluid mechanics, computer-aided design and so on. Fundamentals necessary for the entire book are introduced in Part I, others required for specific machine elements are embedded in relevant chapters. For example, tribology and fluid mechanics are introduced in the chapter on sliding bearings. Such an arrangement helps students effectively bridge the gap between basic scientific knowledge and design activity, assist their better understanding of complex working mechanism of machine elements and ultimately promote their engineering and innovation capability in mechanical design.

This book provides general methodology and fundamentals for the analysis and design of common machine elements, with an aim to help students gain competence in applying these methods and procedures to machine element design. When designing machine elements for important applications, designers should check all important data from standards, design handbooks or industrial manuals. The book uses the International System of Units, with conversion to other unit systems introduced. The book evolved from lecture notes initially prepared for international undergraduates in a mechanical design course, with a hope to increase the competency of mechanical engineering graduates to function well in a globalized profession and to help them communicate effectively and cooperate productively in international collaborative design activities.

Although the author has made every effort to ensure accuracy and conformity with good engineering practice, errors are unavoidable and constructive suggestions and feedback are warmly welcomed.

Finally, I would like to express my appreciation to Dr K. Yahiaoui from University of Wolverhampton for reviewing Chapter 6 and all other reviewers who have contributed to this text. Their reviews greatly enhanced the quality of the book. I am especially grateful to my graduate students for their patiently preparing all the illustrations, to my colleagues for the helpful discussions and to my undergraduate students for their feedback about the book.

Thanks are also due to editorial staff and many other individuals in John Wiley & Sons who have contributed their talents and energy to producing this book.

Special thanks are due to my family for their understanding, support and encouragement during more than 10 years' preparation for the book.

Dalian, China
February 2018

Wei Jiang

About the Companion Website

This book is accompanied by a companion website:

www.wiley.com\go\Jiang\analysis_of_machine_elements

The website includes:

– Solution figures
– Solution of problems

Scan this QR code to visit the companion website.

Part I

Fundamentals of Design and Strength Analysis

1

An Overview of Machine Design

1.1 Introduction

1.1.1 Machines and Machine Elements

A machine is a device that employs power to accomplish a desired function to benefit humankind. It is generally composed of a power source to provide power and movement, and an executive device to fulfil intended function. In between is a transmission system and controller. Power sources can be prime movers, that is, a machine utilizes a natural source of energy to produce power, like an internal combustion engine; or a secondary mover, like an electric motor, which receives energy directly or indirectly from a generator driven by a prime mover [1]. Take a car as an example; power sources, either engines in motor vehicles or batteries in electrical vehicles, provide power to executive devices; for example, wheels, through a transmission system (including couplings, clutches, shafts, powertrain etc.). The control system, that is, steering systems and brakes, control car movement. Lights, meters and windscreen wipers are accessories that facilitate a car running properly.

Machines involve a vast variety of mechanical products in various fields of manufacturing, transportation, aerospace, construction, agricultural, energy and many others. 'Products' covers industrial robots, machine tools and automated assembly systems in manufacturing systems; automobiles, trains, ships and aircraft in transportation vehicles; mobile cranes, dump trucks and concrete mixers in construction equipment. Household appliances, like vacuum cleaners, washing machines and air-conditioning systems are also machines.

A similar concept, that is, mechanism, is a combination of elements formed and connected to transmit motion in a predetermined fashion. Typical mechanisms include linkages, cams and follower systems, gears and gear trains and so on. There is no clear division between mechanisms and machines. If the transmitted forces or power are significant, it is considered a machine; otherwise, it is considered a mechanism [2]. Machinery is a derived term and refers to a grouping of mechanisms and machines [1].

A machine composes individual machine elements properly designed and arranged to work together. Machine elements are the fundamental components of a machine, and are broadly classified as universal elements, such as bolts, keys, splines, pins, belts, chains, gears, bearings and springs that are widely used in different kinds of machines, and special elements such as turbine blades, crankshafts and aircraft propellers, which

Analysis and Design of Machine Elements, First Edition. Wei Jiang.
© 2019 John Wiley & Sons Singapore Pte. Ltd. Published 2019 by John Wiley & Sons Singapore Pte. Ltd.
Companion website: www.wiley.com/go/Jiang/analysis_of_machine_elements

perform specific functions [3]. This book focuses on the analysis and design of universal machine elements.

1.1.2 The Scope of Machine Design

Design is widely considered to be the central or distinguishing activity of engineering [4]. It aims to create and execute a purposeful plan to meet commercial, industrial and social needs. When design is discussed in mechanical engineering domain, especially about mechanical products or machines, it is termed mechanical or machine design.

Machine design is the art of envisioning, creating and developing a brand new or improving on an existing mechanical device for the fulfilment of human needs, with due regard for resource conservation and environmental impact [1, 5]. It is an innovative, iterative, decision making and problem-solving process involving comprehensive utilization of scientific knowledge and creative capability. Designers are required to generate concepts and decide deterministic dimensions for devices or products from limited information, ambiguous and sometimes even partially contradictory requirements to achieve users' objectives while satisfying a set of specified constraints. Therefore, the initial design is usually tentative. With more variables gradually determined, material, geometry, as well as manufacturing and tolerance details are fine-tuned during iteration until final optimum design is achieved [5].

Machine design generally has more than one solution. Since design is a rational process of choosing among design alternatives [6], it greatly depends on the designer's knowledge, previous experience, design method and design philosophies to solve a specific problem. Consider, for example, the design of an automobile. A large number of models are available on the market, and all of them can fulfil the function of transportation. The differences among them are the operation convenience, comfort, aesthetic appearance, cost and so on. These features decide the competitiveness and sales of products.

Analysis and design are indispensable aspects in the process of machine design. Analysis is concerned with predicting the response of an existing or a tentatively designed machine under specified inputs, which is especially important in creative design process; while design attempts to decide the dimensions and shapes of machines to meet performance requirements. Integrating analysis and design skills during the design process distinguishes an outstanding design engineer from a good one [7]. Since design is an evolutionary process, a tentative design is firstly proposed and then analysed to see if it satisfies the given specifications. If not, which is the usual case; the tentative design is revised with changes involving geometry, material and loads, and is analysed again until it satisfies the specified design requirements.

Extensive multidisciplinary skills and knowledge are employed in the design process to convert inadequate, vague requirements into a product that is functional, safe, reliable, manufacturable, competitive and marketable. Engineering science includes mechanics of solids and fluids, materials science, manufacturing processes and so on. Designers are required to use computers and graphics tools to visualize design plans. Since designers often work collaboratively on teams, communication skills, teamwork ability and presentation skills are equally important.

To design a machine or machine element successfully, design engineers not only need to develop competence in understanding and applying scientific knowledge, empirical information, professional judgement and ingenuity in solving practical problems, but also cultivate a strong sense of responsibility and professional work ethic. Mechanical design involves almost all the disciplines of mechanical engineering [8–10]. The extensive knowledge and skills required for a mechanical designer are briefly summarized as:

- Competence in mathematics, statics, dynamics, mechanics of materials, kinematics and mechanisms to facilitate load, stress and strength analyses; an advanced CAD/CAE (computer-aided design or engineering) or FEM (finite element method) technique is preferred;
- Familiar with engineering materials and their properties, materials processing, heat treatments and manufacturing processes;
- Knowledge of tribology, fluid mechanics, heat transfer, electrical and information technology and controls;
- Creativity, complex problem-solving capability and project management skills;
- Competence in graphical representation by sketches, engineering drawing, CAD tools and 3D visualization to convert mental design concepts into technical drawings;
- Both verbal and written communication skills and presentation skills to articulate design projects, describe constraints and limitations and present proposals and technical reports;
- Teamwork and collaboration capability, and a sense of social and ethical responsibilities.

Students are expected to appreciate that machine design is an integration of physical and engineering considerations with social concerns, with an aim to design machines with satisfactory lives and high reliabilities. With the globalization of business world, future design engineers are expected to feel comfortable to work in a vibrant and multicultural environment, learn to satisfy the needs of customers in a competent, responsible, ethical and professional manner and able to communicate complex aspects of design verbally and graphically to other members in both national and international concurrent design teams.

1.2   Machine Design

1.2.1   Machine Design Considerations

An important assessment of design quality is machine's safe and reliable performance of its intended function for the prescribed design life without serious breakdown. Besides, machine design involves a multitude of considerations, as summarized in Table 1.1. It is unrealistic to satisfy all these considerations, as some are seemingly incompatible. Designers are challenged to recognize incongruities and find compromises between these discrepancies. During the design process, the total life cycle of a product, from initial ideation, design, manufacturing, assembly to service and final disposal, should be reviewed, and situations that may practically occur during manufacturing, transporting, storing, installing, servicing and disposal be evaluated. The main concerns selected from Table 1.1 are discussed next and will be addressed throughout the book.

Apart from proper functioning, successful design of competitive machines must prevent premature failures to ensure safe and reliable operation throughout design lifetime. Traditional safety considerations for an element include strength, deflection and

Table 1.1 Machine design considerations.

Functional considerations	Safety and reliability considerations	Manufacturing considerations	Form considerations	Economic, ecological, societal considerations
Functionality	Load/stress/ strength	Materials and properties	Geometry	Marketability
Operation	Deflection/ rigidity	Heat treatments	Dimensions	Cost
Controllability	Friction/wear	Manufacturability	Volume	Recycling
Utility	Corrosion	Assembly and disassembly	Weight	Ecology
Efficiency	Stability	Packaging and transportation	Surface	Ergonomics
Kinematics	Noise and vibration	Maintainability	Styling	Liability
Lubrication	Durability/life		Aesthetics	Cultural/Social impact

stability; while for an element surface, the safety concerns are friction, wear, frictional heat and corrosion. Designers are required to anticipate potentially failure modes under various operating conditions and integrate safety into design process wherever possible. The design philosophy is first to take precaution against failure, but if failure does occur, the design must have a remedy to prevent catastrophic disaster [10].

Manufacturing changes the shapes and sizes of raw materials into the geometry of final elements specified by designers. Currently available manufacturing processes include casting, welding, forming, machining and additive manufacturing. Each process has its own attributes concerning processing power, time, cost and final product qualities. Factors considered while selecting manufacturing process include the geometry and properties of both raw materials and finished elements; and the quality requirements of finished elements, like tolerance, surface finishes, strength and so on. Besides, the number of elements to be produced, the cost, time, energy requirement and environmental impact also need to be considered [8–10].

The decisions on material, manufacturing process and design are interrelated with each other, and therefore should be considered integrally at the early stage of design. These considerations should also cover the whole life of an element, that is, from blank of raw material to manufacturing, heat treatments, assembly, maintenance and final disposal. For example, a large batch production of casted complicated heavy elements should keep the section with uniform thickness to avoid casting shrinkage; the number of machining planes should be reduced to minimize the number of fixtures; if an element needs heat treatments, use fillets rather than sharp corners; to facilitate assembly and disassembly, fittings and locating also need to be paid special attention.

Cost plays an important role in design decision making. Costs spent on product development include expenditures on materials, labour and material processing. Costs of materials and labour increase yearly, while costs of material processing depend on manufacturing processes, machine tools, quantities, required tolerances and so on. The larger the quantities and tolerances, the lower the manufacturing costs. Whenever possible, standard or purchasable products readily available from the market, such as contact bearings, fasteners, keys, motors, pumps and so on, are the first choice. In such cases, all designers need to do is to determine and provide preferred sizes and specifications.

Although, the basic objective of mechanical design is to provide a machine or device that benefits humanity, the consideration of ecological cycle or sustainable growth has become an increasing concern [11]. In the proposed design, recyclable materials are preferred. The selected material processing should have minimal energy consumption and reduced environmental pollution. Efficient material usage is also an important aspect for sustainable development. Light weight design and 3D printing are examples of effective material usage. In short, environment and resource concerns during manufacture, operation and disposal must be seriously taken into consideration.

During machine design, ergonomics or the man-machine relationship has gained increasing attention. Its aim is to ensure that machines are designed to accommodate operators with safety, comfort and efficiency. By incorporating human factors into design, product efficiency and safety can be improved and potential operational problems and product liability can be reduced.

Since designers and manufacturers are legally liable for any damage or harm from products, they have great responsibility to ensure high quality and safe operation of products, and should provide adequate warnings and instructions for use in the product specifications [8].

1.2.2 Machine Design Process

Machine design aims to produce a useful product satisfying the needs of customers. The design of a machine is inherently a complicated and delicate process, involving interactive and iterative procedures, complex calculations and countless design decisions [8]. To develop a safe, efficient and reliable product with excellent functionality and high competitiveness, proper design procedures and approaches should be established and followed. Figure 1.1 outlines a complete product development process currently employed.

The development of new machine from inception through elaboration to termination can be broadly classified into four stages: planning, concept design, detailed design, manufacturing and commercialization [8, 9]. The tasks and focus of each stage are introduced next.

1. Planning

    Design starts with the recognition of needs. The desire or expectation for a new product may be acquired from a target market, from dissatisfaction with existing products or from a particular adverse circumstance; for example, a need for a robot to work in hostile environment.

    The vague or subjective needs are then translated into detailed quantitative specifications to definite the function of the expected product. These specifications are the input and output quantities, such as power, operating speeds, expected life or safety, space or weight limitations.

    Financial investment, estimated price, cost targets, expected profits and other business issues, together with sales operations and time schedules also need to be considered at this stage.

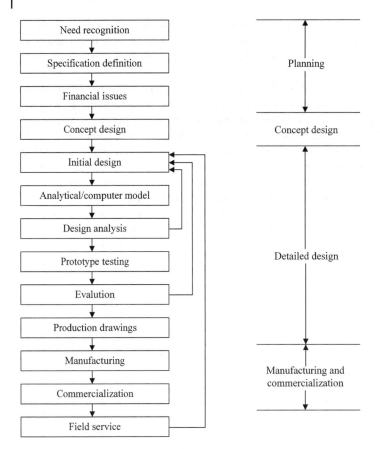

Figure 1.1 Product development process.

2. Concept design

   Based on the simplified assumptions and previous experiences, a designer or a design team may propose several possible alternative concept designs satisfying function and design specifications. These proposed concept designs are investigated, quantified, compared, weighed and ranked in terms of established evaluation criteria, which gives desirable qualitative characteristics of a design according to the function statements and design specifications. The optimum concept design is then decided and will be designed in depth to meet specified criteria of performance, life, weight, cost, safety and others.

3. Detailed design

   Once the optimum concept design is determined, an initial design is proposed and developed into an analytical model or a computer model for pertinent analyses as required, including kinematic, force, strength, rigidity, heat transfer and so on. If analytical results are not satisfactory, designers return to the initial design to revise the previous tentative design. Otherwise, a physical prototype will be built, tested and evaluated. Evaluation criteria can refer to the considerations of machine design listed in Table 1.1. Based on the results of test and evaluation, the tentative design may be judged ready for manufacturing or may be revised and a second analytical model and prototype built and tested. Since design is an evolutionary process, this iterative process is not unusual. After reiterations of analysis, optimization and evaluation, an optimal solution is finally achieved. The initial tentative design is converted into a workable product, and a set of detailed production drawings and design documents are produced to facilitate manufacturing. At this stage, potential manufacturing, assembly and maintenance issues also need to be considered.

4. Manufacturing and commercialization

Once the initial design has been converted into the workable product successfully, practical manufacturing is organized to produce products. A full-scale production can be launched and products can be sold to customers by both domestic and global markets. Commercialization also involves customer service and especially warranty. Finally, service data on failure modes, failure rates, maintenance and safety problems are collected for future product improvement.

In summary, machine design process includes the first three parts of comprehensive product development process. It aims to reach a compromise through repeated revision of initial design to fulfil design function, to satisfy design requirements and practical limitations. The primary focus of this book is on this part of design process, with proper consideration on manufacturing, material selection and heat treatments where necessary.

1.3 Machine Element Design

1.3.1 Machine Element Design Considerations

Machine element design is the design of mechanical elements or components. It is a more focused and an integral part of machine design, consisting of capacity calculation and structural design. Machine element design employs similar knowledge and skills to machine design. Since machine elements are the fundamental components of a machine, each element must be properly designed to facilitate its integration into a machine.

Machine element design must ensure efficient and safe function, with an appropriate combination of size, shape and material to withstand operating loads for the expected service life at minimum costs. The considerations of machine element design are similar to those of machine design, with more concerns on safety, manufacturability, costs, codes and standards.

The safety and reliability of individual machine elements decide the overall safety and life expectancy of a machine. The analysis, prediction and prevention of failure form the basis for the design of machine elements. In the design process, potential failures are prevented by proper material selection, reasonable assumption and simplification of operating loads, correct shape and dimension determination through accurate strength, rigidity and other relevant analyses.

Manufacturing produces elements to the desired dimensional accuracy and surface finish. Since manufacturing process affects final specifications of overall geometry, dimensions, tolerances or surface finishes, proper tolerances and acceptable surface finishes must be specified on design drawings to facilitate manufacturing process selection [8]. Normally, small tolerances and smooth surfaces increase manufacturing cost dramatically. Proper clearances and fits also need to be defined for mating elements.

In general, a compact design is preferred. After calculation, standard sizes close to the minimum acceptable dimension are usually specified for standardized elements, like fasteners, keys and rolling contact bearings. Through the selection of standardized elements, uniformity of practice and reduced cost are achieved.

1.3.2 Common Failure Modes in Machine Elements

A mechanical failure refers to the incapability of a machine or an element to perform its intended function [9]. It may arise from poor design detailing, inadequate material properties, manufacturing deficiencies, hostile service conditions and more often their interactions. A trivial oversight of any of these aspects may cause large detrimental and even catastrophic failure, resulting in serious financial, insurance and legal repercussions [12].

Mechanical failure modes commonly observed in industrial practice include deformation, yielding, fracture, fatigue, pitting and spalling, wear, scoring, scuffing, galling and seizure, corrosion, fretting, creep, buckling and so on.

Deformation includes elastic and plastic deformation, referring to the recoverable and unrecoverable deformation, respectively. When stresses generated in an element exceed yield strength, plastic deformation or yielding occurs. Both elastic and plastic deformation may cause malfunctioning of machine elements. A typical example is shafts supporting a pair of mating gears. The deformation of a shaft may prevent gears meshing properly, leading to interference, impact, wear, noise and vibration.

Fracture may occur in both brittle and ductile material due to either static or fluctuating loads. Ductile rupture occurs when plastic deformation reaches the extreme in ductile materials, leaving a dull, fibrous rupture surface; while brittle fracture happens when elastic deformation achieves the extreme in brittle materials, leaving a granular, multifaceted fracture surface [5, 13]. Fatigue fracture is a sudden and catastrophic failure, taking place due to the initiation and propagation of a microcrack under fluctuating loads over a period of time. The loads causing fatigue failure are usually far below the static failure level.

Surface fatigue failures are usually associated with rolling surface in contact. The repeatedly applied loads produce concentrated cyclic subsurface contact stresses. Micro-cracks initiate slightly below the contact surfaces and propagate until small bits of surface material spontaneously dislodge off the surfaces, producing surface pitting or spalling. Examples of pitting failure are manifested in rolling contact bearings, gear teeth and metal wheels rolling on rails. Pitted surfaces prevent proper function of elements, causing vibration and noise.

Wear is gradual removal of discrete particles from sliding contact surfaces, leading to a cumulative dimensional change on the element profiles. The most common types of wear are abrasive wear and adhesive wear. With an increase of severity of surface damage, adhesive wear is classified as scoring, scuffing, galling and seizure. Adhesive wear is the most common type of wear and the least preventable. The worn surface may impair element profiles, leading malfunction of machines. For instance, wear on gear tooth surfaces may cause intolerable noise and damaging vibration.

Corrosion occurs as a consequence of undesired deterioration of material as a result of chemical or electrochemical interaction with environment. It most happens to the machine elements exposed to corrosive mediums [5].

In summary, failure is the response of a machine or machine element to operating loads and service environments. Imperfection in design, materials and manufacturing may cause failure. Identification of possible failure modes under prescribed operating condition is an essential step in the early stage of machine element design. It is the designer's responsibility to predict, analyse and prevent prospective failure to ensure a successful design of machine elements or machines throughout a design's lifetime.

1.3.3 Design Criteria

Safety is always the paramount criterion in machine design [9], as catastrophic failures result in life losses, property destruction and environmental damage and must be prevented. To provide a safe, reliable and cost-effective machine element, it is essential to establish design criteria against potential failure modes. The design criteria commonly used against the previously discussed failure modes include strength, rigidity, life and wear criteria.

1.3.3.1 Strength Criteria

Strength is the ability to resist loads. It is expressed in terms of ultimate strength, yield strength and fatigue strength [1]. Almost all kinds of failure modes, including yielding, fracture, fatigue, pitting, spalling, wear, scoring, scuffing, galling and seizure, are due to insufficient material strength to withstand loads. Design approaches must satisfy a strength criterion in the form of a stress-strength relationship, either within an element or on the element surface, to ensure safe design.

The strength criterion indicates that the actual stresses σ acting on an element at a critical location under operating conditions must be less than the allowable stress of material $[\sigma]$, expressed as

$$\sigma \leq [\sigma] \tag{1.1}$$

The limiting condition to the right of the inequality depends on material properties. Under a static load, the allowable stress is the material strength divided by the allowable safety factor. The material strength is the yield strength in tension, compression or shear for ductile materials; and ultimate strength for brittle materials. Under a fluctuating load, the material strength is the endurance strength corresponding to the operating conditions. Detailed discussion about endurance strength will be introduced in Chapter 2.

The ability to quantify stress states at a critical location in a machine element is important for assessing failure possibility of an element. Stresses are obtained by simplified calculations according to various loading conditions. For uniaxial loading, the stress calculation is quite straightforward, using stress formula loaded by basic tension, shear or bending. For complex loading, a combined stress theory, either the maximum shear stress theory or the maximum distortion energy theory, is used to calculate multiaxial stresses [14, 15]. For contact loading, the contact stress is calculated by the Hertz formula [16].

Designers must ensure the maximum stresses at critical locations are less than material strength σ_{lim} by a sufficient margin to guarantee adequate safety for an element. This margin is the factor of safety, a reasonable measure of relative safety of a load-carrying element. Thus, an alternate expression for strength criterion by safety factor S is

$$S = \frac{\sigma_{\lim}}{\sigma} \geq [S] \tag{1.2}$$

Safety factors are employed to consider uncertainties and variabilities in loads, material strength, manufacturing qualities and operation conditions, as well as assumption in stress calculations. The selection of an appropriate value of allowable safety factor is based primarily on the these considerations, together with design codes and a designer's previous experience with similar products and conditions [8]. They should be carefully chosen to meet the required reliability at a reasonable cost. The allowable safety factor [S] is generally within the range of 1.25–4 [8–10].

1.3.3.2 Rigidity Criteria

Rigidity is the ability to resist deformation or deflection. It ensures accurate and precise operation of a machine. All machine elements deform under load, either elastically or plastically. Excessive deformation or deflection may cause interference between elements and premature failure due to vibration, wear and fatigue. When deflection is critical to safety or performance of an element the deflection must be analysed to satisfy rigidity criterion. Elastic strain, deflection, stiffness and stability are important considerations for the design of some elements.

Criteria for failure due to excessive deflection are often highly dependent on the application of machine. For example, machine tool frames must be extremely rigid to maintain manufacturing accuracy. Also, in a transmission, the shaft supporting gears must be rigid enough to avoid excessive deflection that may lead to gear disengagements. The following criteria are used to assess the rigidity of an element in different forms of deflection. When an element is subjected to a bending moment M, the deflection y and slope θ should satisfy

$$y \leq [y] \tag{1.3}$$

$$\theta \leq [\theta] \tag{1.4}$$

When an element is subjected to a twisting moment T, the angular deflection φ should satisfy

$$\varphi \leq [\varphi] \tag{1.5}$$

Detailed calculation of deflection y, slope θ and angular deflection φ depends on the cross-section geometry and loads, which can be referred to in *Mechanics of Materials* [14], or by finite element analysis for more complex geometries or loads. The allowable values of deflection $[y]$, slope $[\theta]$ and angular deflection $[\varphi]$ are selected by design requirements.

1.3.3.3 Life Criteria

Normally, it is a requirement that the life L of an element should be longer than the expected life $[L]$. Therefore

$$L \geq [L] \tag{1.6}$$

Detailed applications of life criteria can be found in Chapter 11 for the design of rolling contact bearings.

1.3.3.4 Wear Criteria

Wear is the gradual removal of materials from contact surfaces as the result of relative motion of contacting elements. To against wear, the pressure p between contact surfaces and relative sliding speeds v must satisfy the following criteria,

$$p \leq [p] \tag{1.7}$$

$$pv \leq [pv] \tag{1.8}$$

Wear will be discussed in Chapter 8 for gears, in Chapter 11 for contact bearings and in Chapter 12 for sliding bearings where it will be the prime concern.

Other design criteria, including dynamic and reliability criteria, can be found in relevant references.

Mechanical failures, especially under variable loads, are strongly influenced by the interaction between design, manufacture and materials [12]. Although several design criteria have been established, the basic principles are the same; that is, comparing working conditions with allowable conditions. The allowable conditions largely depend on material properties, while the working conditions rely predominantly on design, manufacturing and operation.

Designers and manufacturers are liable for any damage or harm caused by product defects. Designers are responsible for preventing premature failure by good design, while manufacturers must ensure high quality products by using good quality control and comprehensive testing procedures.

1.3.4 Machine Element Design Process

In principle, the design process is similar for all kinds of machine elements, which intrinsically involves complex calculations, tradeoff decisions and iterative processes, through which an optimum design is obtained. Designs always demand enhanced performance, extended life, reduced weight, lower costs and improved safety [5]. Figure 1.2 presents a flowchart of a machine element design process to facilitate the development of a systematic approach.

The design process of a machine element can be broadly classified into four steps: design task identification, assumptions and decisions, analysis and evaluation and drawings and documentation. The tasks and focus of each step are introduced next.

1. Design task identification
   Broadly, it is required first to identify design requirements and design variables, the interactions of element with surrounding components and select the proper element type according to application and intended function.
2. Assumptions and decisions
   The material and heat treatment need to be selected first so that material property data required for the subsequent calculations can be collected from charts and tables in design handbooks. If required, an initial design for the element is proposed, showing tentative dimensions and shapes that may affect performance or stress analysis.

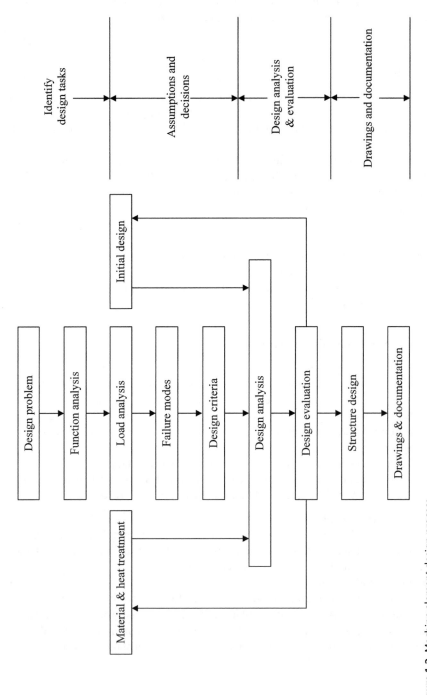

Figure 1.2 Machine element design process.

The actual mechanical system needs to be idealized and simplified to facilitate load analysis. According to the loads and operating conditions, it is important to predict potential failure modes and decide on the corresponding design criteria.

3. Design analysis and evaluation

According to the design criteria, perform strength or rigidity analyses to determine basic geometrical dimensions of machine elements that meet performance requirements against failure. The variability in manufacturing, assembly and service are considered by introducing modifying factors in the analyses. The calculated results need to be rounded up to standard or preferred dimensions.

The performance of machine elements need to be assessed against material limits. A sufficient safety factor should be given to account for uncertainties and imprecision in the calculations due to idealization and simplification in design equations and data approximation. If the performance does not meet design requirements, which is the usual case, revise the previous initial design, dimensions or materials and then iterate the design process until the specified requirements, like desired design lifetime or design safety factor, are satisfied.

4. Drawings and documentation

Once satisfactory results have been obtained, designers need to convert calculated dimensions to detailed drawings, considering manufacturing, assembly, maintenance and standardization. Design is implemented by a complete set of drawings and documentation with design specifications for all elements, including tolerances, fits and finishes. Standards, codes or preferred dimensions are usually incorporated to facilitate purchase and manufacture.

1.4   Materials and Their Properties

The selection of materials is an integral part of machine element design. The material selected for a specific machine element is determined by application requirements or, more precisely, by the requirement of material properties, such as strength, stiffness, reliability and durability. Additional factors, such as recyclability, environmental pollution and costs are also considered while selecting materials.

1.4.1   Types of Materials

Most machine elements are made from metals or metal alloys, such as steels, aluminium alloys and bronzes. Different numbering systems are employed to identify metal materials [8, 9]. Other materials, such as polymers, ceramics and composite materials, are also used in machine elements in various applications.

1.4.1.1   Steels and Alloys

Because of the properties of high strength, high stiffness, durability and relative ease of fabrication, many types of steels are used in machine elements, including carbon steels, alloy steels, stainless steels, structural steels, tool steels and so on.

Steels, including carbon steel and alloy steel, are the most extensively used material for machine elements. Carbon steel is an alloy of iron and carbon with small amounts of manganese, silicon, sulfur and phosphorus. According to the content of carbon, we

Table 1.2 Alloying elements and their effects on microstructure and properties [8–10].

Added elements	Effects by adding alloying elements
Chromium	Forms various carbides of chromium with refined grain structure
	Improves hardness, toughness and wear resistance
	Increases strength at elevated temperatures
Nickel	Dissolved in ferrite without forming carbides or oxides
	Increases strength without decrease ductility
	Improves toughness, hardness and corrosion resistance
Manganese	Added in steels as a deoxidizing and desulfurizing agent
	Dissolved in ferrite and forms carbides
	Increases strength and hardness at proper amount
Silicon	Added in steels as a deoxidizing agent
	Used with manganese, chromium and vanadium to stabilize carbides
	Improves strength, hardness and wear resistance
Molybdenum	Dissolved in ferrite partially and form carbides, contributes to a fine grain size
	Improves hardenability and high-temperature strength
	Increases hardness and toughness
Vanadium	Dissolved in ferrite and easy to form carbides, hence is used in small amounts
	Strong deoxidizing agent and promotes a fine grain size
	Improves strength, toughness, keeps hardness at high temperature
Tungsten	Maintains hardness even at high temperature, widely used in tool steels
	Produces a fine, dense structure
	Similar effect to that of molybdenum, but needs greater quantities
	Increases both toughness and hardness

have low-carbon steels with less than 0.3% carbon, medium-carbon steels with 0.3–0.5% carbon and high-carbon steels with more than 0.5% carbon [9]. Carbon has a great effect on strength, hardness and ductility of any steel. Alloy steels have sufficient quantities of one or more elements other than carbon introduced to modify steel properties. Commonly added alloying elements include chromium, nickel, manganese, molybdenum, vanadium and many others. Their effects on the microstructures and steel properties are listed in Table 1.2.

Alloy steels have two types for special applications, that is, stainless steels and superalloys. Stainless steels contain a minimum of 10% chromium in three main types; austenitic, ferritic and martensitic stainless steels [9]. The most important characteristic of stainless steels is their resistance to corrosive environments.

Superalloys are primarily used for elevated temperature applications, such as in gas turbines, jet engines and heat exchangers. The most attractive properties of superalloys are the high temperature strength and resistance to creep, oxidation and corrosion [10].

Other types of steels are also available, such as structural steels and tool steels. Structural steels are basically low-carbon, hot-rolled steels in the form of sheet, plate and bar, and are usually used in construction and machines. Tool steels, as the designation implies, serve for making tools in manufacturing engineering.

1.4.1.2   Cast Irons and Cast Steels

Cast irons contain iron, carbon (between 2~4%), silicon and manganese [10]. Additional alloying elements are sometimes added to improve material properties. Cast irons have excellent machinability, hardness, wear resistance and damping capability, yet lower tensile strength.

The commonly used cast irons are white cast iron, grey cast iron, ductile and nodular cast iron, malleable cast iron and alloy cast iron. White cast iron is extremely hard, wear-resistant, brittle and hard to machine. Typical applications are found in extrusion dies and railway brake shoes. Grey cast iron has high compressive strength and good wear resistance, yet it is brittle and weak under tension and has poor weldability. Due to low cost and convenient batch casting, grey cast iron is widely used in engine blocks, machine bases and frames. Ductile cast iron, or nodular cast iron, has substantial ductility, improved tensile strength, stiffness and impact resistance. Typical applications include engine crankshafts and heavy-duty gears. Malleable cast iron is annealed white cast irons with high strength and elastic modulus, good machinability and wear resistance. Typical applications are railway equipment and construction machinery. Nickel, chromium and molybdenum are the common elements added in alloy cast iron. The addition of these elements increases strength, hardness, wear resistance and impact resistance [8–10].

Cast steels have the same alloying elements as wrought steels. Since the mechanical properties can be modified by heat treatment, cast steels are used for complex shaped elements requiring a high strength.

1.4.1.3   Nonferrous Alloys

Commonly used nonferrous alloys in machine elements include aluminium alloys, magnesium alloys, nickel alloys, titanium alloys, zinc alloys and copper alloys.

Aluminium alloys have both wrought and cast forms. The attractive properties of aluminium alloys are lightweight, good corrosion resistance, relative ease of forming and machining. They are widely used in structural and mechanical applications without strength requirements. Magnesium alloys are the lightest engineering metals, have a high strength-to-weight ratio and are widely used in aircraft and automotive industries.

Nickel alloys are used in applications requiring corrosion and oxidation resistance, high strength and toughness at temperature extremes as high as 1093°C or as low as −240°C [10]. Examples are turbine engine components, chemical processing equipment and marine components. Commercially available nickel alloys include Monel, Inconel and Hastelloy.

Titanium alloys have excellent corrosion resistance, outstanding strength-to-weight ratios and high-temperature strength. However, titanium alloys are expensive and have poor machinability. Typical applications of titanium alloys include aerospace and aircraft structures, marine components and human internal replacement devices.

Zinc alloys are inexpensive with moderate strength. They have low melting temperature and widely used in castings and zinc galvanizing. Typical zinc die-castings include automotive parts, building hardware and so on.

Copper alloys include brasses and bronzes. When copper is alloyed with zinc, it is called brass. Brass has excellent machinability and corrosion resistance and is used in tubing or piping, screw connections, locks, watches and so on. Bronze refers to copper alloyed with metals such as tin, lead, phosphor, aluminium, silicon, nickel. Wrought bronze alloys include phosphor bronze, leaded phosphor bronze, aluminium bronze and silicon bronze; while cast bronze alloys have tin bronze, leaded tin bronze, nickel tin bronze and aluminium bronze [9]. Bronzes are softer than ferrous alloys but have good strength, machinability and wear resistance. They run well against steel or cast iron when lubricated and can support high loads and operate at high temperatures. These qualities are utilized in sliding bearings and worm gearing where the combination of heavy loads and high sliding velocities places them in analogous situations [10].

1.4.1.4 Polymers

Polymers, usually referred to as plastics, fall into two main groups; that is, thermoplastics and thermosets. Thermoplastics are softened with heat and include acetal, acrylic, acrylonitrile-butadiene-styrene (ABS), nylon, polycarbonate, polyimide and polyvinyl chloride (PVC). They are generally impact resistant. Thermosets are generally heat resistant, including epoxy, phenolic and polyester [9, 10]. Both types are inexpensive, light in weight, resistant to corrosion and wear. Also, they have a low coefficient of friction, low strength and provide quiet and smooth operation.

1.4.1.5 Composite Materials

Composite materials are comprised of matrix materials and reinforcement materials. Each remains distinct and separate from each other. Matrix materials are various plastic resins like nylon, epoxy or polyester and metals, while reinforcements include glass, carbon and SiC (silicon carbide) in the form of continuous fibres, either straight or woven, short chopped fibres and particulates [8–10]. The reinforcement provides stiffness and strength, and the matrix holds the materials together and transfers load to the reinforcement materials.

Unlike isotropic, homogeneous engineering materials that have identical material properties in every direction, the material properties of composites vary with both location and direction. The orientation of multiple laminates can be optimized to reach a desirable high strength-to-weight ratio or high stiffness-to-weight ratio. Therefore, composite materials are becoming increasingly popular in automotive, marine, aircraft and spacecraft applications to realize lightweight design with high structural stiffness and excellent strength performance.

1.4.2 Material Properties

Material properties, including mechanical, physical, chemical, dimensional and processing properties, are important considerations in material selection. Among them, mechanical properties are especially crucial in determining the size and shape of machine elements. They also decide the performance of machine element under operating conditions.

Mechanical properties refer to strength, stiffness, hardness, toughness, ductility, impact, and creep and so on. Material strength considered for design calculation depends largely on the selected materials and loading conditions. For example, yield strengths are mainly applied to ductile materials; ultimate tensile strength for brittle materials; ultimate compressive strength for ceramics and glasses and fatigue strength for cyclic loading. Material stiffness is represented by the elastic modulus or Young's modulus and Poisson's ratio, which is almost the same for all steels [10]. The service

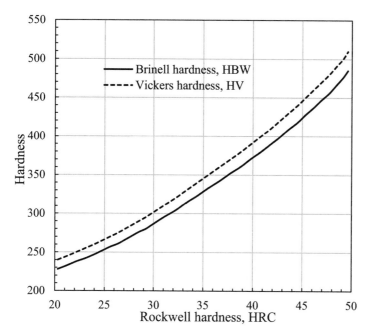

Figure 1.3 Hardness conversions [17].

performance and failure modes have close relations to mechanical properties. For example, wear relates closely to the hardness of a material. The higher the carbon content, the greater the hardness of steels. Commonly used hardness definitions in machine elements are Brinell hardness (HBW or HB), Rockwell hardness (HRA/HRB/HRC, where A, B, C represent different scales) [8] and Vickers hardness (HV). The conversion from HRC to HV and HBW (300 kg load, 10 mm ball) is presented in Figure 1.3.

Physical properties refer to density, electrical, magnetic and optical properties, as well as thermal properties, such as thermal conductivity, thermal expansion, specific heat, melting point, flammability and so on. Chemical properties include composition, corrosion resistance, degradation, oxidation, embrittlement and so on. The consideration of chemical properties closely related to the operating environments of machine elements. Dimensional properties refer to the size, shape, surface finish, tolerances and appearance of raw materials or blank materials, which affect manufacturing costs. Processing properties refer to the materials' manufacturability; that is, capability for machining, casting, forming, welding and so on. Usually, ductile materials can be easily forged, rolled or drawn, while brittle materials should be shaped by other methods.

1.4.3 Heat Treatments

The final material properties are dramatically affected by heat treatments. Heat treatment refers to time-controlled or temperature-controlled heating and cooling processes with an aim to modify materials' mechanical properties. Commonly used heat treatments for steel machine elements include annealing, normalizing, quenching, tempering and case hardening.

During annealing treatment, the material is heated above the upper critical temperature and is held for a designated time until the composition becomes uniform. It is then cooled slowly in a furnace to a temperature below the lower critical temperature and continues to be cooled to room temperature outside the furnace [9]. Annealing produces a soft, ductile, low-strength material, free of significant internal stresses with a refined grain microstructure.

Normalizing is similar to annealing, except the material is heated above the upper critical temperature and cooled in still air to room temperature. Compared with annealing, the rapid cooling in normalizing produces harder steel with higher strength and a coarser grain structure. Normalizing is often used as a final treatment [8].

Through-hardening is accomplished by heating elements above the transformation temperature, followed by rapidly cooling within a quenching medium such as water or oil [9]. In the controlled cooling rate process or quenching, austenite formed above the transformation temperature is transformed into martensite; the hardest, strongest form of steel. The rate of cooling determines the amount of transformation and thus the hardness and strength.

Tempering is usually executed immediately after quenching. It involves reheating the hardened steel to a temperature below the transformation temperature range followed by a desired cooling to ambient temperature. The selection of tempering temperate depends upon the composition and the degree of hardness or toughness required. With increasing tempering temperature, tensile strength and yield strength decrease, whereas ductility improves [9]. The tempering process modifies the steel properties and relieves residual stresses.

Case hardening hardens material surfaces only. Case hardening includes carburizing, nitriding, cyaniding, carbonitriding, induction hardening and flame hardening processes. Carburizing or nitriding is based on the diffusion of additional carbon or nitrogen to some depth on the surface of already machined and heat-treated elements to achieve a high surface hardness. Induction hardening and flame hardening rapidly heat medium carbon or alloy steel element surfaces for a limited time so that a small, controlled depth of material reaches the transformation range then, upon immediate quenching and tempering, only the area above the transformation range produces the high level of martensite required for high hardness. Case hardening produces a great hardness on the outer surface, while retaining ductility and toughness in the core [8, 10].

1.4.4 Material Selection

The selection of materials is an important decision in machine element design. Satisfactory performance of machine elements and machines depends greatly on materials and their properties. Materials selection starts by referring to previous applications or similar experiences, the desirable material properties and the knowledge of physical, economical and processing properties of materials. Here are main points for material selection. A more detailed systemic approach can be found in reference by Ashby [18].

The required material properties are determined by service conditions, potential failure modes and so on. Service conditions could be fluctuating loads, corrosive environments or high temperatures. The required service performance relates to the corresponding material properties. Therefore, an element subjected to a fluctuating load should have high fatigue strength; an element working in a corrosive environment must have corrosion resistance. Analysing potential failure mode is also helpful for establishing required material properties for the element. For example, to resist against excessive wear, hard materials are required; to resist against large deformation, materials with large modulus of elasticity or stiffness are preferred.

Materials may be available in different forms. Proper manufacturing processes, that is, machining, forming, joining, finishing and coating, are required to turn original shaped materials into designed machine elements. Therefore, a material's capability for forming, joining or welding together with the initial material cost and processing costs are important considerations in material selection. Besides, recycling, disposal, legal and health issues are additional considerations [10].

After identifying, selecting, evaluating and ranking candidate materials, combined with making compromises among constraints of properties, availability, machinability and costs, a final selection decision can be made.

1.5   Unit Systems

A unit is a specified magnitude of a physical quantity [10]. Units chosen for any three quantities of force, mass, length, and time are called base units, the fourth unit is the derived unit. If force, length and time are base units, the system is a gravitational system of units, such as the foot-pound-second (fps) system and inch-pound-second (ips)

Table 1.3 Commonly used design variables and their units.

Variables	Symbol	SI units	ips units	fps units
Force	F	N (newton)	lb (pound)	lb
Length	l	m (metre)	in. (inch)	ft (foot)
Mass	m	kg (kilogram)	lb s^2 in.$^{-1}$	slug (lb s^2 ft^{-1})
Time	t	s (second)	s	s
Acceleration	a	m s^{-2}	in. s^{-2}	ft s^{-2}
Angle	θ	rad or degree (°)	rad (radian) or degree (°)	rad or degree (°)
Angular velocity	ω	rad s^{-1}	rad s^{-1}	rad s^{-1}
Energy or work	E, W	J (joule)	lb in.	lb ft
Frequency	f	Hz (Hertz)		
Moment or torque	M, T	N m	lb in.	lb ft
Power	P	N m s^{-1} = W (watt)	lb in s^{-1}	lb ft s^{-1}, hp
Pressure	p	Pa (Pascal)	psi, lb in.$^{-2}$	psf
Stress	σ, τ	MPa	psi, lb in.$^{-2}$	–
Temperature	T	degrees Celsius (°C)	degrees Fahrenheit (°F)	degrees Fahrenheit (°F)
Velocity	v	m s^{-1}	in s^{-1}	ft s^{-1}
Weight	W	N	lb	lb

Table 1.4 Selected conversion relationships.

Quantity	ips to SI conversion	Formula
Force	$1\,\text{lb} = 4.448\,\text{N}$	$1\,\text{N} = 1\,\text{kg}\,\text{m}\,\text{s}^{-2}$
Length	$1\,\text{in.} = 25.4\,\text{mm}$	
Mass	$1\,\text{lb} = 0.454\,\text{kg}$	
Modulus of elasticity	$10^6\,\text{psi} = 6.895\,\text{GPa}$	
Moment or torque	$1\,\text{lb in} = 0.1138\,\text{N m}$	
Power	$1\,\text{hp} = 550\,\text{lb ft s}^{-1} = 745.7\,\text{W (watts)}$	$1\,\text{W} = 1\,\text{J s}^{-1} = 1\,\text{N m s}^{-1}$
Pressure	$1\,\text{psi} = 6895\,\text{Pa}$	$1\,\text{Pa} = 1\,\text{N m}^{-2}$
Stress	$1\,\text{psi} = 6.895 \times 10^{-3}\,\text{MPa}$	$1\,\text{Pa} = 1\,\text{N m}^{-2}$
Work or energy	$1\,\text{lb in.} = 0.1138\,\text{N m}$	$1\,\text{J} = 1\,\text{N m}$

system. If mass, length and time are base units, the system is an absolute system of units, such as the International System of Units (SI) [8]. The unit system used in this book is SI.

In engineering design, any set of calculations must employ a consistent system of units. Table 1.3 lists typical units associated with these systems and their standard abbreviations.

Sometimes a unit has to be converted from one system to another. Their conversion factors are listed in Table 1.4.

1.6 Standards and Codes

A standard is a set of specifications for elements, materials or processes formulated through a cooperative effort among industrial organizations, with an aim to achieve interchangeability, compatibility and uniformity for elements within a country or among cooperating countries [5]. The adoption of standards ensures high design efficiency by avoiding repetitive calculations, providing interchangeable elements on the spot with minimum downtime [1]. It permits convenient manufacture with standard machines and tooling.

Currently used standards are established by organizations or societies of various countries, such as the ANSI by the American National Standards Institute, BS by the British Standards Institute, EN by the European Committee for Standardization, DIN by the German Institute for Standardization, GB by the Standardization Administration of the People's Republic of China and ISO by the International Organization for Standardization. Other commonly used standards also include the AISI developed by the American Iron and Steel Institute, ASME by the American Society of Mechanical Engineers, ASTM by the American Society for Testing and Materials, SAE by the Society of Automotive Engineers and so on.

A code is a set of specifications for analysis, design, manufacture and construction [8]. Codes are legally binding documents, complied by a governmental agency, with an aim to achieve a specified degree of safety, efficiency and performance or quality, and to prevent damage, injury or loss of life [5, 8].

References

1 Hindhede, U., Zimmerman, J.R., Hopkins, R.B. et al. (1983). *Machine Design Fundamentals: A Practical Approach*. New York, NY: Wiley.

2 Norton, R.L. (1999). *Design of Machinery: An Introduction to the Synthesis and Analysis of Mechanisms and Machines*, 2e. McGraw-Hill.

3 Pu, L.G. and Ji, M.G. (2006). *Mechanical Design*, 8e. Beijing: Higher Education Press.

4 Simon, H.A. (1996). *The Sciences of the Artificial*, 3e. Cambridge, MA: MIT Press.

5 Collins, J.A. (2002). *Mechanical Design of Machine Elements and Machines: A Failure Prevention Perspective*, 1e. New York, NY: Wiley.

6 Dym, C.L., Agogino, A.M., Eris, O. et al. (2005). Engineering design thinking, teaching, and learning. *Journal of Engineering Education* 94: 103–120.

7 Katz, R. (2015). Integrating analysis and design in mechanical engineering education. *Procedia CIRP* 36: 23–28.

8 Budynas, R.G. and Nisbett, J.K. (2011). *Shigley's Mechanical Engineering Design*, 9e. New York, USA: McGraw-Hill.

9 Mott, R.L. (2003). *Machine Elements in Mechanical Design*, 4e. Prentice Hall.

10 Juvinall, R.C. and Marshek, K.M. (2011). *Fundamentals of Machine Component Design*, 5e. New York, NY: Wiley.

11 Ramani, K., Ramanujan, D., Bernstein, W.Z. et al. (2010). Integrated sustainable life cycle design: a review. *Journal of Mechanical Design, Transactions of the ASME* 132: 0910041–09100415.

12 James, M.N. (2005). Design, manufacture and materials; their interaction and role in engineering failures. *Engineering Failure Analysis* 12: 662–678.

13 Zhao, S.B. (2015). *Fatigue Design Handbook*, 2e. Beijing: China Machine Press.

14 Gere, J.M. and Timoshenko, S.P. (1996). *Mechanics of Materials*, 4e. CL Engineering.

15 Mendelson, A. (1968). *Plasticity: Theory and Application*, 1e. New York, NY: Collier Macmillan Ltd.

16 Johnson, K.L. (1987). *Contact Mechanics*. Cambridge, UK: Cambridge University Press.

17 ISO 6336–5:2003 Calculation of load capacity of spur and helical gears. Part 5: Strength and quality of materials. International Organization for Standards, 2003.

18 Ashby, M.F. (2005). *Materials Selection in Mechanical Design*, 3e. Oxford: Elsevier Butterworth-Heinemann.

Problems

1 In the context of machine element design, explain the meaning of failure and failure mode.

2 Give a definition of wear failure and list the major subcategories of wear.

3 List three most likely failure modes of a bicycle, and explain why each might be expected.

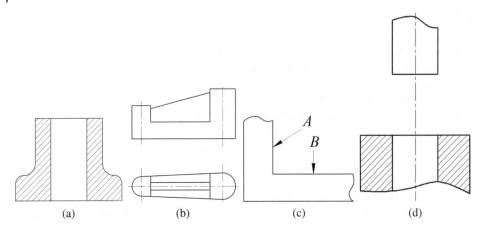

Figure P1.1 Illustration for Problem 4, four designs that may encounter manufacturing difficulties. (a) Casting. (b) Forging. (c) Heat treatment. (d) Assembly.

4 The designs in Figure P1.1 may encounter manufacturing difficulties. Find the problems and revise the designs.

2

Strength of Machine Elements

Nomenclature

b	cylinder length, mm
E	elastic modulus, MPa
F	force, N
K_A	service factor, or application factor
K_σ	combined influence factor
k_f	fatigue stress concentration factor
k_t	geometric stress concentration factor
k_σ	fatigue stress concentration factor in tension
k_τ	fatigue stress concentration factor in shear
m	material constant
N	number of load cycles
N_0	critical number of cycles
N_v	equivalent number of cycles
q	notch sensitivity factor
r	stress ratio
S	safety factor
S_{ca}	calculated safety factor
$[S]$	allowable safety factor
β_q	intensify factor
β_σ	surface condition factor for tension
ε_σ	size factor for tension
μ	Poisson's ratio
ρ	radius of curvature, mm
σ	tensile stress, MPa
σ_{-1}	endurance limit of a material at $r=-1$, MPa
σ_{-1e}	endurance limit of an element at $r=-1$, MPa
σ_{-1Nv}	endurance strength at $r=-1$ and $N=N_v$, MPa

σ_0	endurance limit at $r=0$, MPa
$\sigma_{1,2,3}$	principal stresses, MPa
σ_a	stress amplitude, MPa
σ_b	ultimate, or tensile strength, MPa
σ_{ca}	calculated tensile stress, MPa
σ_e	equivalent stress, MPa
σ_{lim}	material strength, MPa
σ_m	mean stress, MPa
σ_{max}	maximum stress, MPa
σ_{min}	minimum stress, MPa
σ_r	endurance limit at stress ratio r, MPa
σ_{rN}	endurance strength at r and N, MPa
$\sigma_{r\infty}$	endurance limit, or fatigue limit, MPa
σ_s	yield strength in tension, MPa
$[\sigma]$	allowable tensile stress, MPa
τ	shear stress, MPa
τ_{-1}	endurance limit in shear, MPa
τ_{max}	maximum shear stress, MPa
τ_s	yield strength in shear, MPa
$[\tau]$	allowable shear stress, MPa
ψ_σ	mean stress influence factor

Subscripts

-1	completely reversed stress
0	repeated stress
a	stress amplitude
e	material properties of an element
m	mean stress
x,y,z	coordinates
σ	tensile stress, MPa
τ	shear stress, MPa

Analysis and Design of Machine Elements, First Edition. Wei Jiang.
© 2019 John Wiley & Sons Singapore Pte. Ltd. Published 2019 by John Wiley & Sons Singapore Pte. Ltd.
Companion website: www.wiley.com/go/Jiang/analysis_of_machine_elements

To ensure machine elements work safely throughout design life, it is imperative to analyse the strength under service loads, and predict potential failures at early design stage well before the machine is built. Strength analysis requires evaluating each possible failure mode by analysing loads and stresses so as to design an element strong enough against failure. Since loads relate to strength via stress analysis, in this chapter, we start with a brief introduction to loads and stresses, followed by a concise review of static strength due to static loads and a detailed discussion of fatigue strength due to cyclic loads. Contact strength is also covered in this chapter.

2.1 Fluctuating Loads and Stresses

2.1.1 Service Factors and Design Loads

In machines, power is transmitted from power sources, like motors, engines and so on, to executive components. Under stable working condition, nominal loads are applied on the elements. However, the inherent characteristics of primary movers, driven machines and their interactions may cause the actual loads on the machine elements greater than the nominal loads. A service or application factor is introduced to account for this effect. The actual loads, or design loads, are the product of a service factor and nominal loads.

Loads on a driven machine are generally defined in terms of torque or power. The design power is obtained by multiplying a service factor with the nominal power to be transmitted. Service factors are usually selected within the range of 1–5 [1], depending on the power sources, operating conditions and duration of service. Large service factors allow for a high degree of safety to account for uncertainties in design analysis, material properties, manufacturing tolerances, operating environments and so on.

Table 2.1 lists the range of service factors for power transmission elements under various operating conditions. Uniform, light impact, moderate impact, and heavy impact power sources are represented by electric or hydraulic motors, steam engines, multicylinder engines and four or less cylinder engines, respectively. Agitator and light-duty conveyors; machine tools and uniformly loaded conveyors; reciprocating compressors and heavy-duty conveyors and rock crushers and heavy machines are examples representing uniform, light impact, moderate impact and heavy impact driven machines. Small service factors are usually for relative smooth operation, while large service factors for vibratory, shock and pulsation operation. Therefore, the design of gears and couplings selects greater values, while the design of belt drives or worm gearings can select smaller values. Besides, long duration of service requires greater values of service factor.

The service factors listed in Table 2.1 are established after considerable field experience in a particular application and can be used with reasonable accuracy. For specific elements, the precise value of service factor can be found from design handbooks or manufacturer's catalogues [4, 5].

2.1.2 Types of Loads

A load is any external force, torque or moment applied to a machine element. The basic forms of load include tension, compression, shear, torsional shear and friction, as well as

Table 2.1 Service factors for power transmission elements, K_A [2, 3].

Power sources		Driven machines			
		Uniform	Light impact	Moderate impact	Heavy impact
		Agitators, light conveyors, centrifugal compressors	Machine tools, gear pump, mixers	Reciprocating compressors heavy-duty conveyors	Mining machines, heavy machines
Uniform	Electric motor, steam or gas turbines of uniform speed and low torques	1.00~1.10	1.25~1.35	1.50~1.60	≥1.75
Light impact	Electric motor, steam or gas turbines, hydro motor of high and variable torque	1.10~1.25	1.35~1.50	1.60~1.75	≥1.85
Moderate impact	Multi-cylinder internal combustion engines	1.25~1.50	1.50~1.75	1.75~2.00	≥2.00
Heavy impact	Single-cylinder internal combustion engines	≥1.50	≥1.75	≥2.00	≥2.25

torques and moments. To describe a load, information about its magnitude, direction, point of action, duration, frequency and the number of cycles is required. The analysis and design of machine elements involve extensive stress or deflection analysis on the premise of correct load determination.

The accurate determination of realistic operating loads is often a difficult and challenging task in machine element design. When power is transmitted from a mover to the executive components in a machine, nearly all machine elements participate in the activity, either by transmitting power, like gears, or by supporting power transmitting elements, like shafts and bearings. The dynamic effects due to impact or vibration during operation, together with multiple force sources on an element, such as contact, friction, gravity or thermal expansion, make the precise determination of loads even more complicated [1]. Once the loads are acquired, equilibrium analyses combined with vectorial approaches are used to reduce the complex loads into basic loads of tension, shear, moment and so on.

According to the loading history, the load applied to an element could be static, fluctuating, shock or impact and random loads. An ideal static load is applied slowly and is never removed or is removed infrequently. Gravity load is a typical example of static load. The load magnitude, direction and point of action of a static load do not change with time, or change very slowly. A static load can be an axial tension or compression, a shear load, a bending load, a torsional load or any combination of these. Although in engineering practice relatively few elements are subjected to pure static loads, static loads are still the fundamentals to the sizing of machine elements.

Fluctuating or variable loads vary with time during the normal service of an element. They are typically applied for a long enough time so the element experiences many thousands or millions of stress cycles during its expected life [3]. An example is the load acting on a pair of meshing gear teeth. Although every machine load is important, it is estimated that 60–90% of machine elements fail from fluctuating loads due to fatigue [1].

Shock or impact loads are the loads applied suddenly and rapidly, like the force when cars crash. When the applied loads vary irregularly in their amplitudes, such as wind loads acting on wind turbine blades, the loads are called random loads. The latter two types of loads will not be addressed in this book.

2.1.3 Types of Stresses

Once the external load applied to an element has been determined, stresses generated within the element called body stresses can be decided. The calculation of tensile or shear stresses due to tension, shear, toque and bending can be referred to in *Mechanics of Materials* [6]. Detailed stress analyses for specific machine elements will be introduced in the succeeding chapters.

The primary factor considered when specifying stress types is the stress variation. Fluctuating stresses in a machine often take a sinusoidal form because of the nature of rotating machinery. Stress variations are characterized by five parameters; maximum stress σ_{max}, minimum stress σ_{min}, mean (average) stress σ_m, stress amplitude σ_a and stress ratio r, as defined in Figure 2.1. If any two of them are known, the others are readily computed.

The maximum and minimum stresses can be obtained by theoretical stress analysis, finite element analysis or experimental measurement. The mean stress and stress amplitude are computed by

$$\sigma_m = \frac{\sigma_{max} + \sigma_{min}}{2} \tag{2.1}$$

$$\sigma_a = \frac{\sigma_{max} - \sigma_{min}}{2} \tag{2.2}$$

The behaviour of a material under fluctuating stresses is dependent on the manner of stress variation, characterized by the stress ratio, as

$$r = \frac{\sigma_{min}}{\sigma_{max}} \tag{2.3}$$

2.1.3.1 Static Stress
Static stress is generated when an element is subjected to a slowly applied load that is held at a constant value. Figure 2.1a shows a diagram of stress versus time due to static loading. Since $\sigma_{max} = \sigma_{min}$, the stress ratio for static stress is $r = 1$.

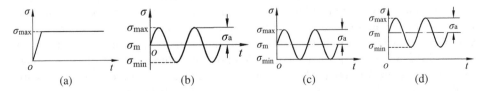

(a) (b) (c) (d)

Figure 2.1 Types of stresses.

Figure 2.2 An example of completely reversed stress.

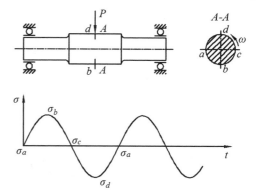

2.1.3.2 Completely Reversed Stress

When a tensile stress is followed by the same level of compressive stress over thousands of cycles, the stress is completely reversed stress. Figure 2.1b shows the diagram of stress versus time for a completely reversed stress. Because $\sigma_{min} = -\sigma_{max}$, the stress ratio for completely reversed stress is $r = -1$ and the mean stress is zero.

An example of a completely reversed stress is a rotating circular shaft loaded in bending, as shown in Figure 2.2. In the position shown, point b at the bottom of the shaft experiences a tensile stress, while point d at the top of the shaft has the same magnitude compressive stress. As the shaft rotates, these two points, as well as all the other points, experience a complete stress cycle during each revolution.

2.1.3.3 Repeated Stress

When a load is applied and removed many times, the stress generated in a machine element is repeated stress. As shown in Figure 2.1c, the stress varies from zero to a maximum value within each cycle. The relations among mean stress, stress amplitude and the maximum stress can be expressed as $\sigma_m = \sigma_a = \sigma_{max}/2$. The minimum stress is zero and the stress ratio is $r = 0$. An example of repeated stress is the stress on the tooth surfaces of a pair of meshing gears.

2.1.3.4 Fluctuating Stress

A fluctuating stress is an alternating stress with a nonzero mean value, as shown in Figure 2.1d. The fluctuating stress can be regarded as the superposition of a completely reversed stress on a static stress. When an axial load is applied to the shaft shown in Figure 2.2, the shaft experiences a fluctuating stress.

2.2 Static Strength

To ensure safety of a machine element under static or steady loads, stresses need to be determined analytically or experimentally and compared with an allowable stress of the material to satisfy strength criteria discussed in Section 1.3.3. Static strength analysis involves the consideration of several factors, including loads, stress states (uniaxial or multiaxial) and material types (ductile or brittle).

2.2.1 Static Strength for Uniaxial Stresses

Uniaxial stresses, such as tensile, compressive, shear, torsional shear or bending stresses, are created by basic axial loads, shear loads, torques or bending moments. They can be calculated by using *Mechanics of Materials* [6]. As long as the maximum stress in an element remains less than the allowable stress, the element is deemed safe, that is

$$\sigma \leq [\sigma] = \frac{\sigma_{\text{lim}}}{[S]} \tag{2.4}$$

Metal behaviour of an element can be ductile or brittle. Most machine elements are made from ductile materials, for example, wrought metals, such as steel, aluminium and copper. Ductile materials are prone to distortion failure and have an identifiable yield strength. The material strength is thus selected as yield strength; that is, $\sigma_{lim} = \sigma_s$.

Brittle materials, such as cast iron, are prone to fracture failure and do not exhibit an identifiable yielding. Brittle materials are stronger in compression than in tension. The material strength is selected as ultimate tensile strength or ultimate compressive strength; that is, $\sigma_{lim} = \sigma_b$.

The allowable stress $[\sigma]$ is determined by material strength divided by an allowable safety factor, which is usually within the range of 1.25–4. The selection of appropriate safety factors depends on the required safety and reliability.

An alternative method to analysis and design machine elements under static stresses is by

$$S = \frac{\sigma_{\text{lim}}}{\sigma} \geq [S] \tag{2.5}$$

That is, the safety factor of machine element under actual stress must be greater than the allowable safety factor.

The discussion here is focus on tensile strength. If shear strength is of the main concern, the tensile stress can be substituted by shear stress in Eqs. (2.4) and (2.5).

2.2.2 Static Strength for Combined Stresses

Combined stresses refer to the stresses that are applied in more than one direction, or the normal stress and shear stress are applied simultaneously. In practical applications, machine elements are usually in biaxial, triaxial or combined stress state rather than uniaxial stress state. It is therefore necessary to establish strength theory to predict failure of machine elements under complex stress states on the basis of material strength data obtained from uniaxial tests. Currently, the most widely used strength theories are the maximum shear stress theory and the maximum distortion energy theory.

2.2.2.1 Maximum Shear Stress Theory

The maximum shear stress theory (or Tresca theory) predicts that a material subjected to a combination of loads will fail whenever the maximum shear stress exceeds the shear strength of the material [6, 7]. The shear strength is usually determined from a standard uniaxial tension test.

In the standard uniaxial tension test on a ductile material, fracture lines are observed at 45° from the axis of tension, and the maximum shear stress at yield is $\tau_{max} = \sigma_s/2$.

For a general stress state, three principal stresses can be determined and ordered such that $\sigma_1 \geq \sigma_2 \geq \sigma_3$. The maximum shear stress is then $\tau_{max} = (\sigma_1 - \sigma_3)/2$. Thus, for a general state of stress, the maximum shear stress theory predicts yield strength as

$$\tau_{max} = \frac{\sigma_1 - \sigma_3}{2} \leq \frac{\sigma_s}{2} \text{ or } \sigma_1 - \sigma_3 \leq \sigma_s \tag{2.6}$$

This implies that the shear yield strength is given by

$$\tau_s = \frac{\sigma_s}{2} \tag{2.7}$$

2.2.2.2 Maximum Distortion Energy Theory

The maximum distortion energy theory predicts that yield occurs when the distortion energy per unit volume reaches or exceeds the distortion energy per unit volume for yield in uniaxial tension of the same material [6, 7]. It is also called von Mises or von Mises–Hencky theory, giving credit to R. von Mises and H. Hencky for the development of it.

When applying this theory, it is convenient to use an equivalent stress σ_e, or von Mises stress, to transform multiaxial stresses into an equivalent uniaxial stress. The equivalent stress is the value of a uniaxial tensile stress that would produce the same level of distortion energy as the actual stresses involved [6, 7]. By the distortion energy theory, the equivalent stress can be derived and expressed by principal stresses as

$$\sigma_e = \frac{1}{\sqrt{2}}[(\sigma_1 - \sigma_2)^2 + (\sigma_2 - \sigma_3)^2 + (\sigma_3 - \sigma_1)^2]^{1/2} \tag{2.8}$$

Thus, the yield strength criterion is expressed as

$$\frac{1}{\sqrt{2}}[(\sigma_1 - \sigma_2)^2 + (\sigma_2 - \sigma_3)^2 + (\sigma_3 - \sigma_1)^2]^{1/2} \leq \sigma_s \tag{2.9}$$

When using *xyz* components of the stress, the von Mises stress and yield strength criterion can be rewritten as

$$\frac{1}{\sqrt{2}}[(\sigma_x - \sigma_y)^2 + (\sigma_y - \sigma_z)^2 + (\sigma_z - \sigma_x)^2 + 6(\tau_{xy}^2 + \tau_{yz}^2 + \tau_{zx}^2)]^{1/2} \leq \sigma_s \tag{2.10}$$

For pure shear in plane problem where $\sigma_x = \sigma_y = 0$, the yield strength can be obtained from Eq. (2.10) as

$$\tau_{xy} \leq \frac{\sigma_s}{\sqrt{3}} \tag{2.11}$$

Thus, the shear yield strength predicted by the maximum distortion energy theory is

$$\tau_s = \frac{\sigma_s}{\sqrt{3}} = 0.577\sigma_s \tag{2.12}$$

Both the maximum shear stress theory and the maximum distortion energy theory can be applied in the analysis and design of a machine element. The maximum shear stress theory gives a simple and moderately conservative approach; while the maximum distortion energy theory provides a more accurate prediction [7].

2.3 Fatigue Strength

Previous discussion is mainly about strength analysis of elements under static loads. More often than not, machine elements are subjected to fluctuating loads and the behaviour of an element under variable loads is entirely different from that under static loads. The failure mode under fluctuating loads is fatigue. The stress-life, strain-life and elastic fracture mechanics methods are currently used to analyse fatigue strength. These methods aim to predict the life in a number of cycles to failure at a specific level of loads. This section will only introduce the stress-life method. The latter two methods will not be covered in this book. Interested readers can read the relevant references [2, 3, 8, 9].

2.3.1 The Nature of Fatigue

Most machines operate under fluctuating loads rather than static loads. Fluctuating loads produce variable stresses, causing fatigue fracture failure after a period of service. Contrary to static fracture failures that happen at high loads and large distortions, fatigue fracture failures are sudden phenomena, usually at relatively low loads, giving no warning in advance and hence are much more dangerous.

Fatigue is a progressive and localized structural damage that occurs when material is subjected to cyclic loading. It starts with a microscopic crack at critical area with high local stresses. The high local stresses at critical area are most likely due to geometrical discontinuities, like keyways, holes and threads; or minute material flaws, like voids and hard precipitated particles or fabrication faults, like tool marks, surface scratches and so on. The microcracks under high cyclic stress may initiate and propagate to a size when the remaining uncracked material is sufficiently weakened and can no longer support the applied load, resulting in a sudden, rapid fracture without warning.

Figure 2.3a shows schematic fracture surface diagram due to fatigue fracture. Fatigue fracture failure usually starts at microcrack A on the surface, propagates across most of the cross section shown by the beach marks at zone B, and finally fracture at zone C. Although a fracture surface feature depends on loading conditions and geometrical discontinuities, it typically consists of a smooth surface fatigue zone B and a rough final fracture zone C. The smooth surface fatigue zone is the bench-marked area, developed by repeated contacting and separating of the mating crack surfaces. The rough

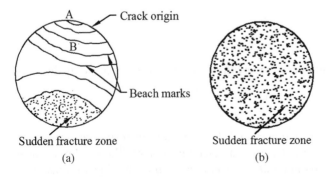

Figure 2.3 Fracture surface diagram.

final fracture zone is the sudden fracture area [10]. Figure 2.3b shows fracture surface diagram due to static fracture for comparison.

The crack initiation and propagation are affected by several conditions. Apart from localized stresses and material microstructures, other factors include tensile residual stresses, temperature, environment and cycling frequency of loads. The estimation of fatigue behaviour requires consideration of these factors, with emphasis on geometries, materials and loading conditions.

2.3.2 Stress-Life Diagrams

The fatigue strength represents a material's capability of withstanding fluctuating loads, which is obtained from a standard fatigue test. In the test, a regular cyclic load with a stress ratio of either -1 or 0 is applied to a smooth or notched specimen on a fatigue testing machine. Different magnitudes of load are applied to generate different stress levels. The number of cycles to failure or fatigue life, corresponding to different stress levels, is recorded. The stress-life diagram (also known as σ-N or S-N curve), which depicts stress versus the number of cycles to failure, is developed as shown in Figure 2.4. From this figure, the fatigue strength of materials can be determined.

In Figure 2.4, point A indicates the ultimate tensile strength of material, which is obtained under static loads. From point A to point B, the number of cycles is rather small, and the stress is almost constant. We can regard this range as a static stress state.

From point B to point C where the number of cycles is less than 10^3, a low cycle fatigue behaviour that is associated with widespread plasticity is observed. This is a low cycle fatigue stage.

Beyond point C is high cycle fatigue stage, where the number of cycles that causes fatigue failure is greater than 10^3. The high cycle fatigue stage can be split into finite life stage and infinite life stage [1]. Within the finite life stage from C to D, the material will fail for a given number of cycles. The stress corresponding to a number of load cycle N is defined as endurance strength, or fatigue strength σ_{rN}. Curve CD can be expressed as

$$\sigma_{rN}^m N = Const \tag{2.13}$$

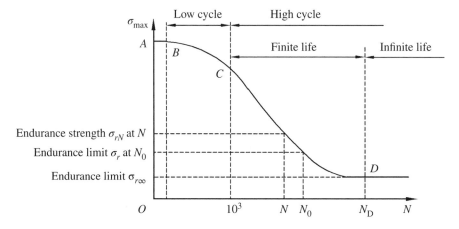

Figure 2.4 Stress-life diagram for ductile materials.

If there is a stress the material can withstand indefinitely without failure, that is, the number of cycles is infinite, this stress is called the endurance limit or fatigue limit $\sigma_{r\infty}$. Beyond point D is an infinite life stage. Therefore,

$$\sigma_{rN} = \sigma_{r\infty} \tag{2.14}$$

Since the number of cycles N_D is normally quite large, a critical number of cycle N_0 is specified, and the corresponding endurance limit at N_0 is termed σ_r. Therefore, from Figure 2.4 and Eq. (2.13), we have

$$\sigma_{rN}^m N = \sigma_r^m N_0 = Const. \tag{2.15}$$

For any number of cycles N, the endurance strength can be derived from (2.15) as

$$\sigma_{rN} = \sigma_r \sqrt[m]{\frac{N_0}{N}} \tag{2.16}$$

where m is material constant depending on stress states and decided by tests. For steels under bending stress, select $m = 9$ [11].

In the case of ferrous metals and alloys, the S-N curve approaches a stress horizontal asymptote after the material has been stressed for a certain number of cycles. The critical number of cycles is defined as $N_0 = 10^6 \sim 10^7$ for steels with hardness no greater than 350HBW, and $N_0 = 10 \times 10^7 \sim 25 \times 10^7$ for steels with hardness greater than 350HBW. The nonferrous alloys do not exhibit a stress asymptote and the critical number of cycles is defined as $N_0 = 25 \times 10^7$ to characterize endurance limits [11].

2.3.3 Endurance Limit Diagrams

2.3.3.1 The Endurance Limit Diagram of a Material

The stress-life diagram represents the endurance strength of materials at different number of cycles under the same stress ratio. The stress-life diagram with different stress ratios are illustrated by similar yet different curves. Since endurance limits vary with different stress ratios in the stress-life diagram, an endurance limit diagram (also called $\sigma_m - \sigma_a$ diagram) shown by the curve ACB in Figure 2.5 is used to evaluate fatigue limits of materials for the same life of, usually $N_0 = 10^7$, under different stress ratios. The $\sigma_m - \sigma_a$ diagram provides a general approach dealing with mean stress σ_m and stress amplitude σ_a for high cycle fatigue design under uniaxial stress fluctuation.

The curve ACB in Figure 2.5 is obtained from fatigue tests. It can be simplified by the Soderberg line AE, Goodman line AB and the lines of AF plus FE or other possible lines. For each simplified criterion, points on or above the respective lines indicate failure. More information about the Soderberg formula and Goodman formula can be found in references [1, 2]. This chapter will give a detailed introduction to the simplified lines of AF and FE.

In Figure 2.5, points A, B, C, D, E represent the endurance limit of completely reversed stress σ_{-1} ($r = -1$), ultimate strength σ_b ($r = +1$), endurance limit of repeated stress σ_0 ($r = 0$), fatigue limit σ_r and yield strength σ_s, respectively. Points on the horizontal axis represent a static stress state, that is, $r = 1$; while points on the vertical axis represent a completely reversed stress state, that is, $r = -1$. Line EF represents the stress state where the endurance limit is the same as the yield strength, implying that the failure mode is yielding. The simplified $\sigma_m - \sigma_a$ diagram AF and FE can be expressed by line equations.

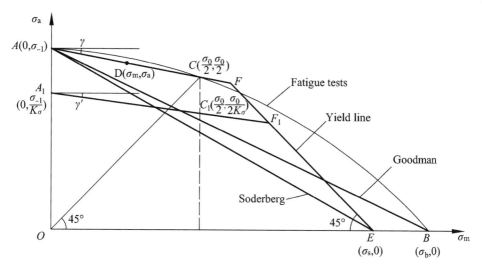

Figure 2.5 Endurance limit diagram.

The line equation for *EF* is

$$\sigma_a + \sigma_m = \sigma_s \tag{2.17}$$

The line equation for *AF* can be obtained by the coordinates of points $A(0, \sigma_{-1})$ and $C(\sigma_0/2, \sigma_0/2)$, as

$$\sigma_{-1} = \sigma_a + \psi_\sigma \sigma_m \tag{2.18}$$

where $\psi_\sigma = \frac{2\sigma_{-1} - \sigma_0}{\sigma_0} = \tan \gamma$ is mean stress influence factor of materials. For carbon steels, select $\psi_\sigma = 0.1 \sim 0.2$; For alloy steels, select $\psi_\sigma = 0.2 \sim 0.3$. For shear stress, select $\psi_\tau = 0.5 \psi_\sigma$.

2.3.3.2 The Endurance Limit Diagram of an Element

The endurance limits of materials are determined by fatigue tests using standard polished specimens. However, in engineering practice, the geometry, dimension and surface conditions of a machine element are usually different from the specimen, leading to a reduced endurance limit that must be modified from published data. Factors affecting endurance limits include stress concentration, size of cross section, surface finish and so on.

Fatigue Stress Concentration Factor, k_f Stress concentration refers to the significant increase of local stresses in the immediate area of geometric discontinuities, such as shaft shoulders, keyways, threads, holes, grooves or notches. The effect of stress increment is measured by geometric or theoretical stress concentration factor k_t, which relates to the geometry of element [8, 10].

Stress concentration reduces fatigue strength of materials or elements. The reduction effect is measured by fatigue stress concentration factor k_f, which is slightly less than the geometric stress concentration factor k_t due to the lessened sensitivity to notches. Their relation is expressed by

$$k_f = 1 + q(k_t - 1) \tag{2.19}$$

Table 2.2 Stress concentration factors for a stepped shaft with a shoulder fillet [12].

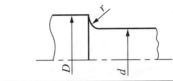

$\dfrac{r}{d}$	Shaft in tension, k_σ $$\sigma = \dfrac{4F}{\pi d^2}$$					Shaft in bending, k_σ $$\sigma = \dfrac{32M}{\pi d^3}$$					Shaft in torsion, k_τ $$\tau = \dfrac{16T}{\pi d^3}$$			
	D/d													
	2.0	1.5	1.1	1.05	1.01	3.0	1.5	1.2	1.05	1.01	2.0	1.33	1.2	1.09
0.04	2.8	2.57	1.99	1.82	1.42	2.4	2.21	2.09	1.88	1.61	1.84	1.79	1.66	1.32
0.10	1.99	1.89	1.56	0.46	1.23	1.80	1.68	1.62	1.53	1.36	1.46	1.41	1.33	1.17
0.15	1.77	1.68	1.44	1.36	1.18	1.59	1.52	1.48	1.42	1.26	1.34	1.29	1.23	1.13
0.20	1.63	1.56	1.37	1.31	1.15	1.46	1.42	1.39	1.34	1.20	1.26	1.23	1.17	1.11
0.25	1.54	1.49	1.31	1.27	1.13	1.37	1.34	1.33	1.29	1.17	1.21	1.18	1.14	1.09
0.30	1.47	1.43	1.28	1.24	1.12	1.31	1.29	1.27	1.25	1.14	1.18	1.16	1.12	1.09

where q is the notch sensitivity factor, ranging from 0 to 1.0, depending on the strength and hardness of material and notch geometry. Since reliable values of notch sensitivity factor are difficult to obtain, this book will use the safest and most conservative value of $k_f = k_t$ [2, 3].

The values of fatigue stress concentration factor vary with axial tension, bending and torsional loading conditions, and designated by k_σ and k_τ for tensile and shear fatigue stress concentration factor, respectively. Abridged geometric stress concentration factors of a stepped shaft with a shoulder fillet are listed in Table 2.2. Detailed information of stress concentration factors with different geometric discontinuities can be found in references [10, 12].

Size Factor, ε_σ or ε_τ Small size elements usually have greater fatigue strength than larger ones with the same surface finish, material and configuration. The larger the element, the greater the statistical probability of flaws existing within an element and the more likely the reduction of fatigue strength. The nonuniformity of material properties due to size is measured by size factor ε_σ for tension and ε_τ for shear. Detailed data of size factor can be found in references [3, 10] and limited data used in this book are presented in Figure 2.6.

Surface Condition Factor, β_σ or β_τ Since fatigue failures originate from relative weak areas on the surface, the surface quality is of particular importance. Poor surface quality, represented by scratches and irregularities, will reduce fatigue strength. This effect is measured by surface condition factor, which can be found in Table 2.3. Fatigue strength of material increases rapidly as surface qualities are progressively improved through machining, grinding and polishing. When experimental data for torsional shear fatigue

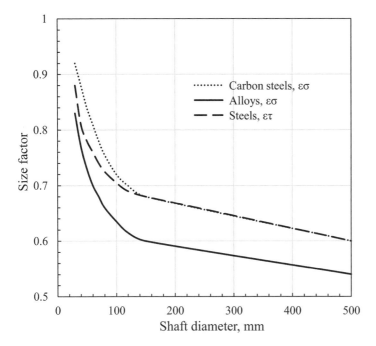

Figure 2.6 Size factors for bending and torsion [10].

Table 2.3 Surface condition factors by different manufacturing methods [13].

Fabrication methods	σ_b/MPa		
	400	**800**	**1200**
Mirror-polished	1	1	1
Fine-ground	0.90	0.90	0.85
Machined	0.80	0.75	0.68
Hot-rolled	0.74	0.50	0.40
As forged	0.55	0.38	0.28

are limited, the surface codification factor for shear β_τ can be selected as approximately equal to that for tension β_σ.

Intensify Factor, β_q Although surface treatment does not have much influence on static strength, it can significantly increase fatigue strength of an element. Surface treatment refers to surface heat treatment and cold work hardening. The former includes flame quenching, induction quenching, carburizing and nitriding; while the latter includes cold rolling, shot peening, hammer peening and so on. The effect of surface treatment on the fatigue strength is measured by intensify factor β_q. Detailed data on intensify

Table 2.4 Intensify factors by various surface treatments [10].

Surface treatments	Specimen	Diameter, d mm^{-1}	Intensify factor, β_q
Induction quenching	Without stress concentration	7~20	1.3~1.6
		30~40	1.2~1.5
	With stress concentration	7~20	1.6~2.8
		30~40	1.5~2.5
Nitriding (depth 0.1~0.4 mm)	Without stress concentration	8~15	1.15~1.25
		30~40	1.10~1.15
	With stress concentration	8~15	1.9~3.0
		30~40	1.3~2.0
Carbonizing (depth 0.2~0.6 mm)	Without stress concentration	8~15	1.2~2.1
		30~40	1.1~1.5
	With stress concentration	8~15	1.5~2.5
		30~40	1.2~2.0
Rolling	Without stress concentration	7~20	1.2~1.4
		30~40	1.1~1.25
	With stress concentration	7~20	1.5~2.2
		30~40	1.3~1.8
Shot peening	Without stress concentration	7~20	1.1~1.3
		30~40	1.1~1.2
	With stress concentration	7~20	1.4~2.5
		30~40	1.1~1.5

factors can be found in reference [10] and abridged data used in this book are listed in Table 2.4.

The effect of these factors can be evaluated by a combined influence factor K_σ, expressed in the formula as [14]

$$K_\sigma = \left(\frac{k_\sigma}{\varepsilon_\sigma} + \frac{1}{\beta_\sigma} - 1 \right) \frac{1}{\beta_q} \qquad (2.20)$$

These discussions relate to the endurance limit of materials subjected to normal tensile stress only, that is, tensile stresses resulting from bending or axial tension. Cases involving fluctuating torsional shear stresses are comparable to normal stresses, and tensile stress can be substituted with shear stress directly.

The actual endurance limit of a machine element will then be

$$\sigma_{-1e} = \frac{\sigma_{-1}}{K_\sigma} \qquad (2.21)$$

Considering these factors, the actual endurance limit diagram of an element is then obtained by moving line ACF proportionally to line $A_1 C_1 F_1$, while keeping EF as it is, as shown in Figure 2.5. Points on lines $A_1 F_1 E$ characterize endurance limits corresponding to a stress ratio between -1 and $+1$.

Again, the actual endurance limit diagram of an element can be expressed by line equations. For line EF_1, the equation is the same as that of line EF, that is

$$\sigma'_{ae} + \sigma'_{me} = \sigma_s \tag{2.22}$$

While in section A_1F_1, the line equation can be obtained by the coordinates of points $A_1(0, \sigma_{-1}/K_\sigma)$ and $C_1(\sigma_0/2, \sigma_0/2K_\sigma)$, as

$$\sigma_{-1} = K_\sigma \sigma'_{ae} + \psi_\sigma \sigma'_{me} \tag{2.23}$$

or

$$\sigma_{-1e} = \frac{\sigma_{-1}}{K_\sigma} = \sigma'_{ae} + \psi_{\sigma e} \sigma'_{me} \tag{2.24}$$

where σ'_{ae} and σ'_{me} are the stress amplitude and mean stress of an element, respectively, and $\psi_{\sigma e} = \frac{1}{K_\sigma} \cdot \frac{2\sigma_{-1}-\sigma_0}{\sigma_0} = \tan \gamma'$ is the mean stress influence factor of an element. These equations are applicable to shear stress by substituting σ with τ.

2.3.4 Fatigue Strength for Uniaxial Stresses with Constant Amplitude

Like static stress state, the strength criterion for constant fluctuating stresses is also $S_{ca} \geq [S]$. The calculated safety factor S_{ca} is defined as the material strength divided by maximum working stress, that is,

$$S_{ca} = \frac{\sigma_{\lim}}{\sigma_{\max}} \tag{2.25}$$

For any number of cycles of N ($10^3 < N < N_0$), the endurance strength is determined by Eq. (2.16) as $\sigma_{rN} = \sqrt[m]{\frac{N_0}{N}}\sigma_r$; while for any stress ratios, the endurance limit is determined by a $\sigma_m - \sigma_a$ diagram.

Generally, when analysing fatigue strength of a machine element, the maximum and minimum stress at critical cross sections are first determined by load variations. These stresses are then converted to mean stress σ_m and stress amplitude σ_a, and the location of working point M (σ_m, σ_a) can be indicated in the $\sigma_m - \sigma_a$ diagram. The maximum working stress is the sum of mean stress σ_m and stress amplitude σ_a. the endurance limit locates on the line A_1F_1E in Figure 2.7. If the location of endurance limit on the line A_1F_1E can be determined, the safety factor can be easily obtained.

The determination of endurance limit on the line A_1F_1E is based on structural constraints and stress variation as loads would fluctuate to cause element failure in service. The typical cases of stress variation include constant stress ratio $r = Const.$, constant mean stress $\sigma_m = Const.$ and constant minimum stress $\sigma_{min} = Const.$ [14], as illustrated in Figure 2.7.

A practical example of a constant stress ratio involves the stresses in a rotating shaft. In the endurance limit diagram, the working point M (or N) indicates the combination of mean stress and stress amplitude representing the critical stress in an element. Connect the origin O and M (or N) and extend the line to intersect with the endurance limit line A_1F_1 or F_1E at M_1 (or N_1), we have

$$\tan \alpha = \frac{\sigma_a}{\sigma_m} = \frac{(\sigma_{\max} - \sigma_{\min})/2}{(\sigma_{\max} + \sigma_{\min})/2} = \frac{1-r}{1+r} = Const. \tag{2.26}$$

Therefore, points on line MM_1 (or NN_1) have the same stress ratio as that of working point of M (or N), that is, the stresses at point M_1 (or N_1) represent the endurance

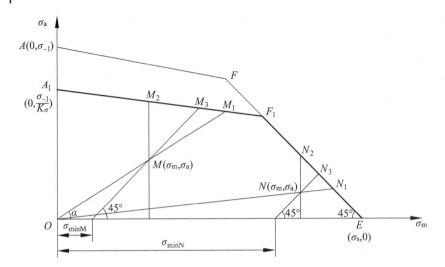

Figure 2.7 Endurance limits for three typical loading cases.

limit at stress ratio r. If the working stress is located within the area of ΔOA_1F_1, the endurance limit can be found by solving two line equations of OM_1 and A_1F_1 to find out the intersection, by

$$\begin{cases} \dfrac{\sigma_a}{\sigma_m} = \dfrac{\sigma'_{ae}}{\sigma'_{me}} \\ \sigma_{-1} = K_\sigma \sigma'_{ae} + \psi_\sigma \sigma'_{me} \end{cases} \tag{2.27}$$

We can obtain the coordinates of intersection point M_1 (σ'_{me}, σ'_{ae}), and then the fatigue strength is expressed as

$$S_{ca} = \frac{\sigma_{\lim}}{\sigma_{\max}} = \frac{\sigma'_{ae} + \sigma'_{me}}{\sigma_{\max}} = \frac{\sigma_{-1}}{K_\sigma \sigma_a + \psi_\sigma \sigma_m} \geq [S] \tag{2.28}$$

A practical example of a constant mean stress is the stresses in a loaded vibrating spring. Points on the vertical lines through the working point of M (or N), that is, MM_2 (or NN_2), represent the same mean stress. Similarly, the fatigue strength for constant mean stress is

$$S_{ca} = \frac{\sigma_{-1} + (K_\sigma - \psi_\sigma)\sigma_m}{K_\sigma(\sigma_m + \sigma_a)} \geq [S] \tag{2.29}$$

A typical example of constant minimum stress is the stresses in a preloaded bolt. Since $\sigma_{min} = \sigma_m - \sigma_a = Const.$, the 45° lines through point M (or N), that is, MM_3 (or NN_3), represent the stresses that have the same minimum stress. Similarly, the fatigue strength is

$$S_{ca} = \frac{2\sigma_{-1} + (K_\sigma - \psi_\sigma)\sigma_{min}}{(K_\sigma + \psi_\sigma)(2\sigma_a + \sigma_{min})} \geq [S] \tag{2.30}$$

In these three cases, if the intersection is on line EF_1, the element will yield rather than fatigue. The strength is then calculated by

$$S_{ca} = \frac{\sigma_s}{\sigma_m + \sigma_a} \geq [S] \tag{2.31}$$

Notes:

1. When it is difficult to determine the possible variation of stresses, use $r = Const.$ to start the calculation.
2. An equivalent set of formulas also holds for cyclic shear stresses. For cases involving fluctuating torsional shear stresses, substitute σ with τ.
3. When the number of cycles is less than the critical number of cycles, that is, $10^3 < N < N_0$, the endurance strength at the number of cycle N is $\sigma_{rN} = \sqrt[m]{\frac{N_0}{N}} \sigma_r$, instead of σ_r. Therefore, σ_{-1} should be substituted with σ_{-1N}.
4. If the endurance limit is located around point F_1 in the $\sigma_m - \sigma_a$ diagram, both fatigue strength and static strength need to be calculated.
5. The fatigue safety factor can be obtained by either a graphical or analytical approach.

In summary, fatigue strength analyses involve establishing a relationship between the fatigue strength of the material or element and the working stress to determine safety factors. The fatigue strength of the material or element is represented by a $\sigma_m - \sigma_a$ diagram; while the working stress is the sum of mean stress σ_m and stress amplitude σ_a.

2.3.5 Fatigue Strength for Uniaxial Stresses with Variable Amplitude

The discussion so far has dealt with fatigue behaviour of an element under uniform stress amplitudes. In real engineering practice, nearly all machine elements are subject to a spectrum of speeds or loads, giving rise to stresses varying in both amplitudes and mean values. Automotive suspension and aircraft structural components are typical examples of elements subject to a spectrum of variable stress amplitudes.

The variations in stress amplitudes make the direct use of standard *S-N* curves inapplicable because these curves are developed from constant stress amplitude operations. Therefore, it becomes necessity to develop a verified theory to predict fatigue strength for spectrum load operations using the constant amplitude *S-N* curves.

The linear cumulative damage rule, proposed by Palmgren of Sweden in 1924 and, independently, by Miner of the United States in 1945, can be used to deal with fatigue strength analysis for elements under variable uniaxial stresses. The rule assumes that any stress amplitude greater than the endurance limit contributes certain fatigue damage to the element. The amount of damage depends on the number of cycles at that stress amplitude and the total number of cycles that would produce failure to a standard specimen at the same stress amplitude. When the total accumulated damage generated by different stress levels reaches a critical value, fatigue failure occurs [15].

2.3.5.1 Linear Cumulative Damage Rule (Miner's Rule)

Assume that σ_1, σ_2, σ_3... are the maximum stress of each cycle; n_1, n_2, n_3... are the number of cycles acting on the element at each stress level σ_i and N_1, N_2, N_3... are the number of cycles to failure at each stress level σ_i acquired from the *S-N* curve.

For stresses greater than the endurance limit, each cycle will cause certain damage. The damage contribution of n_1 cycles at stress σ_1 is assumed to be n_1/N_1; while the damage contribution of n_2 cycles at stress σ_2 is assumed to be n_2/N_2.... Similarly, the damage caused by n_i cycles at stress σ_i is n_i/N_i. For stresses smaller than the endurance limit, it is assumed that they will not cause any damage and can be neglected in the calculation. When the cumulative damage approaches 1 or 100%, fatigue failure ensues. The linear cumulative fatigue damage rule or Miner's rule can be expressed as

$$\frac{n_1}{N_1} + \frac{n_2}{N_2} + \cdots + \frac{n_i}{N_i} + \cdots + \frac{n_z}{N_z} = 1 \tag{2.32}$$

For a more general case,

$$\sum_{i=1}^{z} \frac{n_i}{N_i} = 1 \tag{2.33}$$

2.3.5.2 Prediction of Cumulative Fatigue Damage

We have introduced fatigue prediction for constant stresses amplitude. For variable amplitude stresses, we need first to convert them to constant amplitude stresses and then use the previously introduced method to predict fatigue.

From the *S-N* curve, we have

$$\sigma_i^m N_i = \sigma_{-1}^m N_0 = Const. \tag{2.34}$$

Therefore

$$N_i = N_0 \left(\frac{\sigma_{-1}}{\sigma_i} \right)^m \tag{2.35}$$

Substituting Eq. (2.35) into the linear cumulative-damage rule in Eq. (2.33), we have

$$\sum_{i=1}^{z} n_i \sigma_i^m = N_0 \sigma_{-1}^m \tag{2.36}$$

When Eq. (2.36) is satisfied, it indicates that the element reaches the endurance limit. Assuming the fatigue effect of variable amplitude stresses is equivalent to that of a constant amplitude stress σ_v (usually select $\sigma_v = \sigma_1$) that operates the equivalent number of cycles N_v, we then have

$$\sum_{i=1}^{z} n_i \sigma_i^m = N_v \cdot \sigma_v^m = N_v \cdot \sigma_1^m \tag{2.37}$$

The equivalent number of cycles N_v is then derived as

$$N_v = \sum_{i=1}^{z} n_i \left(\frac{\sigma_i}{\sigma_1} \right)^m \tag{2.38}$$

Corresponding to the equivalent number of cycles N_v, the endurance strength for completely reversed stress σ_{-1Nv} is

$$\sigma_{-1Nv}^m N_v = \sigma_{-1}^m N_0 \tag{2.39}$$

Therefore

$$\sigma_{-1Nv} = \sqrt[m]{\frac{N_0}{N_v}} \sigma_{-1} = \sigma_{-1} \sqrt[m]{\frac{N_0}{\sum_{i=1}^{z} n_i \left(\frac{\sigma_i}{\sigma_1} \right)^m}} \tag{2.40}$$

Therefore, the safety factor and fatigue strength of the element under variable uniaxial stresses is

$$S_{ca} = \frac{\sigma_{\lim}}{\sigma_{\max}} = \frac{\sigma_{-1Nv}}{\sigma_1} = \sigma_{-1m} \sqrt{\frac{N_0}{\sum\limits_{i=1}^{z} n_i \sigma_i^m}} \geq [S] \tag{2.41}$$

2.3.6 Fatigue Strength for Combined Stresses with Constant Amplitude

During machine operation, an element is most likely subjected to a combination of tension, bending and torsion, so both tensile and shear stresses will generate. Under static stress states, yield occurs when the equivalent stress obtained from the maximum distortion energy theory equals the uniaxial yield strength expressed by

$$\frac{1}{\sqrt{2}}[(\sigma_1 - \sigma_2)^2 + (\sigma_2 - \sigma_3)^2 + (\sigma_3 - \sigma_1)^2]^{1/2} = \sigma_s \tag{2.42}$$

where the three principle normal stresses are

$$\sigma_1 = \frac{\sigma}{2} + \frac{1}{2}\sqrt{\sigma^2 + 4\tau^2} \tag{2.43}$$

$$\sigma_2 = 0 \tag{2.44}$$

$$\sigma_3 = \frac{\sigma}{2} - \frac{1}{2}\sqrt{\sigma^2 + 4\tau^2} \tag{2.45}$$

Substituting the three principle normal stresses back into Eq. (2.42), we have

$$\left(\frac{\sigma}{\sigma_s}\right)^2 + \left(\frac{3\tau}{\sigma_s}\right)^2 = 1 \tag{2.46}$$

And from Eqs. (2.12) and (2.46), can be further expressed as

$$\left(\frac{\sigma}{\sigma_s}\right)^2 + \left(\frac{\tau}{\tau_s}\right)^2 = 1 \tag{2.47}$$

This relation was validated by experiment when all the stress components are completely reversed and always in the same phase [16]. By introducing a safety factor, the fatigue strength under completely reversed stress can be evaluated by [17]

$$\left(\frac{S_{ca}\sigma_a}{\sigma_{-1e}}\right)^2 + \left(\frac{S_{ca}\tau_a}{\tau_{-1e}}\right)^2 = 1 \tag{2.48}$$

Here we define $\frac{\sigma_{-1e}}{\sigma_a} = S_\sigma$ and $\frac{\tau_{-1e}}{\tau_a} = S_\tau$, which are associated with completely reversed tensile and shear stress, respectively. Therefore, we have

$$\left(\frac{S_{ca}}{S_\sigma}\right)^2 + \left(\frac{S_{ca}}{S_\tau}\right)^2 = 1 \tag{2.49}$$

Thus

$$S_{ca} = \frac{S_\sigma S_\tau}{\sqrt{S_\sigma^2 + S_\tau^2}} \tag{2.50}$$

When a component is subjected to combined fluctuating stresses, or fluctuating multiaxial stress, the safety factors of tensile and shear stress are first calculated by $S_\sigma = \frac{\sigma_{-1}}{k_\sigma \sigma_a + \psi_\sigma \sigma_m}$ and $S_\tau = \frac{\tau_{-1}}{k_\tau \tau_a + \psi_\tau \tau_m}$, separately, and then use Eq. (2.50) to calculate the safety factor under the combined stresses.

2.3.7 Measures to Improve Fatigue Strength

From Eqs. (2.20) and (2.21), it is understood that the endurance limit of a machine element is influenced by the endurance limit of material and combined influence factor K_σ, or to be more precise, the fatigue stress concentration factor k_σ, size factor ε_σ, surface condition factor β_σ and intensify factor β_q. The measures to improve fatigue strength can therefore be proposed accordingly.

1. Reduce stress concentration
   As discussed previously, the discontinuity in elements along with cyclic stresses may promote the initiation and propagation of fatigue cracks. One of the important discontinuities is geometrical discontinuity. Sudden changes in geometry, like shoulders, keyways, holes, sharp grooves and notches and so on, are geometrical discontinuities where stress concentration occurs. Stress concentration is a highly localized effect and most likely leads to fatigue failure. Care must be taken in the design and manufacture of cyclically loaded elements to reduce stress concentration.
2. Select a material with high endurance limits
   Materials are processed by forging, casting, welding, rolling, extrusion, drawing, heat treatments and so on. During material processing, microscopic surface and subsurface defects may arise, possibly due to the inclusion of foreign materials, voids or crystal discontinuities [5], which may reduce the endurance limits of the material. It is therefore important to select proper composition and process to reduce these defects.
3. Improve manufacturing quality to reduce initial microcracks
   Surface scratches, tool marks and burrs, as well as poor joint design, improper assembly and other fabrication faults, are the regions where microcracks are likely to initiate. Therefore, precise surface geometry and superior surface finish are highly beneficial for fatigue strength.
4. Increase surface strength by surface treatments
   Surface treatments, like case hardening, cold rolling, shot peening and so on, introduce compressive residual stresses at critical stress areas. These compressive residual stresses are beneficial for the improvement of fatigue strength.

2.3.8 Examples of Strength Analyses

Example Problem 2.1
An alloy shaft carrying a maximum stress of $\sigma_{max} = 480\,\text{MPa}$ at critical section must operate at least 10^6 cycles at a stress ratio of $r = 0.25$. The material has properties of $\sigma_{-1} = 480\,\text{MPa}$, $\sigma_s = 800\,\text{MPa}$ and $\psi_\sigma = 0.2$. Select the critical number of cycles as $N_0 = 10^7$ and material constant as $m = 9$. Assume the combined influence factor is $K_\sigma = 1.5$. If the allowable safety factor is 1.5, determine whether the element will meet the strength requirement.

Solution 1: Safety factor is solved by calculation

Steps	Computation	Results	Units
1. Calculate stress amplitudes and mean stress	From Eq. (2.3) $\sigma_{min} = r\sigma_{max} = 0.25 \times 480 = 120\text{MPa}$ Therefore $\sigma_m = \dfrac{\sigma_{max} + \sigma_{min}}{2} = \dfrac{480 + 120}{2} = 300\text{MPa}$ $\sigma_a = \dfrac{\sigma_{max} - \sigma_{min}}{2} = \dfrac{480 - 120}{2} = 180 \text{ MPa}$		
2. Calculate fatigue strength safety factor by infinite life	From Eq. (2.28) $S_{ca} = \dfrac{\sigma_{-1}}{K_\sigma \sigma_a + \psi_\sigma \sigma_m} =$ $\dfrac{480}{1.5 \times 180 + 0.2 \times 300} = 1.45 < [S]$	$S_{ca} = 1.45 < [S]$	
3. Calculate fatigue strength safety factor by finite life	From Eq. (2.16) $\sigma_{-1N} = \sigma_{-1} \sqrt[m]{\dfrac{N_0}{N}} = 480 \times \sqrt[9]{\dfrac{10^7}{10^6}} =$ 619.94 MPa $S_N = \dfrac{\sigma_{-1N}}{K_\sigma \sigma_a + \psi_\sigma \sigma_m} =$ $\dfrac{619.94}{1.5 \times 180 + 0.2 \times 300} = 1.88 > [S]$	$S_N = 1.88 > [S]$	
4. Calculate safety factor by the static strength	From Eq. (2.31) $S_s = \dfrac{\sigma_s}{\sigma_m + \sigma_a} = \dfrac{800}{300 + 180} = 1.67 > [S]$	$S_s = 1.67 > [S]$	

Solution 2: Safety factor is solved by using the $\sigma_m - \sigma_a$ diagram

Steps	Computation	Results	Units
1. Draw $\sigma_m - \sigma_a$ diagram for the material, as shown in Figure E2.1	Draw a line AD from $(0, \sigma_{-1})$ with a slope of $\tan\alpha = -\psi_\sigma = -0.2$. Draw a $\sigma_m - \sigma_a$ diagram for the infinite life of material ADC by σ_{-1}, σ_s, and α. Since $\sigma_{-1N} = \sigma_{-1} \sqrt[m]{\dfrac{N_0}{N}} = 480 \times \sqrt[9]{\dfrac{10^7}{10^6}} = 619.94 \text{ MPa}$ Draw a $\sigma_m - \sigma_a$ diagram for the finite life of material $A_N D_N C$, where $A_N D_N$ is parallel to AD from $(0, \sigma_{-1N})$.	$\sigma_{-1N} = 619.94$	MPa

Steps	Computation	Results	Units
2. Calculate σ_{-1e} and σ_{-1Ne}	The endurance limit of the machine element $$\sigma_{-1e} = \frac{\sigma_{-1}}{K_\sigma} = \frac{480}{1.5} = 320\text{MPa}$$	$\sigma_{-1e} = 320$	MPa
	The endurance strength of the machine element for a finite life $$\sigma_{-1Ne} = \frac{\sigma_{-1N}}{K_\sigma} = \frac{619.94}{1.5} = 413.29\text{MPa}$$	$\sigma_{-1Ne} = 413.29$	MPa
3. Draw $\sigma_m - \sigma_a$ diagram for the element, as shown in Figure E2.1	Draw a $\sigma_m - \sigma_a$ diagram for the infinite life of the element $A_e D_e\,C$ through $(0, \sigma_{-1e})$. Draw a $\sigma_m - \sigma_a$ diagram for the finite life of the element $A_{Ne} D_{Ne}\,C$, where $A_{Ne} D_{Ne}$ is parallel to $A_e D_e$ from $(0, \sigma_{-1Ne})$ The slope of these lines is $$\tan \alpha' = \frac{\psi_\sigma}{K_\sigma} = \frac{0.2}{1.5} = 0.133$$		

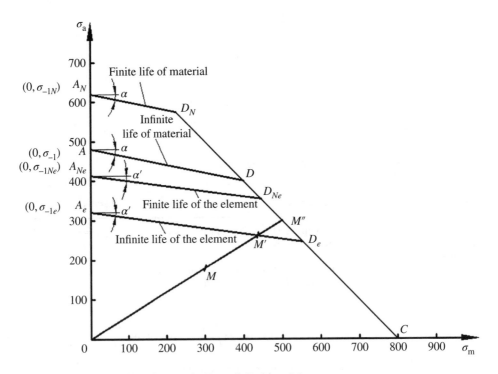

Figure E2.1 Endurance limit diagram for Example Problem 2.1.

Steps	Computation	Results	Units
4. Locate working point and fatigue limit in the $\sigma_m - \sigma_a$ diagram	From Eq. (2.3) $$\sigma_{min} = r\sigma_{max} = 0.25 \times 480 = 120 \text{MPa}$$ Therefore $$\sigma_m = \frac{\sigma_{max} + \sigma_{min}}{2} = \frac{480 + 120}{2} = 300 \text{MPa}$$ $$\sigma_a = \frac{\sigma_{max} - \sigma_{min}}{2} = \frac{480 - 120}{2} = 180 \text{MPa}$$ Locating working point M by (σ_m, σ_a). Connect points O and M and extend the line to intersect with $\sigma_m - \sigma_a$ diagram to obtain M' and M'', which are the endurance limit and yield limit, respectively.		
5. Determine the safety factor	Find the coordinates of points M' (437,262) and M'' (500,300). Safety factor for infinite life $$S = \frac{\sigma'_m + \sigma'_a}{\sigma_m + \sigma_a} = \frac{437 + 262}{300 + 180} = 1.46$$ Safety factor for static strength $$S = \frac{\sigma''_m + \sigma''_a}{\sigma_m + \sigma_a} = \frac{500 + 300}{300 + 180} = 1.67$$	$S = 1.46 < [S]$ $S = 1.67 > [S]$	

Discussion

The results obtained from the $\sigma_m - \sigma_a$ diagram are similar to those from calculation. Using the $\sigma_m - \sigma_a$ diagram, one can easily find out the reason for failure. From the $\sigma_m - \sigma_a$ diagram, the element is prone to yield rather than fatigue; therefore, it is not necessary to analyse fatigue strength.

Example Problem 2.2

An initial design of a shaft is shown in Figure E2.2. The shaft is made of tempered medium carbon steel, with $\sigma_b = 640$ MPa, $\sigma_{-1} = 275$ MPa, $\tau_{-1} = 155$ MPa, $\sigma_0 = 500$ MPa, $\tau_0 = 295$ MPa and rotates at a speed of $n = 20$ rpm. Assume the critical number of cycles is $N_0 = 10^7$ and the material constant of the $S-N$ curve is $m = 9$. The shaft is designed to work 8 hours daily, 300 days yearly for 2.5 years. Operating stresses on the cross section H are: bending stress $\sigma_{bending} = 5.48$ MPa, axial tensile stress $\sigma_{tension} = 0.26$ MPa and torsional shear stress $\tau = 11.0$ MPa. An allowable safety factor is $[S] = 1.6$.

Calculate the endurance strength and check the strength of the shaft at the cross section H.

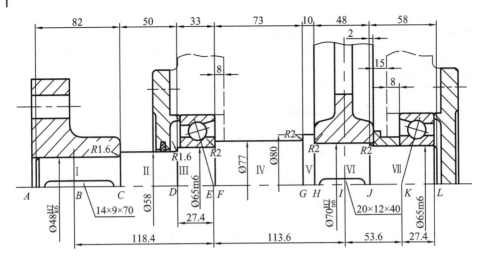

Figure E2.2 An initial design of a shaft.

Solution

Steps	Computation	Results	Units
1. Calculate working stress	The stress at the cross section H: As the shaft rotates, bending stress $\sigma_{bending}$ produced by constant moment M is a completely reversed stress; while axial stress $\sigma_{tension}$ is a static stress. $\sigma_a = \sigma_{bending} = 5.48MPa$ $\sigma_m = \sigma_{tension} = 0.26MPa$ Consider a unidirectional transmission, discontinuous working condition; τ can be regarded as repeated stress. $\tau_a = \tau_m = \dfrac{\tau}{2} = 5.5MPa$	$\sigma_a = 5.48$ $\sigma_m = 0.26$ $\tau_a = \tau_m = 5.5$	MPa MPa MPa
2. Calculate coefficient ψ	$\psi_\sigma = \dfrac{2\sigma_{-1} - \sigma_0}{\sigma_0} = \dfrac{2 \times 275 - 500}{500} = 0.1$ $\psi_\tau = \dfrac{2\tau_{-1} - \tau_0}{\tau_0} = \dfrac{2 \times 155 - 295}{295} = 0.05$	$\psi_\sigma = 0.1$ $\psi_\tau = 0.05$	
3. Calculate endurance strength	The actual number of cycles $N = 8 \times 60n \times 300 \times 2.5 = 7.2 \times 10^6$ since $10^3 < N < N_0 = 10^7$ We need to calculate the strength by finite life. $\sigma_{-1N} = \sigma_{-1}\sqrt[m]{\dfrac{N_0}{N}} = 275 \times \sqrt[9]{\dfrac{10^7}{7.2 \times 10^6}} = 285.2 \text{ MPa}$ $\tau_{-1N} = \tau_{-1}\sqrt[m]{\dfrac{N_0}{N}} = 155 \times \sqrt[9]{\dfrac{10^7}{7.2 \times 10^6}} = 160.7 \text{ MPa}$	$\sigma_{-1N} = 285.2$ $\tau_{-1N} = 160.7$	MPa MPa

Steps	Computation	Results	Units
4. Calculate combined influence factors	From Table 2.2, we can get stress concentration factors by interpolation as $k_\sigma = 2.083, k_\tau = 1.517$ From Figure 2.6, we can obtain the size factor as $\varepsilon_\sigma = 0.78, \varepsilon_\tau = 0.74$ Assuming the shaft is processed by turning, surface quality factors are selected from Table 2.3 as, $\beta_\sigma = \beta_\tau = 0.78$ Select intensity factor $\beta_q = 1.0$. From Eq. (2.20), the combined influence factor for tensile and shear stress are $K_\sigma = \left(\dfrac{k_\sigma}{\varepsilon_\sigma} + \dfrac{1}{\beta_\sigma} - 1\right)\dfrac{1}{\beta_q} = \dfrac{2.083}{0.78} + \dfrac{1}{0.78} - 1 = 2.95$ $K_\tau = \left(\dfrac{k_\tau}{\varepsilon_\tau} + \dfrac{1}{\beta_\tau} - 1\right)\dfrac{1}{\beta_q} = \dfrac{1.517}{0.74} + \dfrac{1}{0.78} - 1 = 2.33$	$K_\sigma = 2.95$ $K_\tau = 2.33$	
5. Calculate safety factors	The fatigue strength calculation needs to consider features such as combined stress state, fluctuating stress and finite life. $S_\sigma = \dfrac{\sigma_{-1N}}{K_\sigma \sigma_a + \psi_\sigma \sigma_m} = \dfrac{285.2}{2.95 \times 5.48 + 0.1 \times 0.26} = 17.61$ $S_\tau = \dfrac{\tau_{-1N}}{K_\tau \tau_a + \psi_\tau \tau_m} = \dfrac{160.7}{2.33 \times \dfrac{11}{2} + 0.05 \times \dfrac{11}{2}} = 12.28$	$S_\sigma = 17.61$ $S_\tau = 12.28$	
6. Calculate safety factor S under combined stress	$S_{ca} = \dfrac{S_\sigma S_\tau}{\sqrt{S_\sigma^2 + S_\tau^2}} = \dfrac{17.61 \times 12.28}{\sqrt{17.61^2 + 12.28^2}} = 10.07 > 1.6$	Safe	

Example Problem 2.3

A rotor shaft has fatigue properties of endurance limit $\sigma_{-1} = 300$ MPa, material constant $m = 9$ and the number of critical cycles $N_0 = 10^7$. In service, the shaft is to be subjected to a spectrum of completely reversed stresses: first 500 MPa for 10^4 cycles, then 400 MPa for 10^5 cycles. After this loading sequence has been imposed, it is desired to change the stress to 350 MPa. Estimate the remaining cycles for the shaft at the final stress level if the duty cycle is to be repeated four times during the life of the shaft.

Solution:

Steps	Computation	Results	Units
1. The number of cycles to failure N_i at each stress level σ_i	From Eq. (2.35) $$N_i = N_0 \left(\frac{\sigma_{-1}}{\sigma_i}\right)^m$$ We have $$N_1 = N_0 \left(\frac{\sigma_{-1}}{\sigma_1}\right)^m = 10^7 \times \left(\frac{300}{500}\right)^9 = 0.01 \times 10^7$$ $$N_2 = N_0 \left(\frac{\sigma_{-1}}{\sigma_2}\right)^m = 10^7 \times \left(\frac{300}{400}\right)^9 = 0.0751 \times 10^7$$ $$N_3 = N_0 \left(\frac{\sigma_{-1}}{\sigma_3}\right)^m = 10^7 \times \left(\frac{300}{350}\right)^9 = 0.2497 \times 10^7$$	$N_1 = 0.01 \times 10^7$ $N_2 = 0.0751 \times 10^7$ $N_1 = 0.2497 \times 10^7$	
2. The number of remaining cycles for 350 MPa	From Miner's rule in Eq. (2.32) $$\frac{n_1}{N_1} + \frac{n_2}{N_2} + \frac{n_3}{N_3} = 1$$ $$4 \times \left(\frac{10^4}{0.01 \times 10^7} + \frac{10^5}{0.0751 \times 10^7} + \frac{n_3}{0.2497 \times 10^7}\right) = 1$$ Therefore $$n_3 = 0.2497 \times 10^7 \times$$ $$\left(\frac{1}{4} - \frac{10^4}{0.01 \times 10^7} - \frac{10^5}{0.0751 \times 10^7}\right) = 4.21 \times 10^4$$	$n_3 = 4.21 \times 10^4$	

2.4 Contact Strength

Previous discussions focus on strength analyses within an element to prevent body failure, such as yielding and fatigue fracture. This section deals with surface strength or contact strength in localized regions, with an aim to prevent surface failure.

2.4.1 Hertzian Contact Stresses

Contact is one of the most common methods of transmitting forces in a machine. When elements make contact with each other, a pair of equal and opposite forces generate according to the action-reaction law. Typical examples are the force transmission between a pair of meshing gears or rolling contact bearings. Theoretically, the contact between curved surfaces of elements is a point or a line. When curved elements are loaded, contact areas deviate elastically from the basic surface curvatures, high contact stresses are correspondingly developed within small contact areas.

Contact stress (also called Hertzian contact stress) refers to the localized stress that develops as two curved surfaces come in contact and deform slightly under imposed

Figure 2.8 Contact stresses on contacting cylinders.

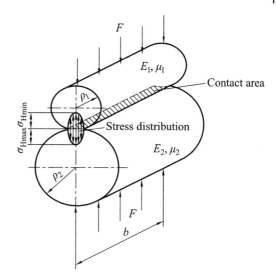

loads. The magnitude of contact stress depends on the material properties, body geometries and applied loads [18]. Generally, small radii lead to small contact areas and, consequently, large contact stress.

Figure 2.8 illustrates the contact area and corresponding contact stress distribution within two cylinders with diameters of ρ_1 and ρ_2, loaded with uniformly distributed force F along cylinder length b. The contact area is a narrow rectangle. The induced contact stress is three-dimensional but may be characterized by stress perpendicular to the plane of contact surface, with parabolic distribution, varying from the periphery of contact surface to the maximum at the centre [1]. Assume the contact is frictionless and the contacting bodies are elastic, isotropic and homogeneous. The maximum contact stress is calculated by the Hertz formula as [14]

$$\sigma_{Hmax} = \sqrt{\frac{F}{b} \cdot \frac{\left(\frac{1}{\rho_1} \pm \frac{1}{\rho_2}\right)}{\pi \left(\frac{1-\mu_1^2}{E_1} + \frac{1-\mu_2^2}{E_2}\right)}} \tag{2.51}$$

where μ, E and ρ are Poisson's ratio, elastic modulus and radius of curvature, respectively. The subscripts 1 and 2 refer to the two cylinders. The negative symbol '−' refers to an internal surface. The Hertz formula is the foundation for the calculation of load carrying capabilities of rolling contact bearings, gears and so on.

2.4.2 Surface Fatigue Failure

In machine elements like rolling contact bearings or mating gear teeth, contact stresses produced by repeated application of loads are cyclic in nature. The cyclic contact stresses, together with material defects, lead to subsurface minute cracks initially. During each cycle of operation, lubricants are forced into the minute cracks under pressure, causing cracks propagate over time and eventually resulting in small bits of material dislodged from the surface, leaving cracks, pits or flaking on the element surfaces, as illustrated in Figure 2.9. This surface damage is called surface fatigue failure.

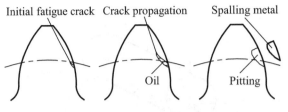

Initial fatigue crack   Crack propagation   Spalling metal

Oil   Pitting

Figure 2.9 Reason for surface fatigue failure.

Surface fatigue failure includes pitting and spalling. Pitting originates with surface cracks and each pit has a relatively small surface area. Spalling originates with subsurface cracks and the spalls are thin 'flakes' of surface material [13]. Both pitting and spalling occur due to insufficient surface strength. It is believed that Hertz stress, the number of cycles, surface finish, hardness, lubrication and temperature all influence surface strength of an element [2]. Consequently, it is advisable to take measures to reduce loads, to increase surface hardness and to improve surface smoothness and lubrication with the ultimate goal of increasing resistance to surface fatigue failure.

References

1 Hindhede, U., Zimmerman, J.R., Hopkins, R.B. et al. (1983). *Machine Design Fundamentals: A Practical Approach*. New York, NY: Wiley.

2 Budynas, R.G. and Nisbett, J.K. (2011). *Shigley's Mechanical Engineering Design*, 9e. New York, USA: McGraw-Hill.

3 Mott, R.L. (2003). *Machine Elements in Mechanical Design*, 4e. Prentice Hall.

4 Wen, B.C. (2015). *Machine Design Handbook*, 5e, vol. 2. Beijing: China Machine Press.

5 Oberg, E. (2012). *Machinery's Handbook*, 29e. New York, NY: Industrial Press.

6 Gere, J.M. and Timoshenko, S.P. (1996). *Mechanics of Materials*, 4e. CL Engineering.

7 Mendelson, A. (1968). *Plasticity: Theory and Application*, 1e. New York, NY: Collier Macmillan Ltd.

8 Parker, A.P. (1981). *Mechanics of Fracture and Fatigue: An Introduction*. London: E&FN Spon.

9 Anderson, T.L. (2004). *Fracture Mechanics: Fundamentals and Applications*, 3e. CRC Press.

10 Zhao, S.B. (2015). *Fatigue Design Handbook*, 2e. Beijing: China Machine Press.

11 Xu, Z.Y. and Qiu, X.H. (1986). *Machine Elements*, 2e. Beijing: Higher Education Press.

12 Pilkey, W.D. (1997). *Peterson's Stress-Concentration Factors*, 2e. New York, NY: Wiley.

13 Juvinall, R.C. and Marshek, K.M. (2011). *Fundamentals of Machine Component Design*, 5e. New York, NY: Wiley.

14 Pu, L.G. and Ji, M.G. (2006). *Mechanical Design*, 8e. Beijing: Higher Education Press.

15 Collins, J.A. (2002). *Mechanical Design of Machine Elements and Machines: A Failure Prevention Perspective*, 1e. New York, NY: Wiley.

16 Du, Q.H. (1963). *Mechanics of Materials*, 2e. Beijing: People's Education Press.

17 Liu, H.W. (1984). *Mechanics of Materials*, 2e. Beijing: People's Education Press.

18 Johnson, K.L. (1987). *Contact Mechanics*. Cambridge, UK: Cambridge University Press.

Problems

Review Questions

1 Distinguish the differences between static loads and variable loads, static stresses and variable stresses. Give an example of variable stresses generated by a static load in an element.

2 Describe fatigue fracture process in a machine element.

3 What is the difference between endurance strength σ_{rN} and endurance limit σ_r?

4 List factors that will affect fatigue strength of a machine element and give the reasons.

5 Describe the linear cumulative damage rule and explain when to use it.

Objective Questions

1 Which of the following is *not* one of factors affecting endurance strength of an element? _____
(a) stress concentration
(b) the size of the element
(c) the magnitude of the load the element carries
(d) materials and their heat treatments

2 Forces F and F_a are static loads. Which of the following in Figure P2.1 represents the stress at point A with a stress ratio of $-1 < r < 1$? _____

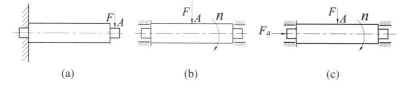

(a)   (b)   (c)

Figure P2.1 Illustration for Objective Question 2.

3 Select the correct statement. _____
(a) Fluctuating stresses are only generated by fluctuating loads.
(b) Static loads cannot generate fluctuating stresses.
(c) Fluctuating stresses are generated by static loads.

(d) Either static loads or fluctuating loads can generate fluctuating stresses.

4 Complete the statement with the correct phrase. _____ of material is chosen for evaluating the fatigue strength of machine elements.
(a) yield strength
(b) endurance limit
(c) ultimate strength
(d) elastic limit

5 Each of the four elements A, B, C, D carries the same maximum stress σ_{max}, with stress ratios of $+1$, 0, -0.5, -1, respectively. Fatigue is most likely to occur on element _____.
(a) A
(b) B
(c) C
(d) D

Calculation Questions

1 An alloy steel element has a mean stress of $\sigma_m = 250$ MPa at its critical section. The stress ratio is $r = 0.25$. Decide the maximum, minimum and stress amplitude of the fluctuating stress. Draw a σ–t curve.

2 The endurance limit of a material is $\sigma_{-1} = 200$ MPa. Assuming the critical number of cycles $N_0 = 5 \times 10^6$, $m = 9$. Estimate endurance strength when N is 60 000, 600 000 and 6 000 000, respectively.

3 A shaft shown in Figure P2.2 carries a maximum working stress $\sigma_{max} = 280$ MPa and minimum stress $\sigma_{min} = -80$ MPa at the shoulder. The mechanical properties of the material are $\sigma_s = 800$ MPa, $\sigma_{-1} = 450$ MPa, $\psi_\sigma = 0.3$. Assuming the combined influence factor is $K_\sigma = 1.62$. If the allowable safety factor is selected as $[S] = 1.3$.
(1) Draw a simplified σ_m–σ_a diagram of the material and the element, and find out the safety factor from the diagram.
(2) Calculate the safety factor of the shaft.
(3) To increase the fatigue strength of the shaft, how should a designer revise the design of the fillet radius?

Figure P2.2 Illustration for Calculation Question 3.

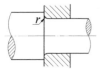

4 A shaft carries a fluctuating load, with the maximum stress $\sigma_{max} = 300$ MPa, the minimum stress $\sigma_{min} = -100$ MPa and constant stress ratio $r = C$. The mechanical properties of the shaft material are $\sigma_s = 800$ MPa, $\sigma_{-1} = 450$ MPa

and $\sigma_0 = 700$ MPa. If the combined influence factor is $K_\sigma = 1.5$ and the allowable safety factor is selected as $[S] = 1.25$, please do the following:

(1) Draw a simplified $\sigma_m - \sigma_a$ diagram of the shaft material;
(2) Find out the safety factor from the diagram;
(3) Calculate the safety factor of the shaft;
(4) Decide the possible failure mode of the shaft;
(5) Propose at least three methods to improve the fatigue strength of the shaft.

5 An element is to be subjected to a spectrum of completely reverse stress. It will be cycled at 500 MPa for 10^4 cycles. Then the stress will be changed to 400 MPa for 10^5 cycles. Finally, the stress will be changed to 350 MPa for 10^5 cycles. The material has endurance limit of $\sigma_{-1} = 300$ MPa, material constant of $m = 9$ and critical number of cycles of $N_0 = 10^7$. Calculate the safety factor.

Part II

Design Applications

3

Detachable Joints and Fastening Methods

Nomenclature

A effective area of contact surface, mm^2

d major diameter, mm

d_0 diameter of bolt shank, mm

d_1 minor diameter, mm

d_2 mean diameter, mm

F external tensile load per bolt, N

F_i load carried by individual bolt in a group, N

F_{max} maximum load in a bolt group, N

F_s shear load acting on each bolt, N

F_Σ total external load, N

f coefficient of friction

h thread height, mm

i the number of contact surface

K_s antiskid factor

k_b stiffness of the bolt

k_m stiffness of the clamped members

L_i distance from y-axis to the bolt axis, mm

L_{min} minimum length of bearing surface, mm

l lead, mm

M overturning moment, N mm

n number of threads, start

p pitch, mm

Q total load in the bolt, N

Q'_p resultant load in the clamped members, N

Q_p preload, N

Analysis and Design of Machine Elements, First Edition. Wei Jiang.
© 2019 John Wiley & Sons Singapore Pte. Ltd. Published 2019 by John Wiley & Sons Singapore Pte. Ltd.
Companion website: www.wiley.com/go/Jiang/analysis_of_machine_elements

r_i	radial distance between centroid and *i*th bolt centre, mm	β	helix angle, °	
		γ	lead angle, °	
S	safety factor	λ	deflection, mm	
S_p	safety factor for crushing	σ	tensile stress, MPa	
T	torque, N mm	σ_{ca}	calculated tensile stress, MPa	
T_1	frictional torque in the thread, N mm	σ_p	bearing stress, MPa	
		σ_s	yield strength, MPa	
T_2	frictional torque at bearing surface, N mm	$[\sigma]$	allowable tensile stress, MPa	
		$[\sigma_p]$	allowable bearing stress, MPa	
W	effective section modulus, mm³	τ	shear stress, MPa	
z	number of bolts in a joint group	$[\tau]$	allowable shear stress, MPa	
α	thread angle, °	φ_v	equivalent frictional angle, °	

3.1 Introduction

3.1.1 Applications, Characteristics and Structures

Machine joints are used to connect two or more components together. They form indispensable parts in machine constructions, as they facilitate manufacturing and assembly, accommodate shipping and handling, permit disassembly for repair, replacement and maintenance. Virtually all machines comprise an assemblage of individual parts, separately manufactured and joined together by various fastening methods [1]. For example, a jumbo jet such as Boeing's 747 uses about 2.5 million fasteners [2].

Machine joints can be broadly classified as detachable and permanent joints. Detachable joints can be disassembled without damaging any element in the connection. Connections using threaded fasteners, keys and splines are typical examples of detachable joints. Permanent joints refer to connections like riveting, welding, bonding (brazing, soldering, adhesive bonding) and so on that cannot be disassembled without damaging elements of the joint.

As joints and connections cause geometrical and material discontinuities, high local stresses and potential failures, the number of joints should be reduced [1–4]. As a matter of fact, joints remain the weakest link in a machine and their high safety concern presents an incessant challenge to designers and engineers. It is therefore necessary to have a thorough understanding of performance and careful analysis of joints under all conditions of service, especially in cars, aeroplanes, steam and gas turbines and so on where mechanical reliability and human safety are vital.

This chapter will discuss in detail the analysis and design of conventional standard threaded fasteners. The detachable fastening methods for shaft and hub, as well as permanent connection methods will be introduced in the following chapters.

3.1.2 Selection of Fastening Methods

The selection of fastening methods depends upon many factors. The first is whether the joint is to be permanent or detachable. The selection of permanent joints is mainly due to the consideration of manufacturing and assembly costs; while the selection of detachable joints considers far more factors, including structure, assembly, transportation, maintenance and so on. Normally, permanent joints are less expensive than detachable joints.

Fastening methods use such devices as bolts, screws, nuts, keys, pins, rivets, welds and adhesives for various applications. Typical application scenarios of the individual devices are another consideration. Usually, threaded fasteners, welds, rivets, adhesive bonds are used to connect plates; threaded fasteners and welds can join rods; keys and splines can connect shafts with hubs; pins, setscrews are chosen for retaining shaft-mounted components; interference fit for mounting rolling contact bearings; couplings and clutches link shafts together [1–4]. In general, the use of threaded fasteners remains the basic and the most widely used fastening method in the design and construction of a machine.

Other considerations include size, thickness, geometries and weight of the components to be joined. And the most important factor is the loading conditions, which greatly influence the strength of joints.

3.2 Screw Threads

3.2.1 Types of Screw Threads

Screw threads are helical ridges formed by cutting or cold forming a groove onto the surface of a cylindrical bar, producing a screw, bolt and stud; or internally in a cylindrical hole, fabricating a nut [5]. Different types of screw threads are classified according to various criteria, such as function, profile, pitch, thread position, directions and starts.

Figure 3.1a is basically the same for both Unified (inch series) and ISO (metric) threads [3], used on screws for fastening. Square, Acme and buttress threads shown from Figure 3.1b–d are power screws. Among them, the square thread has the greatest strength and efficiency, yet is more difficult to fabricate because of the 0° thread angle. Acme and buttress profiles have small thread angles making them easier to manufacture. Acme profile is selected to carry bidirectional loads, while buttress is for unidirectional loading.

For most standard screw threads, at least two pitches are available; that is, coarse series and fine series. Coarse threads are recommended for ordinary applications, especially where rapid assembly or disassembly is required. Fine threads have better capability of resisting loosening from vibrations because of their smaller lead angle. They have a smaller thread depth and larger root diameter that provide higher static tensile strength. Fine threads are used in automobiles, aircraft and other applications that are subject to vibration. Extra-fine threads may be used for more precise adjustments or thin-wall tubing applications.

According to the thread position, external threads are the threads on a screw, while internal threads are the threads on a nut or threaded hole. Threaded fasteners work by assembling the matching external and internal threads together.

There are also left-handed and right-handed threads, which are based on the direction of helix line. Threads are usually made right-handed unless otherwise indicated. That is, if the bolt is turned clockwise, the bolt advances towards the nut.

Screws can be multiple-threaded by having two or more threads cut beside each other. Multiple threads have the advantages of smaller thread height and increased lead for

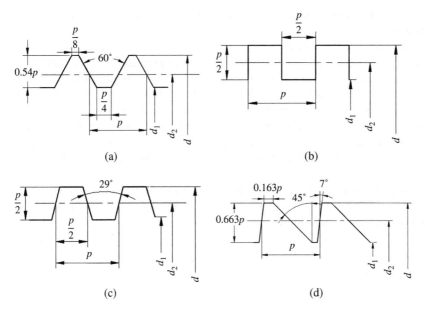

Figure 3.1 Typical types of screw threads. (a) Unified and M profile. (b) Square. (c) Acme. (d) Buttress. Source: Adapted from Juvinall and Marshek, 2011, Figure 10.4, p. 415. Reproduced with permission of John Wiley & Sons, Inc.

fast advancement of nuts. Standardized products such as screws, bolts and nuts all have single threads.

3.2.2 Standards and Terminology

The geometry and terminology of screw threads, illustrated in Figure 3.2, is explained as follows:

1. Major (or nominal) diameter d: the largest diameter of a screw thread.
2. Minor (or root) diameter d_1: the smallest diameter of a screw thread.
3. Pitch diameter (or mean diameter) d_2: the diameter where the width of the thread and groove are equal.
4. Pitch p: the axial distance between corresponding points on adjacent threads. The pitch in Unified Inch standard is the reciprocal of the number of threads per inch.
5. Lead l: axial distance the mating thread (or nut) will advance in one revolution. For a single-threaded screw $l = p$; for a multiple-threaded screw, $l = np$, while n is the number of threads.
6. Thread height h: the radial distance between the major diameter and minor diameter.
7. Thread angle α: the angle between the flanks of adjacent threads measured in an axial plane.
8. Lead angle γ: the angle between the perpendicular of the screw axis and the rise of the thread, $\tan\gamma = l/\pi d_2$.

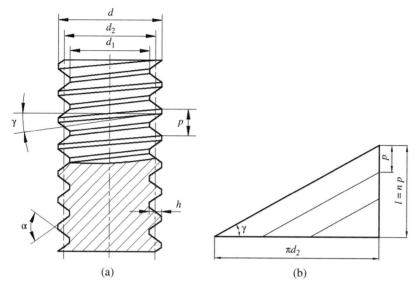

Figure 3.2 Geometry and terminology of external screw thread.

9. Helix angle β: the angle between the screw axis and the rise of the thread.
10. Starts n: the number of threads.

Two identification codes are used to identify screw threads; that is, Unified and ISO. Most fasteners in the US, Canada and the UK use the Unified Inch profile according to the ANSI standard, while Europe and China use the ISO (metric) standard.

Two Unified thread series are in common use, that is, UN and UNR, with the difference that a root radius is used in the UNR series. Unified threads are specified by a short-hand designation in the sequence of nominal major diameter, the number of threads per inch, thread series and tolerance classes. For example, 1/4-28UNF-2A identifies an external thread with a nominal major diameter of 1/4 in., unified fine thread series with 28 threads per in., class 2 fit and right-handed. Detailed dimensions of coarse thread UNC (Unified National Coarse) and fine thread UNF (Unified National Fine) within the UN series can be found in reference [6].

Metric threads are designated by M ('for metric') followed by a major diameter and a pitch in millimetres separated by the symbol '×'. A right-handed thread is assumed unless the designation is followed by -LH. Thus, M24×1.5-LH is a left-handed thread with a nominal major diameter of 24 mm and a pitch of 1.5 mm. Standard sizes for selected metric threads are given in Table 3.1.

Table 3.1 Basic dimensions of selected metric screw threads [3].

Nominal diameter d (mm)	10	12	14	16	18	20	22	24
Pitch p (mm)	1.5	1.75	2	2	2.5	2.5	2.5	3
Minor diameter d_1 (mm)	8.16	9.85	11.6	13.6	14.9	16.9	18.9	20.3

3.3   Threaded Fastening Methods

3.3.1   Types of Threaded Fastening Methods

Threaded fastenings are the basic assembly method in the design and construction of machines. Threaded fasteners join the matching external and internal threads together by the following typical methods.

1. Bolted joints

    A bolted joint is designed to let a bolt to pass through holes in mating members and secured by tightening a nut from the opposite end of the bolt [2]. Both ordinary bolted joints (Figure 3.3a) and precision bolted joints (Figure 3.3b) are available. Ordinary bolted joints have bolt shanks inserted in the holes with a clearance, while precision bolted joints have bolt shanks fitting into reamed holes without appreciable clearance. Both of them are applied to join relative thin members by using thorough holes.

2. Stud joints

    A stud is a stationary bolt attached permanently to one of the members to be joined. The mating member is then placed over the stud and a nut is tightened to clamp the members together, as indicated in Figure 3.3c. A stud joint applies when one of jointed components is too thick to drill a body size hole. The joint can be disassembled frequently by removing the nut and washer, without damaging the threaded member.

3. Cap screw joints

    A cap screw fastening is designed to let a cap screw to insert through a hole in one member to be joined and into a threaded hole in the mating member [2], as shown in Figure 3.3d. The threaded hole may be formed by tapping or by the cap screw itself. It applies when one of jointed components is too thick to drill a body size hole or the space is too small to mount a nut. Such joints cannot be disassembled frequently.

4. Setscrew joints

    A headless setscrew is designed to be inserted into a tapped hole to bear directly on the mating element, as shown in Figure 3.3e. Setscrews depend on compression to develop clamping force to lock components into place. They are inexpensive and adequate for light service. However, they should not be used in vibration applications where loosening would impose a safety hazard.

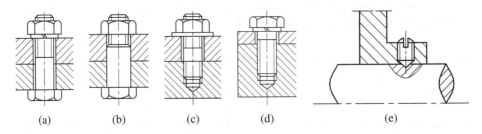

(a)   (b)   (c)   (d)   (e)

Figure 3.3 Types of threaded fastening methods.

Many other types of threaded fastening methods are available in various applications, for example anchor bolts to fix machines on the foundation, or eyebolts to lift machine components and so on.

3.3.2 Threaded Fasteners

Threaded fasteners are standard and commercially available products used to connect two or more elements. The possibility for different combinations of material grades, thread dimensions, tolerance grade and manufacturing methods forms a large variety of threaded fasteners. These combinations are considered in American National Standards, British Standards, Chinese Standards, ISO Standards and so on [6, 7]. The use of standard products offers the advantages of interchangeability and low cost, and guarantees they are indispensable devices throughout the industrialized world.

Threaded fasteners perform the function of locating, clamping, adjusting and transmitting force from one machine element to another [5]. They form a big family, including bolts, studs, machine screws, setscrews, nuts and so on. A bolt is a headed, threaded fastener designed to connect two unthreaded components with the aid of a nut, as shown in Figure 3.3a, b. Bolts have standard thread length and total length. Various standard head styles and thread configurations are readily available. A stud is a headless fastener threaded on both ends and is usually screwed permanently into a tapped hole (Figure 3.3c).

A screw (machine screw or cap screw) is a headed, threaded fastener designed for the assembly of two components, one of which contains its own internal thread as shown in Figure 3.3d. Screws generally have several head (see Figure 3.4) and tip configurations, and are tightened by a screwdriver into tapped holes.

Headless setscrews are designed to bear directly on the mating part by being inserted into tapped holes to prevent relative motion (see Figure 3.3e). Figure 3.5 shows several types of points of setscrew. The setscrew transmits torque by the friction between the point and the mating part or by the resistance of the material in shear.

Different types of nuts and locknuts are available. The purpose of using nuts is to make the threads deflect to distribute the load of bolt more evenly to the nut. The material of nut must be selected carefully to match that of the bolt. Figure 3.3a–c shows common hexagonal nuts. Jam nuts have reduced thickness and usually used together with hexagonal nuts for loosing prevention. Locknuts and lock washers are used for locating the element axially on a shaft, as shown in Figure 10.8b.

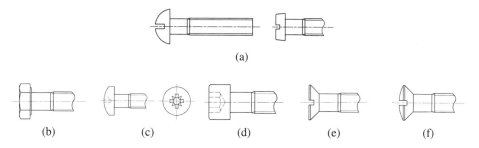

Figure 3.4 Typical cap screw heads. (a) Slotted round-head screw. (b) Hex-head cap screw. (c) Phillips round-head screw. (d) Hexagonal socket-head screw. (e) Slotted flat-head screw. (f) Slotted oval-head screw.

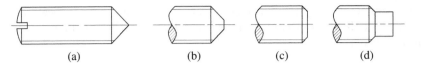

Figure 3.5 Typical points of setscrews: (a) cone point, (b) cup point, (c) flat point and (d) full-dog point.

Washers are used under a bolt head or under both a bolt head and nut. The basic type of washer is a plain flat washer, which distributes clamping forces over a wide area. When loaded, a split helical lock washer will deform axially, generating axial forces on the fastener to prevent loosing, as used in Figure 3.3a, d.

This book only presents typical types of threaded fasteners. Detailed information for almost endless threaded fasteners can be found in various standards, design handbooks [6, 7] and on the Internet.

3.3.3 Tightening Torque and Preloading

The purpose of a bolt fastening is to clamp two or more elements together. For most applications, a bolted joint is tightened during assembly before it starts to carry operating loads. While tightening, the bolt head is usually held stationary and the nut is twisted. The nut moves along the screw and, when resistance is encountered, axial force will generate in the thread [5]. This axial force is called preload, initial tension or pretension. It is this force that clamps two or more members together. The consequence of tightening process is that the bolt is preloaded in tension, while the clamped elements in compression.

The value of preload must be properly controlled to enhance resistance to potential fastener loosening and fatigue failure. Too small a preload may cause leakage, while too high a preload may twist off bolts or screws. The maximum preload is taken to be 75% of the proof load [2, 4], which is the product of the proof strength times the tensile stress area of the bolt or screw.

The required preload can be controlled by tightening torque T, which can be estimated by [8]

$$T = T_1 + T_2 \approx 0.2 Q_p d \tag{3.1}$$

where T_1 and T_2 are frictional torque in the thread and at the bearing surface, respectively. Q_p is the preload and d is the major diameter of the thread.

In practice, tightening torques are controlled or monitored approximately by a built-in dial that indicates the torque in a torque wrench. A more accurate approach to determine the value of preload is by bolt elongation measurement, especially for high reliability design.

3.3.4 Fastener Loosening and Locking

Twisting a nut stretches the bolt and produces clamping force within the connected members. However, the initial tension or preload may be lost gradually due to wear, creep, impact, vibration or corrosion during operation. When the initial preload is lost, the in-service threaded fasteners will loosen, causing separation of the connected members and resulting in malfunctioning of machines. Periodically retightening is a

Figure 3.6 Locking devices.

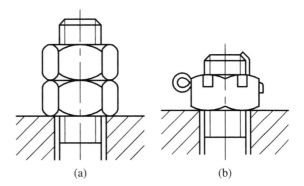

(a)   (b)

convenient and effective method to re-establish a proper preload to prevent excessive loosening.

Several methods are available to restrain a nut from becoming loose on a bolt. The first is by increasing supplementary friction, such as using jam nuts or double nuts in Figure 3.6a, or using a split helical lock washer in Figure 3.3a. The second is by using special devices, as illustrated in Figure 3.6b, by inserting a split cotter pin through the cross-hole drilled in bolt passing through the slotted nut and in Figure 10.8b by locknut and lock washer. The third is to permanently damage screw threads by brazing, soldering, punching or gluing.

3.4 Force Analysis of Multiply Bolted Joints

In common practice, a group of bolts are placed in a specified pattern to form a multiply bolted joint to carry various external loads, as illustrated in Figure 3.7. The external loads are shared among these bolts. Since the materials, dimensions, preloads and so on are usually identical for each bolt in a multiply bolted joint, it is therefore important to identify the bolt that carries the largest load and ensure its safety. This section will analyse force distribution among bolts within a group under typical loading conditions.

3.4.1 Multiply Bolted Joints Subjected to Symmetric Transverse Loads

If a transverse load is applied symmetrically to a multiply bolted joint, as illustrated in Figure 3.8a, it is usually acceptable to assume that the loads are uniformly distributed

Figure 3.7 Loads carried by multiply bolted joints.

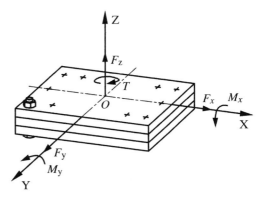

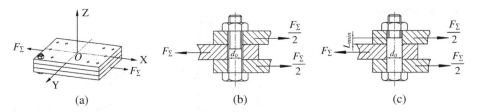

Figure 3.8 Multiply bolted joints subjected to symmetric transverse loads.

among all the bolts. The transverse load can be carried by two kinds of bolted joints, that is, ordinary bolted joints or precision bolted joints.

In an ordinary bolted joint there is a clearance between the hole and bolt, as shown Figure 3.8b. The transverse load is thus carried by friction between the joint interfaces and ensured by the clamping action of the bolt. Assuming each bolt has the same preload, then we have

$$f \cdot Q_p \cdot z \cdot i \geq K_s F_\Sigma$$

Therefore

$$Q_p \geq \frac{K_s F_\Sigma}{fzi} \tag{3.2}$$

where
K_s – antiskid factor to account for reliability, usually select as $K_s = 1.1 - 1.3$;
f – coefficient of friction. For dry cast iron and steel machined surface, select
  $f = 0.1 - 0.2$; For coarse surface without machining, select $f = 0.3 - 0.45$.

In a precision bolted joint, the bolt provides precise alignment of mating members, that is, the diameter of the hole and bolt shank are exactly the same, as shown in Figure 3.8c. The transverse load is thus carried by shearing and bearing of the bolt shank. Assuming each bolt is subjected to an identical load, the load each bolt carries is then

$$F_s = \frac{F_\Sigma}{z} \tag{3.3}$$

3.4.2 Multiply Bolted Joints Subjected to a Torque

When a multiply bolted joint carries a torque in a jointed plane, the connected members tend to rotate around the centroid of bolt group. Both ordinary bolted joints and precision bolted joints can be used in such a situation to carry the torque.

When using ordinary bolted joints, the torque is carried by frictional force between the connected members, as shown in Figure 3.9a. Assuming each bolt carries the same preload, from the equilibrium condition of torque, we have

$$fQ_p r_1 + fQ_p r_2 + \dots + fQ_p r_z \geq K_s T$$

Therefore,

$$Q_p \geq \frac{K_s T}{f \sum_{i=1}^{z} r_i} \tag{3.4}$$

where r_i is radial distance from the centroid to the centre of ith bolt.

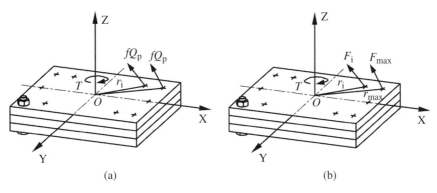

Figure 3.9 Multiply bolted joints subjected to a torque.

When using precision bolted joints, the torque is carried by shearing and bearing of bolt shank, as shown in Figure 3.9b. Assuming each bolt has the same stiffness, the shear load each bolt subjected to is proportional to the distance from the centroid to the centre of each bolt. It thus gives

$$\frac{F_1}{r_1} = \frac{F_2}{r_2} = \ldots = \frac{F_i}{r_i} = \ldots = \frac{F_{max}}{r_{max}}$$

The force taken by each bolt depends upon its radial distance from the centroid; that is, the bolt farthest from the centroid takes the greatest load, while the nearest bolt takes the smallest load. From the equilibrium condition, we have

$$F_1 r_1 + F_2 r_2 + \ldots + F_z r_z = T$$

Combining these two equations, the maximum load a bolt subjected to is

$$F_{max} = \frac{T r_{max}}{\sum\limits_{i=1}^{z} r_i^2} \tag{3.5}$$

3.4.3 Multiply Bolted Joints Subjected to a Symmetric Axial Load

If a total external axial load F_Σ is symmetrically applied to a multiply bolted joint, as illustrated in Figure 3.10, it is usually acceptable to assume that the external load is uniformly

Figure 3.10 Multiply bolted joints subjected to a symmetric tension load.

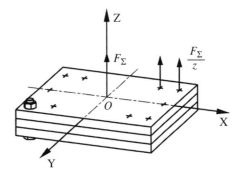

distributed among all the bolts. Therefore, the external load each bolt subjected to is

$$F = \frac{F_{\Sigma}}{z} \tag{3.6}$$

The total load a preloaded bolt carries will be discussed in detail in Section 3.5.3.3.

3.4.4 Multiply Bolted Joints Subjected to an Overturning Moment

Figure 3.11a shows an overturning moment M acting on a multiply bolted joint. Assume the overturning moment M is in the symmetrical plane XOZ, vertical to the jointed plane. Before the bolted joint takes the overturning moment, all the bolts are preloaded and elongated equally and the foundation or the connected member is compressed uniformly. When the overturning moment M is applied to the multiply bolted joint, the bolts on the left side of y-axis will be stretched further and the foundation will be relaxed; while on the right side, the bolts will be pressed, and the foundation will be compressed further. According to the equilibrium condition, the total moment generated by the load in each bolt F_i should be balance with the external moment M. Therefore

$$F_1 L_1 + F_2 L_2 + \ldots + F_z L_z = M$$

The force taken by each bolt depends on the distance from its centre to the y-axis; that is, the bolt farthest from the y-axis takes the largest load, while the nearest bolt takes the smallest load. We can therefore write

$$\frac{F_1}{L_1} = \frac{F_2}{L_2} = \ldots = \frac{F_i}{L_i} = \ldots = \frac{F_{max}}{L_{max}}$$

Solving these equations simultaneously, the external force on the most heavily loaded bolt in the group is

$$F_{max} = \frac{M L_{max}}{\displaystyle\sum_{i=1}^{z} L_i^2} \tag{3.7}$$

For the connected member, the contact surfaces must be free from crushing on the right and free from separation on the left, satisfying the following equations, respectively.

$$\sigma_{pmax} \approx \frac{z Q_p}{A} + \frac{M}{W} \le [\sigma_p] \tag{3.8}$$

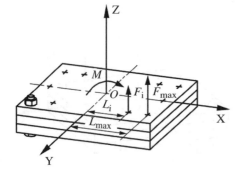

Figure 3.11 Multiply bolted joints subjected to an overturning moment.

and

$$\sigma_{p\min} \approx \frac{zQ_p}{A} - \frac{M}{W} > 0 \tag{3.9}$$

where W is effective section modulus of the contact surface and A is the effective area of the contact surface.

In real engineering practice, the total load a multiply bolted joint is subjected to is usually the combination of the previously discussed cases. The basic approach to obtain the load on each bolt is decomposition and superposition. The complex external load is first decomposed into simple loading cases and the load each bolt carries under each simple loading condition is obtained by previous analyses. The vector sum of these results forms the total load a bolt carries. Since all the bolts within a group are identical, we only need to consider the bolt carries the maximum load. When the bolt that carries the maximum load is identified, the strength of bolt can be determined by the methods to be introduced next.

3.5 Strength Analysis

3.5.1 Potential Failure Modes

According to the previous force analysis, it is found that although multiply bolted joints can carry various external loads, each individual bolt actually carries either tension loads or shear loads. Correspondingly, they are termed tension bolts or shear bolts.

A tension bolt carries external axial loads, preload or the combination of both. Potential failure modes for a tension bolt may be elastic or plastic deformation. Machinery constantly operate dynamically; accordingly, tension bolts usually have a small dynamic load superimposed on a much larger static preload. The fluctuating load will cause fatigue fracture failure in fasteners. A shear bolt is subjected to a transverse shearing load. The potential failure modes may be the crushing or shearing of bolt shanks.

Other failure modes can also be observed depending on operating conditions, for example, fretting fatigue due to small amplitude cyclic relative motions at the interface, corrosion fatigue in corrosive environments, creep and thermal relaxation at elevated temperatures in jet engines and nuclear reactors, wear in movable joints and so on.

Statistically, the distribution of typical bolt failure is about 15% at the fillet under head, 20% at the end of thread and 65% at the first thread engaged in a nut [9, 10]. These are the locations of high stress concentration.

3.5.2 Strength Analysis for Shear Bolts

Shear bolts, usually used in precision bolted joints, are used to carry transverse loads, as illustrated in Figure 3.8c. Since there is no clearance between the bolt shank and the inner surface of hole, the transverse load is carried by bearing between the shank and cylindrical hole and shear at the shank cross section. Assume that the bearing stress is uniformly distributed over the projected contact area of the bolt shank, the crushing strength is then [11]

$$\sigma_p = \frac{F_s}{d_0 L_{\min}} \leq [\sigma_p] \tag{3.10}$$

The shearing strength at the bolt shank cross section is calculated by

$$\tau = \frac{F_s}{\frac{1}{4}\pi d_0^2} \leq [\tau] \tag{3.11}$$

3.5.3 Strength Analysis for Tension Bolts

Tension bolts, usually used in ordinary bolted joints, are assembled with or without pretension. Tension bolts with pretension are capable of carrying not only static and dynamic axial loads but also transverse loads; while tension bolts without pretension carry axial loads only.

3.5.3.1 Tension Bolts Subjected to Axial Loads Only

A tension bolt without tightening is capable of carrying static axial loads only. One of its limited applications is in a hoisting hook. The strength of bolt is calculated by

$$\sigma = \frac{F}{\frac{1}{4}\pi d_1^2} \leq [\sigma] \tag{3.12}$$

3.5.3.2 Preloaded Tension Bolts Subjected to Transverse Loads

The purpose of using bolts is to clamp two or more elements together. The clamping load stretches the bolt by twisting the nut until the bolt elongates approaching the proof strength. If the nut does not loosen, this bolt tension remains as the preload. The tensile stress caused by the initial preload Q_p on the bolt is

$$\sigma = \frac{Q_p}{\frac{1}{4}\pi d_1^2} \tag{3.13}$$

The torsional shear stress due to frictional torque in the thread T_1 generated by tightening can be obtained from [8]

$$\tau = \frac{T_1}{\frac{1}{16}\pi d_1^3} = \frac{Q_p \tan(\gamma + \varphi_v)\frac{d_2}{2}}{\frac{1}{16}\pi d_1^3}$$

Selecting the average value of γ, ϕ_v and d_2 for the commonly used bolts within the range of M10–M64, we have [8]

$$\tau = \frac{Q_p \tan(\gamma + \varphi_v)\frac{d_2}{2}}{\frac{1}{16}\pi d_1^3} = 0.5\frac{Q_p}{\frac{1}{4}\pi d_1^2} = 0.5\sigma$$

Since the bolt is subjected to both tensile and shear stresses, the equivalent stress is calculated by the maximum distortion energy theory as

$$\sigma_{ca} = \sqrt{\sigma^2 + 3\tau^2} \approx 1.3\sigma$$

Therefore, the calculated tensile stress due to preloading is

$$\sigma_{ca} = \frac{1.3Q_p}{\frac{1}{4}\pi d_1^2} \tag{3.14}$$

Although tightening torque generates combined tensile and shear stresses in a bolt, we can simplify the analysis by calculate the tensile stress only, while increasing the value by 30% to account for the effect of torsional shear stress.

When preloaded bolts carry transverse loads, friction between joint members resists the external transverse loads. The bolt is subjected to preload only, not affected by the transverse loads. The magnitude of preload is determined by Eq. (3.2), and the strength for a preloaded tension bolt subjected to a transverse load is thus

$$\sigma_{ca} = \frac{1.3 Q_p}{\frac{1}{4}\pi d_1^2} \le [\sigma] \tag{3.15}$$

3.5.3.3 Preloaded Tension Bolts Subjected to Combined Preload and Static Axial Loads

When an external static axial load is applied to a preloaded tension bolt, the bolt is subjected to a combined preload and static axial loads. As shown in Figure 3.12, before the joint carries an external load, the tightening torque initially applied to the nut or on the head of bolt produce a preload load Q_p in the bolt. The preload Q_p exerts tension on the bolt and stretch the bolt by a deflection λ_b; while the same amount of force Q_p on the clamped members compress the members by a deflection λ_m. The stiffness of a bolt is the ratio of the applied force to the corresponding deflection, expressed as

$$k_b = \tan \theta_b = \frac{Q_p}{\lambda_b} \tag{3.16}$$

Similarly, the stiffness of the connected members is

$$k_m = \tan \theta_m = \frac{Q_p}{\lambda_m} \tag{3.17}$$

When an external static axial load F is applied to a preloaded tension bolt, the bolt will stretch further by $\Delta\lambda$. Thus, the total deflection of the bolt is $\lambda_b + \Delta\lambda$ and the total deflection on the clamped members is $\lambda_m - \Delta\lambda$. The initial preload and the applied axial load add to give resultant forces of Q and Q'_p on the bolt and the clamped members, respectively. Thus, only part of the applied force is carried by the bolt. The amount is dependent on the relative stiffness of the bolt and the clamped members.

Figure 3.12 Force-deflection relationships of bolt and connected members.

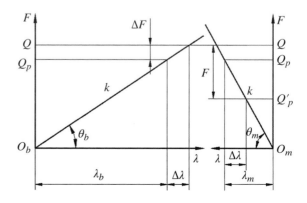

According to the force equilibrium and force-deflection relationships, the forces in the bolt and in the clamped members are calculated by

$$\begin{cases} \Delta F = \Delta \lambda k_b \\ F - \Delta F = \Delta \lambda k_m \end{cases}$$

From these equations, we then have

$$\begin{cases} \Delta F = \dfrac{k_b}{k_b + k_m} F \\ F - \Delta F = \dfrac{k_m}{k_b + k_m} F \end{cases}$$

Therefore, the total load in the bolt is

$$Q = Q_p + \frac{k_b}{k_b + k_m} F \tag{3.18}$$

The total load in the bolt includes the initial preload due to tightening, and a partial of subsequently applied operating force, which tends to separate the clamped members. And the resultant load in the clamped members is

$$Q'_p = Q_p - \frac{k_m}{k_b + k_m} F \tag{3.19}$$

where $\frac{k_b}{k_b + k_m}$ is the relative stiffness of the joint. Since gaskets are usually inserted at the interface between bolted members, the relative stiffness depends on the material of gaskets. For a metal gasket, it is approximately 0.2–0.3; for leather 0.7; for asbestos, 0.8 and for rubber, 0.9.

The total load in the bolt depends greatly on the relative stiffness. If $k_b \gg k_m$, we have $Q \approx Q_p + F$; while for $k_b \ll k_m$, we have $Q \approx Q_p$. Therefore, when a bolted joint carries a relatively large load, high stiffness gaskets should be used.

To prevent the separation of joint, the resultant load in the clamped members Q'_p should be greater than zero, that is,

$$Q'_p \geq 0 \tag{3.20}$$

For ordinary bolted joints subjected to a stable load, select $Q'_p = (0.2\text{–}0.6) \, F$; for ordinary bolted joints subjected to a variable load, $Q'_p = (0.6\text{–}1.0) \, F$; for a sealing case, $Q'_p = (1.5\text{–}1.8) \, F$ and for anchor bolt joints, $Q'_p \geq F$ [8].

The static strength of a preloaded bolt subjected to a combined preload and static axial loads is then

$$\sigma_{ca} = \frac{1.3Q}{\frac{1}{4}\pi d_1^2} \leq [\sigma] \tag{3.21}$$

3.5.3.4 Preloaded Tension Bolts Subjected to Combined Preload and Variable Axial Loads

When a preloaded tension bolt is subjected to a variable load fluctuating between zero and an upper extreme F, the total load in a bolt varies from Q_p to Q, as indicated in

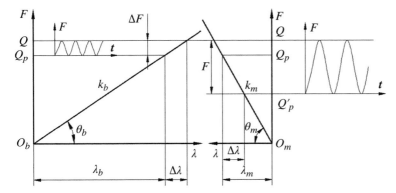

Figure 3.13 Preloaded tension bolt subjected to combined preload and variable axial load.

Figure 3.13. The alternating stress experienced by the bolt is then

$$\sigma_{max} = \frac{Q}{\frac{1}{4}\pi d_1^2} \quad \text{and} \quad \sigma_{min} = \frac{Q_p}{\frac{1}{4}\pi d_1^2} \tag{3.22}$$

Therefore, the minimum value of stress keeps constant and the amplitude of stress is calculated by

$$\sigma = \frac{\sigma_{max} - \sigma_{min}}{2} = \frac{k_b}{k_b + k_m} \cdot \frac{2F}{\pi d_1^2} \tag{3.23}$$

From the endurance limit diagram and methods introduced in Chapter 2, the fatigue safety factor can be obtained by duplicating Eq. (2.30) as

$$S_{ca} = \frac{2\sigma_{-1} + (K_\sigma - \psi_\sigma)\sigma_{min}}{(K_\sigma + \psi_\sigma)(2\sigma_a + \sigma_{min})} \geq S \tag{2.30}$$

If the design is not satisfactory, additional bolts and/or a different sized bolt may be called for.

3.5.4 Measures to Improve Fatigue Strength of Bolted Joints

The fatigue strength of a multiply bolted joint depends on the strength of each individual bolt in the group. Many factors affect bolt fatigue strength, including the stress amplitude, the distribution of load among thread teeth, stress concentration, mechanical properties and processing methods. The following will analyse typical factors affecting fatigue strength and propose measures for improvement.

1. Reduce cyclic stress amplitudes on the bolt
   When the minimum stress keeps constant, stress amplitude greatly affects fatigue strength [12]. For a preloaded bolt, when the external operating load varies between 0 and F, the total load acting on the bolt varies from Q_p to Q, as illustrated in Figure 3.13. The load variation in the bolt will reduce if the stiffness of the bolt is reduced or the stiffness of the clamped members is increased, as can be noticed from Eq. (3.18).

However, reducing the stiffness of the bolt or increasing the stiffness of the clamped members under a constant preload will inevitably reduce the resultant load in the clamped members and consequently weaken the sealing effect. Therefore, it is preferable to increase the preload simultaneously so that the resultant load in the clamped members does not change too much. While increasing the preload, the preload must be controlled to prevent twisting off the bolt while tightening.

2. Reduce stress concentration

    Stress concentration is an important factor affecting fatigue strength. Stress concentration often happens where geometrical discontinuity appears. For a bolt, the screw end, the connection of bolt head and shank and the area where the cross-section changes are the places stresses concentrate. Proper modification of standard bolts, like increased fillet radius under the head, can effectively reduce stress concentration.

3. Adopt proper manufacturing methods

    Rolled threads are preferred compared with cut or grounded threads, as rolling processes cause work hardening of materials and generate a favourable grain structure and compressive residual stresses, which benefit the improvement of fatigue strength. Furthermore, proper heat treatment like nitriding and shot peening will also increase fatigue strength of bolts.

3.6 Design of Bolted Joints

3.6.1 Introduction

More often than not, multiply bolted joints involve several bolts placed in a specified pattern to improve strength and stability of a connection. The design of multiply bolted joints consists of two tasks: one is to decide the number of bolts in a pattern and their layout; the other is to specify the dimension and material of the fasteners. The former requires structural design, while the latter needs force and strength analysis.

3.6.2 Materials and Allowable Stresses

A wide variety of materials are used for threaded fasteners. The selection of materials for threaded fasteners is normally based on the requirements of strength, weight, corrosion resistance, magnetic properties, life expectancy and costs. The most widely used materials are carbon steels and alloy steels. Stainless steels and nickel-based superalloys, such as Inconel and Hastelloy, are used for fasteners working in corrosive or high temperature environments. Aluminium, bronze and brass threaded fasteners are used for applications where corrosion resistance and good thermal and electrical conductivity are required. Nylon and plastics are both suitable for applications with more economical considerations than strength requirements.

The strength of steels for bolts and screws is used to determine property classes. Three strength ratings are involved; namely, tensile strength, yield strength and proof strength. The proof strength is defined as the stress at which the bolt or the screw would undergo permanent deformation, usually 90–95% of the yield strength [4]. The value depends on the material, heat treatment and other factors.

Recommended materials for property class 4.6–5.8 are low or medium carbon steels; for property class 8.8–9.8, low carbon alloy steel and medium carbon steel, heat treated

Table 3.2 Mechanical properties of fasteners – bolts, screws and studs.

Property class	Size range, inclusive	Minimum proof strength, MPa	Minimum tensile strength, MPa	Minimum Yield strength, MPa	Material	Head Marking
4.6	M5–M36	225	400	240	Low or medium carbon	4.6
4.8	M1.6–M16	310	420	340	Low or medium carbon	4.8
5.8	M5–M24	380	520	420	Low or medium carbon	5.8
8.8	M16–M36	600	830	660	Medium carbon, Q&T	8.8
9.8	M1.6–M16	650	900	720	Medium carbon, Q&T	9.8
10.9	M5–M36	830	1040	940	Low-carbon martensite, Q&T	10.9
12.9	M1.6–M36	970	1220	1100	Alloy, Q&T	12.9

Source: Budynas and Nisbett, 2011, Table 8.11, p. 435. Reproduced with permission of McGraw-Hill.

by quenching and tempering; for property class 10.9, low and medium carbon alloy steels by quenching and tempering and for property class 12.9, alloy steel by quenching and tempering [2, 7].

Table 3.2 details mechanical properties of different property classes for metric fasteners. Property classes are identified by a numerical code system ranging from 4.6 to 12.9 to represent the strength. The numeral before the decimal point is the tensile strength divided by 100 ($\sigma_b/100$), while the numeral following the decimal point is 10 times the ratio of yield strength to tensile strength ($10\sigma_s/\sigma_b$). For example, property class

4.6 implies that the tensile strength is 400 MPa, and the ratio of yield strength to tensile strength is 0.6. Nuts are graded to mate with their corresponding class bolts.

The American Society for Testing Materials (ASTM), the Society of Automotive Engineers (SAE) and the Standardization Administration of China (SAC) have established similar standard specifications for materials and strength levels for threaded fasteners. Designers can refer to relevant standards or design handbooks while designing [6, 7, 13].

The allowable tensile stress is decided by

$$[\sigma] = \frac{\sigma_s}{S} \tag{3.30}$$

where yield strength σ_s is selected from Table 3.2, and safety factor S is selected as 4–1.3 for M6–M60 carbon steel bolts and 5–2.5 for M6–M60 alloy steel bolts, respectively. When preloaded bolted connections carry variable loads, safety factors may be twice the values of those for static loading [7].

The allowable bearing stress is calculated by

$$[\sigma_p] = \frac{\sigma_s}{S_p} \tag{3.31}$$

where the safety factor for crushing S_p is 1.25 for steel and 2.5 for cast iron [8].

The allowable shear stress is calculated by [8]

$$[\tau] = \frac{\sigma_s}{2.5} \tag{3.32}$$

3.6.3 Design Criteria

For a tension bolt that is subjected to a preload, or a static axial load or a combination of both, it must meet the strength requirement by $\sigma \leq [\sigma]$, as expressed in Eqs. (3.12, 3.15, 3.21), depending on the loading conditions. When a tension bolt carries a fluctuating load, fatigue strength $S \leq [S]$ should also be guaranteed, as expressed by Eq. (2.30). Besides, to ensure safe and proper functioning of bolted joints, the preload induced by bolt tightening must be great enough to prevent joint separation during operation.

The design criterion for a shear bolt is to guarantee crushing strength $\sigma_p \leq [\sigma_p]$ expressed by Eq. (3.10), and shear strength $\tau \leq [\tau]$ by Eq. (3.11).

3.6.4 Design Procedure and Guidelines

Although multiply bolted joints usually work together in practical engineering, the failure of a single fastener in a group can be destructive or even catastrophic. Designers must select and decide the type, material, property class and size of standard fasteners to ensure all of them will most adequately suit the application. The following provides the procedure and guidelines for bolted joint design:

1. Decide the pattern of bolt layout and the number of bolts in the group. Ideally, it is better to ensure each bolt within the group is uniformly loaded, or to minimize the maximum load a bolt carries.
2. Analyse operating loads on the bolted joint. Dissolve the loads into simple load conditions; that is, tensile loads, shear loads, torques and moments.

3. Determine the load each bolt subjected to according to the bolted joint type (ordinary or precision bolted joints), assembly condition (with preload or without preload) and loading condition (transverse or axial loading, static or dynamic loading). The basic approach is first to determine the forces that act on each bolt by each applied load separately. Then, superpose the forces vectorially to identify the bolt that carries the greatest load in the group.
4. Determine the critical section dimension, usually the minor diameter d_1, according to the design criteria.
5. Decide the size of bolts, nuts, washers and so on according to design handbooks and manufacturers' catalogues.

3.6.5 Structural Design

Structural design determines bolt layout pattern and bolt number, following these guidelines:

1. The bolt layout is usually designed as a simple, symmetric geometry, such as a square, rectangle or circle.
2. Select an even number of bolts in a group to facilitate manufacture and to ensure equal loading. Each bolt should have the same size (both diameter and length), thread series (coarse or fine), material and property class.
3. Ensure rational arrangement of bolts. Sufficient room should be provided for a wrench between adjacent bolts or between the bolt and the edge of jointed members, as indicated in Figure 3.14a.
4. The coarse surface to be joined by a bolt should be machined, as shown in Figure 3.14b, to avoid additional induced bending stress during operation.

3.6.6 Design Cases

Example Problem 3.1
A hydro-cylinder with an inner diameter of $D = 160$ mm is shown in Figure E3.1. Oil pressure is $p = 1\,\mathrm{N\,mm^{-2}}$. The cylindrical barrel and the flange are connected by eight uniformly distributed bolts. The relative stiffness of the bolted connection is $k_b/(k_b + k_m) = 0.3$. The applied preload is $Q_p = 2.5F$ (F is the operating force) and the allowable stress of bolt material is $[\sigma] = 120\,\mathrm{N\,mm^{-2}}$.

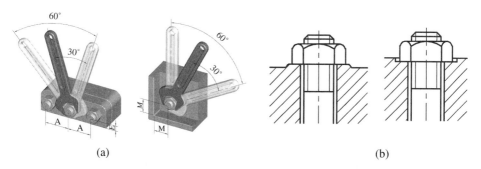

(a)   (b)

Figure 3.14 Examples of structural design considerations.

Decide:

(1) The total tensile load carried by each bolt;

(2) The resultant load on the clamped members;

(3) The minor diameter required for the bolts.

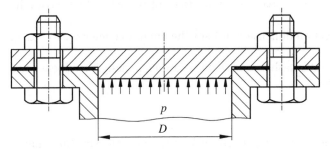

Figure E3.1 Illustration for Example Problem 3.1.

Solution:

Steps	Computation	Results	Units
1. The total tensile load carried by each bolt	The external load is shared by each bolt, therefore $$F = \frac{\pi D^2 p}{4z} = \frac{\pi \times 160^2 \times 1}{4 \times 8} = 2513N$$ The total tensile load carried by each bolt: $$Q = Q_p + \frac{k_b}{k_b + k_m}F = 2.5F + 0.3F = 2.8F =$$ $2.8 \times 2513 = 7036.4N$	$Q = 7036.4$	N
2. The resultant load on the clamped members	$$Q_p' = Q_p - \left(1 - \frac{k_b}{k_b + k_m}\right)F =$$ $Q_p - 0.7F = 1.8F = 1.8 \times 2513 = 4523.4N$	$Q'_p = 4523.4$	N
3. The minor diameter required for the bolt	For the bolt to have enough strength, it must satisfy $$\sigma = \frac{1.3Q}{\frac{\pi}{4}d_1^2} \le [\sigma]$$ Therefore, $$d_1 \ge \sqrt{\frac{4 \times 1.3Q}{\pi[\sigma]}} = \sqrt{\frac{4 \times 1.3 \times 7036.4}{3.14 \times 120}}$$ $= 9.85mm$	$d_1 = 9.85$	mm

Example Problem 3.2

A sliding bearing (see Chapter 12) is bolted to a smooth floor. The bearing is supposed to carry a load of $F = 5000\,N$ applied at the middle of the bearing width. The angle between the force and the horizontal line is $\alpha = 45°$. Detailed dimensions are shown in

Figure E3.2. The coefficient of friction between the bearing bottom surface and the floor is 0.16. The allowable bearing stress between the bearing and floor is $[\sigma_p] = 125$ MPa. Select proper bolts for the connection.

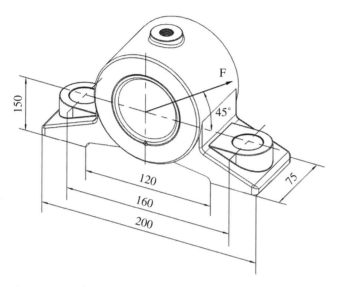

Figure E3.2 A sliding bearing bolted to a smooth floor by a group of two bolts.

Solution:

Steps	Computation	Results	Units
1. Structural design	Propose the number of bolts and pattern. According to the structure, two identical bolts are used and a tightening torque is also applied.		
2. Force analysis	(1) Determine the forces on the bolted joint. Resolve load F into vertical and horizontal components, which are the axial and shear force on the joints, respectively.		
	Axial force $$F_{\Sigma v} = F\sin 45^\circ = 5000 \times \sin 45^\circ = 3536\text{N}$$	$F_{\Sigma v} = 3536$	N
	Shear force $$F_{\Sigma h} = F\cos 45^\circ = 5000 \times \cos 45^\circ = 3536\text{N}$$	$F_{\Sigma h} = 3536$	N
	Overturning moment to be resisted by the bolted joints $$M = F_{\Sigma h} \times 150 = 530400\text{N}\cdot\text{mm}$$	$M = 530\,400$	N·mm
	(2) Determine the force on each individual bolt under the axial force $$F_a = \frac{F_{\Sigma v}}{z} = \frac{3536}{2}N = 1768N$$	$F_a = 1768$	N

Steps	Computation	Results	Units
	(3) Compute the force on the bolts to resist the bending moment from		
	$$F_{max} = \frac{ML_{max}}{\sum\limits_{i=1}^{z} L_i^2} = \frac{530400 \times 80}{80^2 + 80^2} N = 3315N$$		
		$F = 5083$	N
	The total axial working force on the left bolt		
	$$F = F_a + F_{max} = 1768 + 3315 = 5083N$$		
	(4) The connection has tendency of sliding relative to each other under the load $F_{\Sigma h}$. To prevent such situation, the following condition should be met.		
	$$f\left(zQ_p - \frac{k_m}{k_b + k_m}F_{\Sigma v}\right) \geq K_s F_{\Sigma h}$$		
	Select $\dfrac{k_b}{k_b + k_m} = 0.2$, so		
	$\dfrac{k_m}{k_b + k_m} = 1 - \dfrac{k_b}{k_b + k_m} = 0.8$, select antiskid factor		
	$K_s = 1.2$, The preload of each bolt is	$Q_p = 14\,674$	N
	$$Q_p \geq \frac{1}{z}\left(\frac{K_s F_{\Sigma h}}{f} + \frac{k_m}{k_b + k_m}F_{\Sigma v}\right) =$$		
	$$\frac{1}{2} \times \left(\frac{1.2 \times 3536}{0.16} + 0.8 \times 3536\right)N = 14674N$$		
	(5) The total force the left bolt subjected to is		
	$$Q = Q_p + \frac{k_b}{k_b + k_m}F = 14674 + 0.2 \times 5083 =$$		
	$15691N$	$Q = 15\,691$	N
3. Determine the diameter of the bolts	Select the bolt property class as 4.6. From Table 3.2, the yield strength of material is $\sigma_s = 240\,MPa$.		
	Assuming the safety factor is $S = 1.5$. The allowable stress of bolt material is		
	$$[\sigma] = \frac{\sigma_s}{S} = \frac{240}{1.5} = 160MPa$$		
	The required minor diameter of the bolt will be		
		M16	mm
	$$d_1 \geq \sqrt{\frac{4 \times 1.3Q}{\pi[\sigma]}} = \sqrt{\frac{4 \times 1.3 \times 15691}{3.14 \times 160}} = 12.7mm$$		
	From Table 3.1, select major diameter $d = 16\,mm$ ($d_1 = 13.6 > 12.7\,mm$).		

Steps	Computation	Results	Units
4. Check the connection	(1) To prevent crushing failure of the right bolt $$\sigma_{pmax} = \frac{1}{A}\left(zQ_p - \frac{k_m}{k_b + k_m}F_{\Sigma v}\right) + \frac{M}{W}$$ $$= \frac{1}{75 \times (200 - 120)} \times (2 \times 14674 - 0.8 \times 3536)$$ $$+ \frac{530400}{\frac{75}{12 \times \frac{200}{2}} \times (200^3 - 120^3)}$$ $$= 5.77MPa < [\sigma_p]$$ (2) To prevent leakage of the left bolt $\sigma_{pmin} =$ $$\frac{1}{A}\left(zQ_p - \frac{k_m}{k_b + k_m}F_{\Sigma v}\right) - \frac{M}{W} = 3.07MPa > 0$$	$\sigma_{pmax} < [\sigma_p]$ $\sigma_{pmin} > 0$	

References

1 Collins, J.A. (2002). *Mechanical Design of Machine Elements and Machines: A Failure Prevention Perspective*, 1e. New York, NY: Wiley.

2 Budynas, R.G. and Nisbett, J.K. (2011). *Shigley's Mechanical Engineering Design*, 9e. New York, NY: McGraw-Hill.

3 Juvinall, R.C. and Marshek, K.M. (2011). *Fundamentals of Machine Component Design*, 5e. New York, NY: Wiley.

4 Mott, R.L. (2003). *Machine Elements in Mechanical Design*, 4e. Prentice Hall.

5 Hindhede, U., Zimmerman, J.R., Hopkins, R.B. et al. (1983). *Machine Design Fundamentals: A Practical Approach.* New York, NY: Wiley.

6 Oberg, E. (2012). *Machinery's Handbook*, 29e. New York, NY: Industrial Press.

7 Wen, B.C. (2015). *Machine Design Handbook*, 5e, vol. 2. Beijing: China Machine Press.

8 Pu, L.G. and Ji, M.G. (2006). *Mechanical Design*, 8e. Beijing: Higher Education Press.

9 Xu, Z.Y. and Qiu, X.H. (1986). *Machine Elements*, 2e. Beijing: Higher Education Press.

10 Pilkey, W.D. (1997). *Peterson's Stress-Concentration Factors*, 2e. New York, NY: Wiley.

11 Gere, J.M. and Timoshenko, S.P. (1996). *Mechanics of Materials*, 4e. CL Engineering.

12 Parker, A.P. (1981). *Mechanics of Fracture and Fatigue: An Introduction.* London: E&FN Spon.

13 Mechanical properties of fasteners – Bolts, screws and studs. GB/T 3098.1-2010, Standardization Administration of the People's Republic of China, Beijing, 2011.

Problems

Review Questions

1 Why is preloading important for a bolted connection? How should one control the applied preload?

2 Why is loosening prevention important for a bolted connection? Give examples to describe measures to resist loosening.

3 A rigid coupling uses property class 5.8 ordinary bolted joints to transmit torque T_1. Currently, the coupling is required to transmit a higher torque T_2, but the number of bolts and their diameters cannot be changed. Propose three methods for the coupling to transmit a higher torque T_2.

4 Please discuss the differences and relations between the load carried by a bolt and the load carried by a multiply bolted joint.

5 How could one improve the static strength and fatigue strength of a bolted connection?

Objective Questions

1 Two components are to be connected by bolts. One of components is relatively thick and the connection is disassembled frequently. Which of the following is the best choice_____?

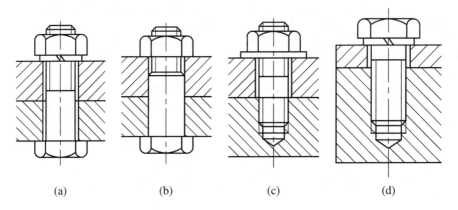

(a)   (b)   (c)   (d)

Figure P3.1 Illustration for Objective Question 1.

2   A preloaded ordinary bolted joint carries axial loads. The preload is Q_p, the external load is F. The total load the bolt carries is _____.
   (a) $= Q_p + F$
   (b) $< Q_p + F$
   (c) $> Q_p + F$
   (d) $= Q_p + F/2$

3   A preloaded tension bolt is subjected to a variable axial load. Which of the following measures can improve the fatigue strength of the bolt? _____.
   (a) Increasing the stiffness of the bolt k_b and reducing the stiffness of the connected member k_m.
   (b) Reducing the stiffness of the bolt k_b and increasing the stiffness of the connected member k_m.
   (c) Increasing the stiffness of the bolt k_b and the stiffness of the connected member k_m.
   (d) Reducing the stiffness of the bolt k_b and the stiffness of the connected member k_m.

4   The dimension of the mating nut is selected according to which parameter of the bolt? _____
   (a) d
   (b) d_2
   (c) d_1
   (d) d/p

5   When multiply bolted joints carry a transverse load or a torque, the load the bolt is subjected to _____.

Figure P3.2 Illustration for Objective Question 4.

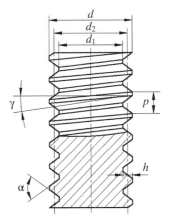

(a) must be a shear load
(b) must be a tensile load
(c) could be either a shear load or a tensile load
(d) both a shear load and a tensile load

Calculation Questions

1 A preloaded bolt carries a fluctuating axial load varying from 0 to F. Assuming the criterial area of the bolt is A_c, the stiffness of the bolt and the connected member are k_b and k_m, respectively. Decide on the stress amplitude.

2 A flange is attached to a pressurized cylinder by 16 identical bolts. The fluid pressure is cycled from 0 to 1.2 MPa. Let $D = 500$ mm, the resultant load on the clamped members must be kept as $Q'_p = 1.5F$. The allowable stress of bolt material is $[\sigma] = 160$ MPa. Determine the size of the bolts.

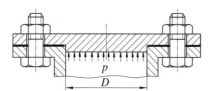

Figure P3.3 Illustration for Calculation Question 2.

3 A flange and pressure vessel is connected by 12 M16 ($d_1 = 13.6$ mm) bolts. The inner diameter of the pressure vessel is $D = 250$ mm. The allowable stress of the bolt is $[\sigma] = 120$ N mm^{-2}. The resultant load on the connected member is $Q'_p = 1.5F$ (F is the operating load). Decide on the maximum pressure the bolted joint can carry.

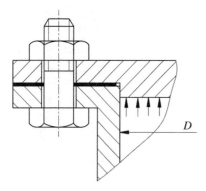

Figure P3.4 Illustration for Calculation Question 3.

4 Two half couplings are connected by eight M16 bolts distributed around the bolt circle with a diameter of $D = 130$ mm. The total length of the bolt is 70 mm, including the screw length of 28 mm. The shank diameter is $d_0 = 17$ mm. The detailed dimensions of structure are shown in Figure P3.5, where $d_1 = 55$ mm, $L_1 = L_2 = 100$ mm and $b_1 = b_2 = 25$ mm. The two half couplings are made of grey iron HT200, with the allowable bearing stress as $[\sigma_{p1}] = 100$ MPa. The bolts are made of low carbon steel Q235, with the allowable shear stress $[\tau] = 90$ MPa, allowable bearing stress $[\sigma_{p2}] = 300$ MPa and allowable tensile stress $[\sigma] = 150$ MPa.
  1. Decide the maximum torque that can be transmitted by the coupling;
  2. Assuming the coefficient of friction is $f = 0.15$ and antiskid factor $k_s = 1.2$, if ordinary bolted joints are selected to transmit the same amount of torque, decide on the required minor bolt diameter.

Figure P3.5 Illustration for Calculation Question 4.

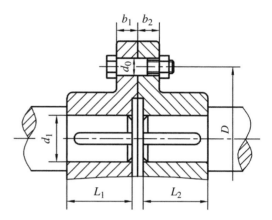

Design Problems

1 Compare the three design layouts in Figure P3.6 using precision bolted joints, where $L = 200$ mm and $a = 50$ mm.

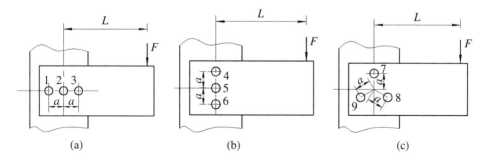

(a)   (b)   (c)

Figure P3.6 Illustration for Design Problem 1.

Structure Design Problems

1 Correct the mistakes in Figure P3.7.

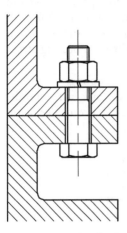

Figure P3.7 Illustration for Structure Design Problem 1.

2 Figure P3.8 shows a cap screw joint. Correct the mistakes in the figure by drawing a new one.

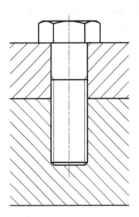

Figure P3.8 Illustration for Structure Design Problem 2.

3 Draw a setscrew in Figure P3.9 to connect the hub with the shaft.

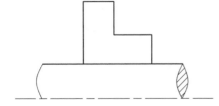

Figure P3.9 Illustration for Structure Design Problem 3.

CAD (Computer-Aided Design) Problems

1   Write a flow chart for a bolt design process.

2   Design an interface similar to Figure P3.10, and complete the Example Problem 3.2.

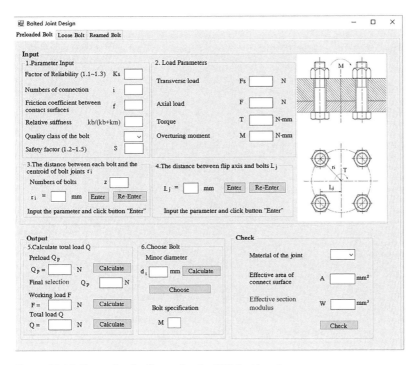

Figure P3.10 Illustration for Illustration for CAD Problem 2.

4

Detachable Fastenings for Shaft and Hub

Nomenclature

b key width, mm
d shaft diameter, mm
F force acting on a key, N
h key height, mm
L length of a key, mm

l working length of a key, mm
T torque, N mm
σ_p bearing stress, MPa
$[\sigma_p]$ allowable bearing stress, MPa
τ shear stress, MPa
$[\tau]$ allowable shear stress, MPa

4.1 Keys

4.1.1 Applications, Characteristics and Structure

When power is to be transmitted to, or supplied from a rotating shaft, it is necessary to attach power transmission elements, such as pulleys, sprockets or gears, to the shaft. The relative rotation between the shaft and the attached elements are prohibited. Detachable retention devices, such as keys, splines and pins, are commonly used to secure the connection between the hub of power transmission elements and the shaft.

Analysis and Design of Machine Elements, First Edition. Wei Jiang.
© 2019 John Wiley & Sons Singapore Pte. Ltd. Published 2019 by John Wiley & Sons Singapore Pte. Ltd.
Companion website: www.wiley.com/go/Jiang/analysis_of_machine_elements

Keys are elements enabling transmission of torque and power between power transmitting elements and shafts. A key is first installed into longitudinal groove cut into a shaft, called a keyseat. The hub of a power transmitting element with a similar groove, usually called a keyway or a keyseat, is then slid over the key. The key seats in the groove of the shaft and the hub so that exactly half the height of the key bears on the shaft and the other half bears on the hub [1]. Keys are demountable, facilitating assembly and disassembly of a shaft system.

4.1.2 Types of Keys

Several types of keys are available, including parallel keys, Woodruff keys, taper keys and gib-head keys. Parallel keys are named because the top, bottom and the sides of key are parallel. The square key, as shown in Figure 4.1, is the most common type. It is also available in rectangular size, called a flat key, which is usually used for either large shafts or small shafts where a shorter height can be tolerated.

Parallel keys include three types, as shown in Figure 4.1b. Type A has rounded ends, which is installed in a profiled keyseat fabricated by an end mill with a diameter equal to the width of key. The resulting keyseat is flat-bottomed with a sharp, square corner at its end. Type B has square ends, which is installed in a sled runner keyseat produced by a circular milling cutter having a width equal to the width of key. As the cutter begins or ends the keyseat, it produces a smooth radius, with less stress concentration than a profiled keyseat [1]. Type C has a square and a rounded end and is often used to connect shaft and hub at the end of shaft [2]. Parallel keys are used for transmitting unidirectional torques where periodic withdrawal of hub members may be required [3]. Keyed components generally have a slip fit onto the shaft to facilitate assembly and disassembly.

The guided key and feather key are used when the connected hub components are required to slide freely along the shaft. While using a guided key in Figure 4.2a, it is common practice to screw the guided key 2 to the shaft 4 to prevent backlash between the key and keyseat. When the sliding distance is large, feather key 2 in Figure 4.2b is fixed on the hub 1 and move together with the hub along the shaft 3.

Woodruff keys are mainly used for angular location on shaft ends, as shown in Figure 4.3. A circular groove holds the key in place while the hub is slid over the shaft. Woodruff keys can yield better concentricity after assembly, however, due to the greater depth of keyseat, Woodruff keys are usually used for light duty applications.

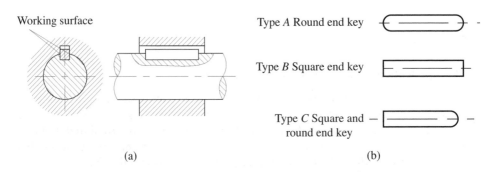

(a)   (b)

Figure 4.1 Different types of parallel keys.

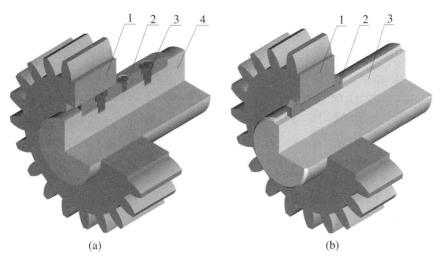

(a)   (b)

Figure 4.2 Guided key and feather key.

Figure 4.3 Woodruff key.

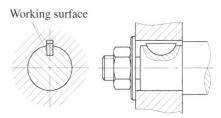

Taper keys (Figure 4.4a) have a slope of 1 : 100 between the top surface of key and the mating hub surface, with round ends, square ends and a gib-head end. The installation of a round end key is similar to that of a parallel key, that is, installing the key first and then driving the hub over the key; while square-end and gib-head keys are designed to be inserted from the end of the shaft after the hub is in position.

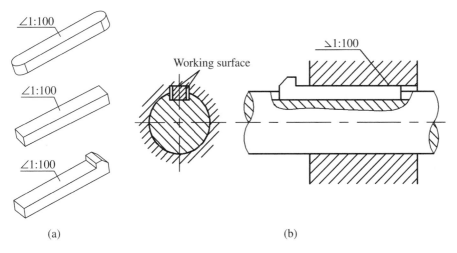

(a)   (b)

Figure 4.4 Taper keys and gib-head keys.

When taper keys, including gib-head keys in Figure 4.4b, are installed in place, they are normally driven tightly and rely on the top and bottom working surfaces to transmit torque and power. In contrast to parallel keys, there is clearance on both sides. When overload or the shaft and hub move relative to each other, the side surface can work like parallel keys to transmit torque. Therefore, taper keys work for heavy duty service to transmit unidirectional, reversing or vibrating torques and in applications where periodic withdrawal of the key may be necessary [3].

Taper keys are used in pairs to transmit heavy unidirectional torques. Two pairs of taper keys arranged at 120° interval are used to transmit bidirectional torques. Because of the eccentricity caused by taper keys, they are best suited for large diameter shafts where the adverse effect of eccentricity can be neglected.

4.1.3 Strength Analysis

The size of key for a particular application is usually selected after the shaft diameter has been specified. The standard sizes for width and height, as functions of shaft diameter, are listed in abridged form in Table 4.1. The length of key is selected according to the hub width and the torsional load to be transmitted. Keys normally extend along the full width of hub and, for good stability, hub widths are commonly $1.5d$ to $2d$, where d is the diameter of mating shaft.

Keys are used as detachable fasteners for the connection of shafts and hubs and are subjected to shearing and compressive bearing stresses. Therefore, potential failures for keys include shearing across the shaft and hub interface and crushing failure due to bearing stress between the sides of key and the shaft or hub.

Figure 4.5 shows that a torque in the shaft creates a force on the lower left side of the key. The key in turn exerts a force on the right side of the hub keyseat. The reaction force of the hub react back on the upper right side of the key. These forces on the key directly shears the key over its cross section. The force is commonly assumed to be uniformly distributed over the surfaces and is the quotient of torque and shaft radius. To ensure safety, the shear stress must satisfy,

$$\tau = \frac{2T}{bld} \leq [\tau] \tag{4.1}$$

The failure in crushing is related to the compressive stress on the side of key, the side of shaft keyseat or the side of hub keyseat; whichever is the weakest. The area in the bearing is the same for either of these zones, expressed as $l \times h/2$. Failure occurs on the surface with the lowest allowable stress. The crushing strength is

$$\sigma_p = \frac{4T}{hld} \leq [\sigma_p] \tag{4.2}$$

Table 4.1 The dimension of parallel key versus shaft diameter [5], mm.

Shaft diameter d	>17–22	>22–30	>30–38	>38–44	>44–50	>50–58	>58–65	>65–75	>75–85
Width b × Height h	6 × 6	8 × 7	10 × 8	12 × 8	14 × 9	16 × 10	18 × 11	20 × 12	22 × 14
Length L		14–70	18–90	22–110	28–140	36–160	45–180	50–200	56–220	63–250
Length series	14, 16, 18, 20, 22, 25, 28, 32, 36, 40, 45, 50, 56, 63, 70, 80, 90, 100, 110, 125, 140, 180, 200, 220, 250								

(a)

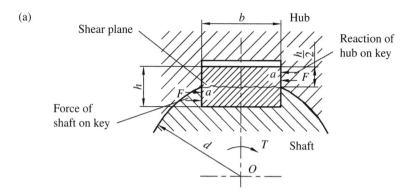

(b)

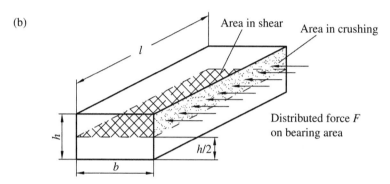

Figure 4.5 Forces acting on a key.

where
l – working length of key, mm. For Type A key, $l = L - b$; Type B, $l = L$; Type C,
$l = L - b/2$.

If either shear or crushing strength is insufficient, double keys oriented at 180°
intervals from one another along the shaft circumference may be used to carry the
load. Considering the nonuniform load distribution, the strength calculation should
introduce a factor of 1.5 instead of 2.

Keys are often made from cold-drawn low-carbon steel, high-carbon steel or heat-
treated alloy steels for higher strength requirements. The allowable bearing stresses
$[\sigma_p]$ for key, hub and shaft made from steels or cast iron are between 60–150 and
30–80 MPa, respectively. Larger values are selected for static loads, while smaller values
are for impact loads [5]. The allowable shear stress can be roughly estimated as half of
the allowable bearing stress by the maximum-shear-stress theory, expressed as

$$\tau = 0.5 \ \sigma_p \tag{4.3}$$

Example Problem 4.1

A steel shaft is to transmit a stable torque of 1000 N m to the steel spur gear through a
key. The diameter of the shaft connecting the gear is $d = 80$ mm and the width of hub is
$w = 120$ mm, as shown in Figure E4.1. Select a proper key to transmit the torque. If the
transmitted torque is doubled, modify the design.

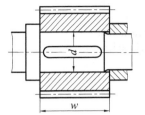

Figure E4.1 Illustration for Example Problem 4.1.

Steps	Computation	Results	Units
1. Select the type of key	Select Type *A* parallel key	Type *A*	
2. Decide the size of the key	Decide the size of key $b \times h$ from Table 4.1. Since $d = 80$ mm, select $b \times h$ as 22×14; According to the width of the hub, select the length of the key from the same table as $L = 110$ mm. Therefore, the working length of the key is $l = L - b = 110 - 22 = 88$ mm.	$b = 22$ $h = 14$ $L = 110$ $l = 88$	mm mm mm mm
3. Select the allowable stress	The failure mode of key connection is the crushing of the weakest material. For a steel gear and shaft under stable loads, select the allowable bearing stress as $[\sigma_p] = 80$ MPa.	$[\sigma_p] = 125$	MPa
4. Calculate the crushing strength of the key	From Eq. (4.2) $$\sigma_p = \frac{4T \times 10^3}{hld} = \frac{4 \times 1000 \times 10^3}{14 \times 88 \times 80} = 40.6 \text{ MPa}$$ Since $\sigma_p < [\sigma_p]$, the selected key is acceptable.	Select a key with $b \times h \times L$ as $22 \times 14 \times 110$	mm
5. Design modification	If the transmitted torque is doubled, then $$\sigma_p = \frac{4T}{hld} = \frac{4 \times 2000 \times 10^3}{14 \times 88 \times 80} = 81.2 > [\sigma_p]$$ The following design modification are suggested: 1) Since the actual stress is close to the allowable stress, select Type *B* parallel key can increase the working length of the key that transmits torque. The strength is $$\sigma_p = \frac{4T}{hld} = \frac{4 \times 2000 \times 10^3}{14 \times 110 \times 80} = 64.9 < [\sigma_p]$$ 2) Select two Type *A* keys to carry the load. The strength is $$\sigma_p = \frac{4T}{hld} = \frac{4 \times 2000 \times 10^3}{14 \times 88 \times 1.5 \times 80} = 54.1 < [\sigma_p]$$	Select a type B key with $b \times h \times L$ as $22 \times 14 \times 110$ or Select two Type *A* keys with dimensions $22 \times 14 \times 110$	mm mm

4.2 Splines

Splines are standardized products. They perform the same function as keys in transmitting torque between a shaft and mating elements. Splines are used when multiple keys are required or when a high torque needs to be transmitted.

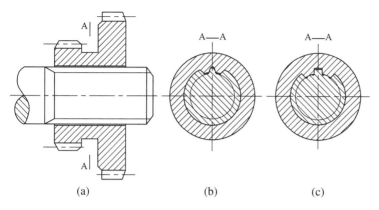

(a)   (b)   (c)

Figure 4.6 Types of splines.

A spline can be regarded as multiple keys machined into a shaft as an external spline, with corresponding grooves machined on the bore of hub of a power transmitting element as an internal spline [1], as shown in Figure 4.6a. Mating splines in the shaft and hub provides the strongest joint connection for torque transmission [6]. Either involute or straight-sided profiles are available, as illustrated in Figure 4.6b and c, respectively. Involute splines are machined by standard hobs, with pressure angles of 30° or 45°. Straight-sided splines depend on the major diameter fit to produce accurate concentricity between the shaft and mating hub; while involute splines rely on tooth profile to centre the shaft and are thus preferred.

Splines are machined according to the standards and usually contain 4, 6, 10 or 16 teeth. Therefore, splines have higher load carrying capacity, greater fatigue strength over keys as more teeth share loads. A spline is made with a loose slip fit to allow for axial motion between a shaft and mating elements [7]. Since splines are integral with shafts, they are more accurately centred and guided. However, the manufacturing cost is high.

The potential failure mode of a spline is crushing on the working surface. The strength analysis of spline is similar to that of a key. Detailed analysis can be found in reference [5].

4.3 Pins

Pins are simple, inexpensive and standardized fasteners. They are used to fasten machine elements together, to keep them aligned (Figure 4.7a) or to prevent rotational motion between shaft and mounted elements (Figure 4.7b). According to their functions, pins are classified as Dowel pins, shear pins and many others.

Dowel pins, including cylindrical pins (Figure 4.7a) and taper pins (Figure 4.7b), are often used to precisely locate elements to facilitate machining or assembly. To insert a pin, the parts are first joined and secured, and a hole is then drilled and reamed to tight tolerances. Pins are held in place by interference fit and must be driven or pressed out for removal. Cylindrical pins confront difficulty in providing prolonged precise positioning if repeated disassembly occurs. A taper pin may be used instead to attain repeatable assembly quality. Taper pins are sized according to the diameter at the large end.

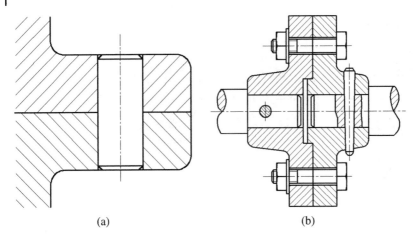

           (a)   (b)

Figure 4.7 Types of pins.

Taper pins can also join a shaft and the hub of mounted power transmission elements. The pin prevents rotational motion between the shaft and mounted elements when a torque is transmitted. The torque transmission capacity of connection is limited by the shear strength of pin. Pins are suitable for transmitting low to medium torques.

A shear pin serves as a safety or protective component, usually in couplings or clutches, as shown later in Table 13.3. Pins are purposely made of relatively weak materials or with deliberately small diameters to ensure the pins will break if loads exceed acceptable operating limits, thus protecting critical or expensive components of a machine. Therefore, the diameter of a pin is determined according to the overload when shear occurs.

Many other types of pins are used in various machines. For example, the split cotter pin prevents the nut from turning on the bolt in Figure 3.6b. The clevis pin has a ridge at one end and is kept in place by a cotter pin inserted through a hole in the other end, as shown in the chain joint connections in Figure 7.3.

References

1 Mott, R.L. (2003). *Machine Elements in Mechanical Design*, 4e. Prentice Hall.
2 Pu, L.G. and Ji, M.G. (2006). *Mechanical Design*, 8e. Beijing: Higher Education Press.
3 Oberg, E. (2012). *Machinery's Handbook*, 29e. New York: Industrial Press.
4 Gere, J.M. and Timoshenko, S.P. (1996). *Mechanics of Materials*, 4e. CL Engineering.
5 Wen, B.C. (2015). *Machine Design Handbook*, 5e, vol. 2. Beijing: China Machine Press.
6 Juvinall, R.C. and Marshek, K.M. (2011). *Fundamentals of Machine Component Design*, 5e. New York: Wiley.
7 Budynas, R.G. and Nisbett, J.K. (2011). *Shigley's Mechanical Engineering Design*, 9e. New York: McGraw-Hill.

Problems

Review Questions

1 Explain how parallel keys and taper keys work.

2 What are the potential failure modes of parallel keys?

3 What measure should be taken if a key does not have sufficient strength?

Objective Questions

1 The dimension of the cross section of a key is selected by _____.
 (a) the transmitted torque
 (b) the transmitted power
 (c) the length of the wheel hub
 (d) the diameter of the mating shaft

2 The length of a key is selected by _____.
 (a) the transmitted torque
 (b) the transmitted power
 (c) the length of the wheel hub
 (d) the diameter of the mating shaft

3 Which of the following is a taper key in Figure P4.1? _____.

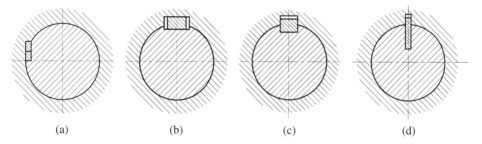

(a)   (b)   (c)   (d)

Figure P4.1 Illustration for Objective Questions 3 and 4.

4 Which represents a parallel key in Figure P4.1? _____

5 A parallel key is supposed to transmit a maximum torque of T in Figure P4.2. If a torque of $1.5\,T$ is to be transmitted, it is better to _____.

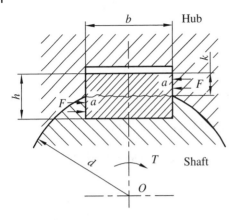

Figure P4.2 Illustration for Objective Question 5.

(a) install a pair of keys
(b) increase the width of the key to $1.5b$
(c) increase the diameter of the shaft to $1.5d$
(d) increase the height of the key to $1.5h$
Note: parameters of d, b and h can be found in Figure P4.2.

Calculation Questions

1   Figure P4.3 shows a rigid flanged coupling made of medium carbon steel, with an allowable bearing stress of $[\sigma_p] = 100$ MPa. The maximum transmitted stable torque is 1200 N m. Select the key type, determine the main dimensions of the key and check its strength. The dimensions in the figure are $d_1 = 55$ mm, $L_1 = L_2 = 84$ mm.

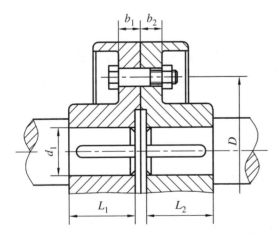

Figure P4.3 Illustration for Calculation Question 1.

2   The cone clutch shown in Figure P4.4 transmits a power of $P = 2.5$ kW. Shaft 4 with a diameter of $d = 52$ mm rotates at a speed of $n = 250$ rpm, subjected to an impact loading. The material for the coupling is cast steel, with an allowable bearing stress as 75 MPa. The length of the hub 3 is $L_1 = 120$ mm. Other dimensions are $S = 10$ mm, $L_2 = 40$ mm. Select the Type A parallel key with dimensions of $16 \times 10 \times 140$ mm. Check the strength of the key.

Figure P4.4 Illustration for Calculation Question 2.

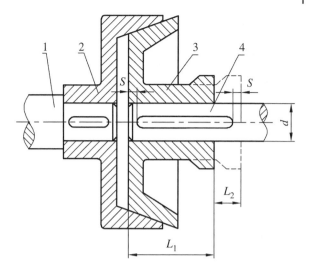

Design Problems

1. The main tasks in key design include:
   (1) select the length of key according to the length of hub
   (2) select the type of key according to their applications
   (3) select the dimension of key cross section according to the diameter of mating shaft
   (4) check the strength of key

   The normal design procedure is _____.
   (a) $2 \rightarrow 1 \rightarrow 3 \rightarrow 4$
   (b) $2 \rightarrow 3 \rightarrow 1 \rightarrow 4$
   (c) $1 \rightarrow 3 \rightarrow 2 \rightarrow 4$
   (d) $3 \rightarrow 4 \rightarrow 2 \rightarrow 1$

2. A cast steel V-belt sheave installed at the end of a shaft by a key in Figure P4.5. The sheave diameter and mating shaft diameter are $d_d = 250$ mm and $d = 45$ mm, respectively. The width of hub is $L = 65$ mm. The belt drive works steadily. Decide the type and dimension of key and the maximum torque the shaft can transmit.

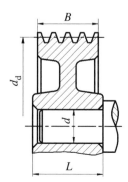

Figure P4.5 Illustration for Design Problem 2.

3 A coupling and gear is connected with a shaft by keys, as shown in Figure P4.6. The shaft transmits a torque of $T = 1000\,\text{N m}$, with slight impact. The shaft, gear and coupling are made of medium carbon steel, cast steel and grey iron, respectively, with corresponding allowable stresses of 120, 110 and 55 MPa, respectively. The shaft diameters connecting the coupling and gear are $d_1 = 55\,\text{mm}$, $d_2 = 85\,\text{mm}$. The width of the coupling and gear are $L_1 = 116\,\text{mm}$, $L_2 = 90\,\text{mm}$, respectively. Select the type and dimension of keys and check their strength.

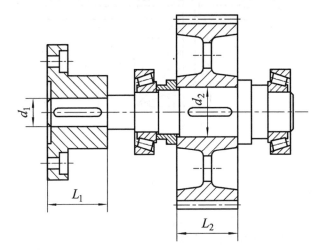

Figure P4.6 Illustration for Design Problem 3.

Structure Design Problems

1 To facilitate assembly and disassembly of a taper key, we usually fabricate a slope of $1:100$ _____.
(a) at the bottom of keyway on the shaft
(b) on the top side of key
(c) at the bottom of keyway on the hub
(d) both (b) and (c)

2 In a gear reducer shown in Figure P4.7, gears 2 and 3 are installed on the shaft II. These two gears should be installed on _____.

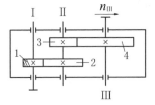

Figure P4.7 Illustration for Structure Design Question 2.

(a) the two keys on the same generatrix
(b) the same key
(c) the two keys located at an interval of 180° circumferentially
(d) the two keys located at an interval of 120° circumferentially

3 When using a pair of keys is inevitable, two parallel keys are arranged at 180° interval; two taper keys at 120° interval; and two woodruff keys on the same line, as shown in Figure P4.8. Please explain the reason.

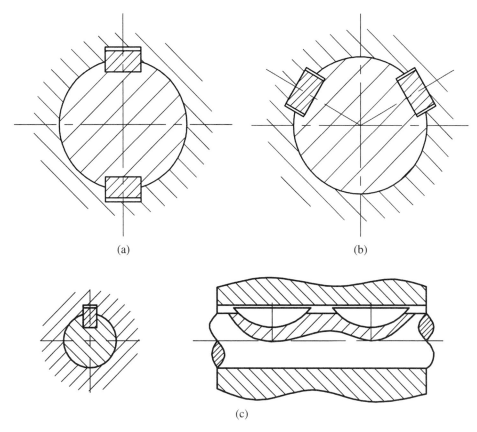

(a)   (b)

(c)

Figure P4.8 Illustration for Structure Design Question 3.

4 In a safety coupling shown in Figure P4.9, either one or two pins can be used. Analyse which design is better.

Figure P4.9 Illustration for Structure Design Question 4.

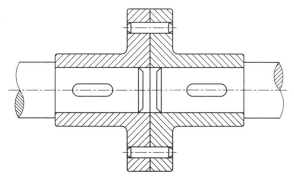

5

Permanent Connections

Nomenclature

d	diameter of a rivet, mm
F	tensile or transverse load, N
F_s	shear load, N
k	leg length, mm
l	the length of weld, mm
M	moment, N·mm
m	throat length, mm
t	plate thickness, mm

w	rivet spacing or pitch, mm
σ	tensile stress, normal stress, MPa
$[\sigma]$	allowable tensile stress, MPa
σ_p	bearing stress, MPa
$[\sigma_p]$	allowable bearing stress, MPa
τ	shear stress, MPa
$[\tau]$	allowable shear stress, MPa

Subscripts

1,2...i	plate number
x, y, z	coordinates

5.1 Riveting

5.1.1 Applications, Characteristics and Structure

As permanent joints, riveted joints are widely used in mechanical products like aircraft, automotive, as well as in the construction of bridges, boilers, buildings, due to simple structure, moderate strength, small assembly time and low production cost. Particularly, riveted joints are capable of joining thin components or dissimilar materials. Nevertheless, the increased weight and protruding rivet heads are the obvious disadvantages associated with riveted joints. With the rapid development of high-strength steel bolts and the improvement of welding processes, the usage of rivets has shown a sharp decline in recent decades [1–3].

Riveted joints are a common means of fastening rolled steel plates, sheet metals and other relatively thin components to form a permanent joint. Figure 5.1 shows typical riveted joints. In Figure 5.1a, two plates are lap jointed by a row of rivets; in Figure 5.1b and Figure 5.1c, two plates are butted together with one strap or with top and bottom straps by two rows of rivets, respectively.

5.1.2 Types of Rivets

A conventional rivet is a short metal pin with a preformed head and a shank to be formed into a second head after assembly [4]. During assembly, rivet holes are drilled

Analysis and Design of Machine Elements, First Edition. Wei Jiang.
© 2019 John Wiley & Sons Singapore Pte. Ltd. Published 2019 by John Wiley & Sons Singapore Pte. Ltd.
Companion website: www.wiley.com/go/Jiang/analysis_of_machine_elements

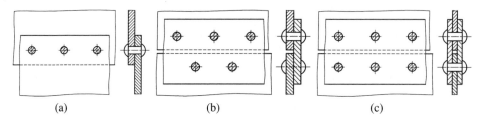

Figure 5.1 Typical riveted joints.

Table 5.1 Types of rivets and rivet joints.

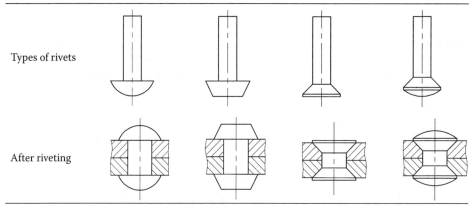

or punched first. Rivets are then inserted through holes of various thicknesses of plates, and clinched in place by high speed automatic machinery to form a permanent joint [4]. Table 5.1 presents some of conventional rivets with different heads and permanent rivet joints after riveting.

5.1.3 Strength Analysis

While assembling rivet joints, compressive loads associated with forming the second head generate between contact surfaces. If external transverse loads applied to a riveted joint exceed the friction between contact surfaces, the rivet joint may fail. Common failure modes include the pure prevention of the shearing of rivets, the crushing of rivets or plates, the rupture of connected plates due to pure tension and the edge shearing or tearing of the margin.

Considering the potential failure modes of riveted joints, the analysis and design of a rivet joint includes the prevention of the shearing of rivets, the crushing of rivet-plate interface and the weakening effect of rivet holes on the connected plates due to removal of material. The failure due to edge shearing or tearing can be avoided by spacing the rivets at least twice the rivet's diameter away from the edge [4]. Hence, this type of failure is usually prevented by proper structural design.

The analysis of tensile and transverse shear stresses in rivets is comparable to that introduced for bolts. Referring to Figure 5.2 showing the case of a single transverse row of rivets, the shear strength of a single rivet is

$$\tau = \frac{4F}{\pi d^2} \leq [\tau] \tag{5.1}$$

where $[\tau]$ is allowable shear stress for rivets.

Figure 5.2 Forces on rivets.

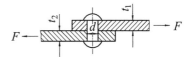

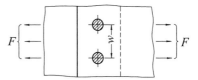

Assume bearing stress is uniformly distributed over the projected contact area of the cylindrical surface of rivet, the crushing strength is calculated by

$$\sigma_p = \frac{F}{dt_{min}} \leq [\sigma_p] \tag{5.2}$$

where t_{min} is the minimum thickness of riveted plates and $[\sigma_p]$ is the allowable bearing stress for plates or rivets, whichever is smaller. It is standard practice that the nominal diameter of rivet d rather than the diameter of hole is used to calculate bearing stress, although the rivet expands and nearly fills up the hole after assembly.

The tensile strength of plate at the section containing rivet holes is given by

$$\sigma = \frac{F}{(w-d)t_{min}} \leq [\sigma] \tag{5.3}$$

where F is taken as load per rivet and w is rivet spacing or rivet pitch, and $[\sigma]$ is the allowable stress of plates.

Rivets are usually made from ductile materials, such as carbon steel, aluminium and brass. The allowable shearing stress $[\tau]$, bearing stress $[\sigma_p]$ and tensile stress $[\sigma]$ depend on the ultimate strengths of rivet material and plate material, and the selected safety factors. They can be found in design handbooks [5]. In general, a rivet cannot provide as strong an attachment as a bolt or screw of the same diameter.

5.1.4 Design of Riveted Joints

Riveted joint design involves the selection of rivet diameter, rivet spacing and the number of rivets so that the joint is strong enough against potential failure modes. Ideally, the joint should be designed to achieve equal strength against the three previously introduced failure modes. The procedure and guidelines for riveted joint design are similar to bolted joint design and are summarized as:

1) According to the load and specific requirements, decide the pattern of rivet layout and the number of rivets. Preferably, it is better to ensure each rivet is uniformly loaded. Also, the spacing between rivets must be sufficient for driving tools; usually three times the rivet diameter is considered as minimum spacing.
2) Analyse the loads acting on the riveted joints. Determine the load each rivet is subjected to.
3) Select the initial rivet type and diameter from design handbooks and manufacturer' catalogues, calculate shear and bearing stress to ensure that all stresses are below allowable values.

Example Problem 5.1

A riveted joint shown in Figure E5.1 connects two plates with a thickness of $t = 10$ mm by rivets with the diameter of $d = 20$ mm. The width the connected plate is $l = 180$ mm. The allowable shear stress of rivets is $[\tau] = 120$ MPa, the allowable bearing stress of rivets is $[\sigma_p] = 250$ MPa and the allowable tensile stress of plates is $[\sigma] = 150$ MPa. Determine the maximum allowable tensile force F the rivet joint can carry.

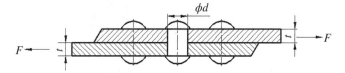

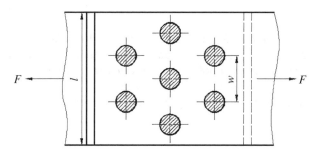

Figure E5.1 Illustration for Example Problem 5.1.

Solution:

Steps	Computation	Results	Units
1. Force decided by shear strength of rivets	From Eq. (5.1), $F_1 = 7 \times \dfrac{\pi d^2}{4}[\tau] = 7 \times \dfrac{\pi \times 20^2}{4} \times 120 = 263760$ N	$F_1 = 264$	kN
2. Force decided by crushing strength between rivets and plates	From Eq. (5.2), we have $F_2 = 7dt[\sigma_p] = 7 \times 20 \times 10 \times 250 = 350000$ N	$F_2 = 350$	kN
3. Force decided by tensile strength of the plates	In considering tensile stresses in the plate, two possibilities must be examined. If the plate fails in tension at cross section A–A in Figure E5.2, the force causes the tensile failure of the plate is obtained from Eq. (5.3) as $F_3 = (l - 2d)t[\sigma] = (180 - 2 \times 20) \times 10 \times 150 = 210000$ N If the plate fails in tension at cross section B–B in Figure E5.2, rivets on cross section A–A must either fail in shear or in crushing. The force causes the tensile failure of the plate cross section B–B is obtained from Eq. (5.3) as $F_{\text{tension}} = (l - 3d)\, t[\sigma] = (180 - 3 \times 20) \times 10 \times 150 =$ 180000N		

Steps	Computation	Results	Units
	The force causes the shear failure of the rivets at cross section $A–A$ is obtained from Eq. (5.1) as $$F_{shear} = 2 \times \frac{\pi d^2}{4}[\tau] = 2 \times \frac{\pi \times 20^2}{4} \times 120 = 75360 \text{N}$$ The force causes the crushing failure of the rivets at cross section $A–A$ is obtained from Eq. (5.2) as $$F_{crushing} = 2dt[\sigma_p] = 2 \times 20 \times 10 \times 250 = 100000 \text{ N}$$ The force required for tensile failure at $B–B$ and simultaneous shear of rivets at cross section $A–A$ is $$F = F_{tension} + F_{shear} = 180000 + 75360 = 255360 \text{ N}$$ The load required for tensile failure at $B–B$ and simultaneous crushing of rivets at cross section $A–A$ is $$F = F_{tension} + F_{crushing} = 180000 + 100000 = 280000 \text{N}$$ Therefore, the maximum allowable load for the joint is 210 000 N.	$F_3 = 210$	kN

Figure E5.2 Solution for Example Problem 5.1.

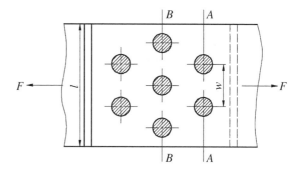

5.2 Welding

5.2.1 Applications, Characteristics and Structure

Welded joints are permanent joints formed by various welding processes involving metallurgical bonding of metals. One of the most widely used joining processes is fusion welding, which fuses the base metal to form welds. During the welding process, local heat is applied to the work pieces to increase temperature high enough to melt the metal.

The three major types of fusion welding processes are gas welding, arc welding and high-energy beam welding. Gas welding mainly refers to oxyacetylene welding. Depending on the method of supplying filler materials and of shielding the molten weld metal from atmosphere, arc welding includes shielded metal arc welding (SMAW), gas metal arc welding (GMAW), gas tungsten arc welding (GTAW) and submerged arc welding (SAW). High-energy beam welding includes electron beam welding and laser beam welding [7].

Welding is a reliable, efficient and economical joining process for connecting metal sections, which is widely used in most engineering fields, including the shipbuilding, power and oil industries. Welded joints are superior to riveted joints for strong strength, light weight and low cost. Nevertheless, welding may generate unacceptable residual stresses and thermal distortions if the welding process is not properly controlled.

5.2.2 Types of Welded Joints and Types of Welds

Welded joints refer to the relationship between connected parts. Basic welded joint designs in fusion welding include butt, lap, T-, edge and corner joints [7]. Welds are named after the edge geometry of the parts to be jointed. Basically, two types of weld seams are available. The first is butt welds, which are used to join components within a plane. The second is fillet welds, which are used to connect components in different planes. They are usually designed as equal-leg right angles. Representative weld joints and welds are presented in Table 5.2.

5.2.3 Strength Analysis

Loads acting on a weldment may be uniformly or eccentrically applied to the weld. Similar to the previous treatment of loads on bolts and rivets, complex loads on a weldment

Table 5.2 Types of welded joints and welds.

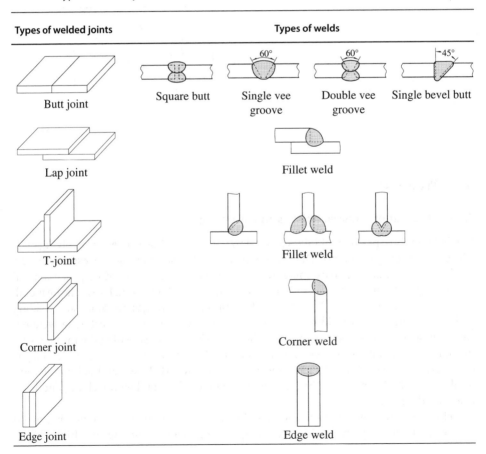

| Types of welded joints | Types of welds |

Figure 5.3 Loads on a butt welded joint.

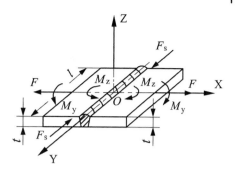

are first decomposed into typical loads, that is, tension or compression forces, torques or moments, as shown in Figure 5.3. The loads on each weld under each type of load are then analysed separately and combined vectorially to determine the maximum load. The common potential failure mode under these loads is fracture.

Assume stresses are uniformly distributed at weld cross sections. The following section introduces strength analysis of butt welds and fillet welds.

5.2.3.1 Butt Welds

For a single V-groove butt weld loaded by either tensile or compressive force F, the weld strength is analysed against the average normal stress as

$$\sigma = \frac{F}{lt} \leq [\sigma] \tag{5.4}$$

Similarly, the shear strength due to shear load F_s against the average shear stress is

$$\tau = \frac{F_s}{lt} \leq [\tau] \tag{5.5}$$

And the strength due to moment loading M_y or M_z is calculated by

$$\sigma = \frac{6M_y}{lt^2} \leq [\sigma] \tag{5.6}$$

or

$$\sigma = \frac{6M_z}{l^2 t} \leq [\sigma] \tag{5.7}$$

5.2.3.2 Fillet Welds

When a fillet weld joint carries a transverse load, the weld seam along the horizontal plane is subjected to shear stress, while along the vertical plane this is a tensile (or compressive) stress, as illustrated in Figure 5.4. The size of a fillet weld is defined as leg length k, and, although not necessary, the two legs are usually of the same length. When calculating transverse shear stresses and axial stresses, throat length m is assumed to be in a 45° orientation from the intersection of the plates to the straight line connecting the ends of the two legs. For the usual convex weld with equal legs, $m = 0.707\,k$. The throat area used for stress calculations is then the product of ml, where l is the length of weld.

Since fillet welds usually fail in the throat section, the throat stresses are thus usually used in design. The stress and strength at the throat section of a fillet weld can be approximately calculated as

$$\frac{F}{\sin 45°\,kl} = \frac{F}{0.707kl} \leq [\tau] \tag{5.8}$$

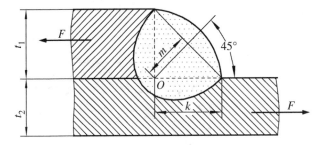

Figure 5.4 Forces on a fillet weld.

The allowable stress of weld relates to the weld material, parent material and welding processes. If the allowable stress of parent material is $[\sigma]$, for a weld seam, the allowable tensile stress can be selected as approximately $(0.9–1)[\sigma]$; the allowable compressive stress $[\sigma]$ and the allowable shear stress $(0.5–0.65)[\sigma]$ [5].

These calculations neglect weld induced residual stresses, which may sometimes have adverse effects on the strength of weldments. Although not rigorously correct, these convenient procedures are considered to be justified for estimation. Detailed stress and strength calculation for butt and fillet welds subjected to torsional, bending or fatigue loading, as well as strength analyses for other weld configurations can be referred to in relevant design handbooks [5].

5.2.4 Design of Welded Joints

The design of welded joints requires consideration of the loads carried by welded joints, the types of weld materials and parent materials and joint geometries. Welded joint design usually follows the following procedure and guidelines:

1) Propose welded joint geometry and design the members to be joined;
2) Select parent and weld materials and decide their allowable stresses;
3) Determine the magnitude and the direction of force on the weld under each load, combine the forces vectorially at the point of weld where the force appears to be maximum;
4) Analyse stresses the weld is subjected to and compare them with the allowable stresses of materials to determine the required dimensions.

Example Problem 5.2
The steel strap with thickness of 12 mm is to be welded to a rigid frame to carry a static load of $F = 20$ kN. The strap and frame are welded together by fillet welds along two sides, each of which has a length of $l = 40$ mm, as shown in Figure E5.3. Select steel for strap, with yield strength of $\sigma_s = 350$ MPa. The strap is to be welded with an E70 electrode. With a safety factor of $S = 3$, decide the size of weld leg.

Figure E5.3 Illustration for Example Problem 5.2.

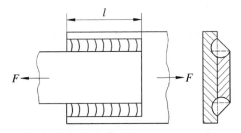

Solution:

Steps	Computation	Results	Units
1. Determine the yield strength in shear	Using the maximum distortion energy theory, we have the yield strength in shear as $$\tau_s = \frac{1}{\sqrt{3}}\sigma_s = \frac{1}{\sqrt{3}} \times 350 = 202 \text{ MPa}$$	$\tau_s = 202$	MPa
2. Determine the allowable shear stress	$$[\tau] = \frac{\tau_s}{S} = \frac{202}{3} = 67.3 \text{ MPa}$$	$[\tau] = 67.3$	MPa
3. Determine the required length of the weld leg	According to Eq. (5.8), we have $$k \geq \frac{F}{0.707 \sum l[\tau]} = \frac{20000}{0.707 \times 2 \times 40 \times 67.3} = 5.25 \text{ mm}$$ Select the value as 6 mm.	$k = 6$	mm

5.3 Brazing, Soldering and Adhesive Bonding

5.3.1 Applications, Characteristics and Structure

Brazing and soldering are similar to welding yet accurately belong to bonding [2]. Brazing and soldering use heat to melt filler materials that flow into the space between parts to be joined. They differ from welding in that the temperature is always below the melting point of base metal. Brazing heats the base metal to a specified temperature above 450°C; while soldering is similar to brazing except at a temperature below 450°C and has a relatively low strength. In both processes, the filler materials act like molten metal glue that solidifies immediately upon cooling. Common filler materials for brazing are alloys of copper, silver or nickel, while most solders are tin-lead alloys. Brazed joints require little or no finishing and can braze dissimilar metals and even nonmetals [2]. Soldering has wide applications in plumbing and in joining electrical and electronic parts.

Adhesive bonding is one of permanent joints, using polymeric adhesives to join plates or tubes. Compared with fasteners, the use of adhesives not only significantly reduces weight, but also eliminates stress concentration associated with drilled holes. Besides, adhesive bonded joints can effectively seal against leakage as long as the liquid or gas does not react with the adhesive [2]. And finally, adhesives can be used to join dissimilar materials or relative thin components. Nevertheless, it is more temperature sensitive than mechanical fasteners. Safety and environmental factors are important concerns in plumbing and in adhesive bonding.

5.3.2 Types of Adhesive and Their Selection

Adhesives may be classified in a variety of ways depending on their chemical composition (e.g. epoxies, polyurethanes, polyimides), physical form (e.g. paste, liquid, film, tape), adhesion method (e.g. reactive, nonreactive), origin (e.g. natural, synthetic) or load carrying capability (e.g. structural, semistructural, nonstructural) [1, 8].

Structural adhesives are strong adhesives that are normally used to carry, especially, shear loads. Most structural adhesives are thermosetting, becoming irreversibly hardened upon being cured. Typical examples include epoxies, urethanes, anaerobics, acrylic and cyanoacrylates.

Epoxies are the most versatile and widely used structural adhesives, with a large variety of formulations and properties available. One-part epoxies require heat curing; while two-part types cure at room temperature but require premixing. Urethanes are similar to epoxies in great versatility, relatively high strength, good toughness, flexibility and impact resistance. Anaerobics cure in the absence of oxygen. Acrylic adhesives are preferred for many metals and plastics under ordinary industrial conditions as they are tolerant of dirty surfaces. Cyanoacrylates are particularly appropriate for fast curing requirement [2, 3]. Most structural adhesives can carry significant stresses.

Nonstructural adhesives provide cost effective means required for assembly of finished products where failure would be less critical. They have some strength at normal application. However, at elevated temperature, their properties decrease rapidly. Typical examples include contact adhesives and pressure sensitive adhesives. They are generally intended for light duty applications [1].

Although almost any solid materials can be bonded with a suitable adhesive, the selection of proper adhesive depends on operating environment and applications. Since exposure to water, solvents, ultraviolet light and high temperature can significantly degrade adhesive performance, environmental limitations and effects must be recognized while selecting adhesives. Other special properties, like rust-proofing, insulating, electrical conducting, transparent and super high or super low temperatures, should also be considered for specific applications.

5.3.3 Analysis and Design of Adhesive Joints

Adhesive joints function as one component when carrying loads. Normally, adhesive joints have larger load carrying capacities for tension and shear than for peer and cleavage. When adhesive joints are loaded with tension or shear (Figure 5.5a), stresses are uniformly distributed over the bonded area, with minor stress concentration at the edge of contact. However, when adhesive joints take peel or cleavage (Figure 5.5b), peel stress and cleavage stress can be quite large and often account for debonding failure.

The design of adhesive joints need to consider their load carrying capability. Since adhesive bonds perform better when loaded in shear, design the joint to carry shear load instead of peel load. Wherever possible, bond to multiple surfaces in preference to a single surface, as bonding to multiple surfaces permits loads to be carried predominantly in shear [1].

The adhesive, bonded materials and their interaction influence the load carrying capacity of bonded joints. The mechanical properties of adhesives are affected by adhesive composition, curing conditions (including curing temperature, pressure and time), surface treatments, and operation environment. Detailed data can be found in design

Figure 5.5 Forces on adhesive bonded joints.

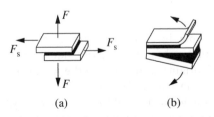

(a)   (b)

handbooks [5, 9]. Usually, adhesives with adequate ductility are preferred for reducing stress concentration and increasing toughness to resist debond propagation [1].

Since peel stresses focus at bond ends, increasing bonded area or utilizing rivets at bond ends can effectively reduce debonding failure and significantly improve durability and reliability of adhesive joints.

References

1 Budynas, R.G. and Nisbett, J.K. (2011). *Shigley's Mechanical Engineering Design*, 9e. USA: McGraw-Hill.
2 Juvinall, R.C. and Marshek, K.M. (2011). *Fundamentals of Machine Component Design*, 5e. Wiley.
3 Mott, R.L. (2003). *Machine Elements in Mechanical Design*, 4e. Prentice Hall.
4 Hindhede, U., Zimmerman, J.R., Hopkins, R.B. et al. (1983). *Machine Design Fundamentals: A Practical Approach*. New York: Wiley.
5 Wen, B.C. (2015). *Machine Design Handbook*, 5e, vol. 2. Beijing: China Machine Press.
6 Oberg, E. (2012). *Machinery's Handbook*, 29e. New York: Industrial Press.
7 Kou, S. (2003). *Welding Metallurgy*, 2e. Wiley.
8 Pu, L.G. and Ji, M.G. (2006). *Mechanical Design*, 8e. Beijing: Higher Education Press.
9 Petrie, E.M. (2007). *Handbook of Adhesives and Sealants*, 2e. New York: McGraw-Hill.

Problems

Review Questions

1 Compare permanent connections of riveted joints, welded joints and bonded joints, and analyse their advantages and disadvantages.

2 If two sheets of thin plates with a thickness of 0.2 mm are to be connected, which of the following methods are feasible; bolting, riveting, welding or bonding? List any other joining methods you think suitable.

Objective Questions

1 A welded seam carries a variable tension and compression load. From the load carrying capacity aspect, which of the following in Figure P5.1 is the best design? _____

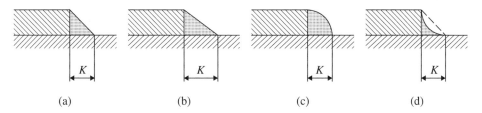

(a)　　　　　(b)　　　　　(c)　　　　　(d)

Figure P5.1 Illustration for Objective Question 1.

2   In Figure P5.2, a butt weld carries a moment M_z and tension F, while the other loads are negligible. If $[\sigma]$ is the allowable stress of the weld, which of the following can be used to calculate the strength of weld seam? _____

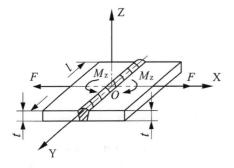

Figure P5.2 Illustration for Objective Question 2.

(a) $\sigma = \dfrac{6M_z}{lt^2} \le [\sigma]$

(b) $\sigma = \dfrac{6M_z}{lt^2} + \dfrac{F}{lt} \le [\sigma]$

(c) $\sigma = \dfrac{6M_z}{l^2 t} \le [\sigma]$

(d) $\sigma = \dfrac{6M_z}{l^2 t} + \dfrac{F}{lt} \le [\sigma]$

Calculation Questions

1   Two plates are joined by rivets. The dimensions shown in Figure P5.3 are $t = 10$ mm, $d = 10$ mm, $w = 70$ mm, $l = 100$ mm, $e = 15$ mm and $c = 35$ mm. The allowable tensile stress of the plates is $[\sigma] = 200$ MPa. The allowable bearing stress between the rivet and plate is $[\sigma_p] = 400$ MPa and the allowable shear stress of the rivet is $[\tau] = 180$ MPa. Calculate the maximum load that can be carried by the rivet joint.

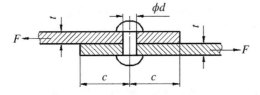

Figure P5.3 Illustration for Calculation Question 1.

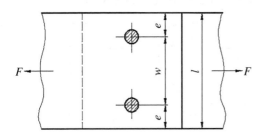

2 Two identical low carbon steel plates with a thickness of $t = 10$ mm and width of $l = 180$ mm are going to be joined by the welding process, as shown in Figure P5.4. Both butt joint and lap joint are proposed. The allowable tensile stress of the weld is $[\sigma] = 200$ MPa and the allowable shear stress is $[\tau] = 150$ MPa. Decide the maximum load F each of joint can carry.

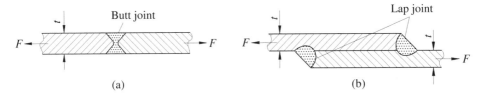

Figure P5.4 Illustration for Calculation Question 2.

Design Problems

1 Two rolled carbon steel plates with a dimension of $l \times t = 200$ mm $\times$ 12 mm are joined by rivets shown in Figure P5.5. The diameter of the rivets is $d = 20$ mm and the pitch is $w = 60$ mm. The allowable tensile stress of the plate is $[\sigma] = 200$ MPa. The allowable shear stress of the rivet is $[\tau] = 150$ MPa. The allowable bearing stress between the rivet and plate is $[\sigma_p] = 300$ MPa. Check the strength of the rivet joint if a tension force of $F = 300$ kN is applied. If the strength is deficient, how can the design be improved?

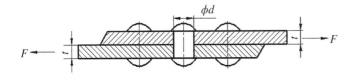

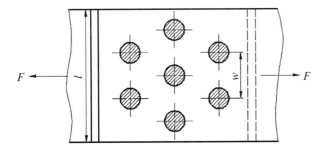

Figure P5.5 Illustration for Design Problem 1.

2   Two identical low carbon steel plates with same thickness are to be joined by welding process. A butt joint and lap joint are proposed, as shown in Figure P5.4. The welded joint is supposed to carry a static tension load of F. The allowable tensile stress of the weld is $[\sigma] = 200$ MPa and the allowable shear stress is $[\tau] = 150$ MPa. Which design is better?

Structure Design Problems

1   Compare the two designs in Figure P5.6. Assume the dimensions of plates and rivets, the number of rivets and the materials of plates and rivets are exactly the same in the two layouts.

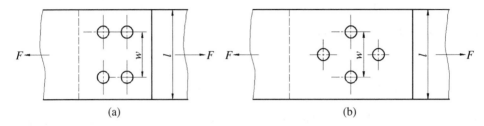

(a)   (b)

Figure P5.6 Illustration for Structure Design Problem 1.

6

Belt Drives

A mechanical drive, sometimes called a power transmission, is usually used to transmit power and motion from one rotational element to another. Mechanical drives include belt drives, chain drives, gear drives, wormgear drives and so on. They can be broadly classified as flexible and non-flexible or rigid power transmission drives. In a flexible drive, there is an intermediate, flexible element such as a belt or a chain between the driving and driven shafts. The rotary motion of the driving shaft is first converted into translatory motion of the belt or chain and then converted back into the rotary motion of the driven shaft. In a rigid drive, such as in a gear drive, the rotary motion of the driving shaft is directly converted into the rotary motion of the driven shaft by the direct contact of pinion and gear. This chapter will discuss in detail the features, as well as analysis and design of belt drives. The analysis and design of other mechanical drives will be introduced in the succeeding chapters.

Analysis and Design of Machine Elements, First Edition. Wei Jiang.
© 2019 John Wiley & Sons Singapore Pte. Ltd. Published 2019 by John Wiley & Sons Singapore Pte. Ltd.
Companion website: www.wiley.com/go/Jiang/analysis_of_machine_elements

Nomenclature

A	cross sectional area of belt, mm^2
a	centre distance, mm
b_d	datum width of V-belt groove, mm
b_p	pitch width, mm
D	sheave or pulley diameter, mm
d_d	datum diameter, mm
d_p	pitch diameter, mm
E	equivalent elastic modulus of belt, MPa
F_0	initial tension, N
F_1	tight tension, N
F_2	slack tension, N
F_c	centrifugal tension, N
F_e	effective tension, N
F_{ec}	critical or maximum effective tension, N
F_f	frictional force, N
F_Q	force acting on the pulley or sheave shaft, N
f	coefficient of friction
f_v	equivalent coefficient of friction of V-belt
h	belt height, mm
i	speed ratio
K_A	service factor
K_L	correction factor for belt length
K_α	correction factor for contact angle
L_d	belt datum length, mm
N	normal force, N

n	rotational speed of a pulley or sheave, rpm
P	transmitted power, kW
P_0	basic power rating of a single standard V-belt, kW
ΔP_0	basic power rating increment of a single standard V-belt, kW
P_{ca}	design power, kW
P_r	actual power rating of a single V-belt, kW
Q	radial force acting on the pulley or sheave, N
q	mass of belt per unit length, kg m^{-1}
r	pulley radius, mm
v	belt speed, m s^{-1}
v_1	linear speed of the small pulley or sheave, m s^{-1}
z	number of belt
α	contact angle, ° or rad
ε	slip ratio
φ	wedge angle, °
σ_1	tensile stress in tight side, MPa
σ_2	tensile stress in slack side, MPa
σ_b	bending stress in a belt, MPa
σ_c	centrifugal stress in a belt, MPa
σ_{max}	maximum stress in a belt, MPa
$[\sigma]$	allowable stress of a belt, MPa

Subscripts

1	driving pulley or sheave
2	driven pulley or sheave

6.1 Introduction

6.1.1 Applications, Characteristics and Structures

A belt drive represents one of the major types of flexible drive. They are extensively used in automotive or industrial drives to reduce a higher rotational speed of one part to a lower value over a considerable distance.

Figure 6.1 shows a power transmission system consisting of a belt drive, a gear drive and a chain drive. The power required by the driven machine is provided by an electric motor. Since the electric motor usually operates at a high speed and delivers a low torque, which is not appropriate for the driven machine, a power transmission system is employed. In a power transmission system, belt drives are usually used at the first stage from the power source. Gear reducers, together with other drives with properly designed speed ratios, are used between the belt drive and the driven machine to satisfy the speed requirement of the driven machine.

Figure 6.1 A typical power transmission system.

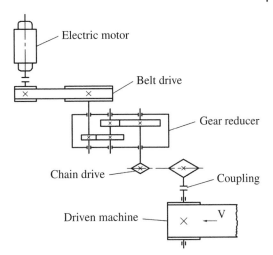

Since belt drives are used to transmit motion over comparatively long distances, the tolerance on the centre distance is not as critical compared with gear drives, and the layout of driving and driven shafts has considerable flexibility. Although a belt may slip when it overloads, it provides a measure of safety for the driver and driven machines. Belt drives also have advantages of smooth operation, simple structure and shock absorption. In many cases, their use simplifies machine design and substantially reduces manufacturing and maintenance costs [1].

However, except for timing belts, the speed ratio of a belt drive is inaccurate due to inevitable elastic creep between belt and pulleys. Belt slippage may also shorten belt life. Besides, the power and torque transmission capacities are limited by the coefficient of friction and interfacial pressure between belt and pulleys [2]. To ensure proper functioning, a belt must be mounted on pulleys with proper initial tension, and belt-tensioning devices are essential to maintain the initial tension in belt drives.

A typical belt drive consists of a driving pulley 1, a driven pulley 3 and a continuous belt 2 (Figure 6.2). A belt is a flexible band passing around two wheels that transmits motion and power from one to another. Wheels with a flat profile are called pulleys, while wheels with grooves are termed sheaves [3]. When a belt drive is used for speed reduction, which is the usual case in industry, the small driving pulley is mounted on the high-speed shaft while the large driven pulley is connected to the driven machine shaft. The speed reduction will cause proportional increase of torque. Since belts are virtually endless loops, machines must have provision to allow belt adjustment and replacement.

Figure 6.2 shows two typical layouts of belt drives, that is, an open-belt drive (Figure 6.2a) and cross-belt drive (Figure 6.2b). An open-belt drive is used when the driven pulley is to be rotated in the same direction as the driving pulley. A cross-belt drive is adopted when the driven pulley is to be rotated in the opposite direction to the driving pulley.

6.1.2 Types of Belts

Because of cost and efficiency, several types and sizes of belts are available for different load levels. There are primarily four types of belt configurations, that is, flat, V-, round

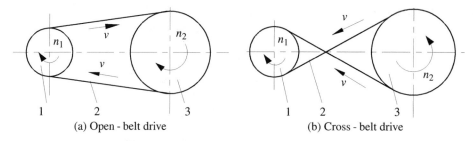

(a) Open - belt drive   (b) Cross - belt drive

Figure 6.2 Typical belt drive layouts.

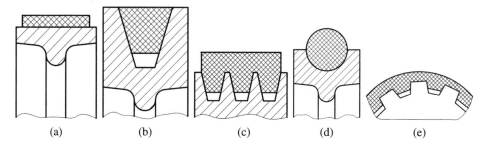

Figure 6.3 Types of belts.

and timing (or synchronous) belts, each with its individual characteristics. The cross sections of various belts are illustrated in Figure 6.3.

Flat belts are the simplest and least expensive type, with narrow rectangular cross sections (Figure 6.3a). They are designed to run on cylindrical pulleys. The contact surface between the belt and pulley is the working surface. The driving force is limited by frictional force on the working surface. Flat belts are satisfactory at high speeds and relatively low powers. When high powers are to be transmitted, flat belts become overly large and cannot compete with V-belts.

V-belts are generally endless, having trapezoidal cross sections with standard lengths (Figure 6.3b). Grooved pulleys, or sheaves, are used for V-belts. The shape of cross section causes the V-belt to wedge tightly into the V-shaped groove and the two contact sides are working surfaces. A ribbed belt combines a flat belt with V-belts, as shown in Figure 6.3c. The flat belt section carries loads, while the V-belt section provides grip on the sheave. Ribbed belts improve load carrying capacity and avoid high bending stresses.

Round belts have a circular cross section, as shown in Figure 6.3d, and are limited to light duties, mainly used in machinery for the clothing industry and in household appliances. They may be purchased in various lengths or cut to length and are joined either by a staple, a metallic connector, gluing or welding.

Timing or synchronous belts are endless flat belts with a series of evenly spaced teeth on the inner side of circumference, designed to engage with mating toothed wheels or sprockets to minimize slippage, as shown in Figure 6.3e. They are used for synchronized power transmission to ensure that the driven sheave always rotates at a constant speed ratio to the driving sheave. The power transmission capacity of a timing belt is limited by the tensile strength of belt and the shear strength of cogs. Timing belts possess high-speed characteristics of flat belts with power capacity approaching that of chains.

Figure 6.4 V-belt terminology.

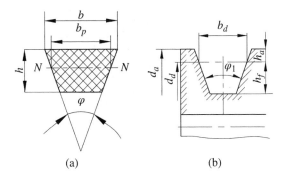

(a)　　　　　　　(b)

6.1.3 V-Belts

6.1.3.1 Terminology and Dimensions of V-Belts

Among the previously mentioned belt types, perhaps V-belts have the widest industrial applications. V-belt configurations have been standardized and widely tested for reliability and life. The datum width system is adopted by belt industry for the configuration of both standard and narrow V-belts [4]. Typical cross sections of a V-belt and sheave groove are shown in Figure 6.4. The pitch line, which is designated as N-N in Figure 6.4a, is the neutral axis of the belt section. The width of pitch line, or the pitch width, b_p, keeps constant as the belt bends around sheaves. The datum width of V-belt groove, b_d, is measured at points where the neutral axis of belt contacts the groove (Figure 6.4b). Therefore, the datum groove width b_d is identical with the pitch width b_p; that is, $b_d = b_p$. As a result, the datum diameter d_d measured at the groove width b_d is identical to the pitch diameter d_p at pitch width b_p. Since the sheave diameter D, commonly called the pitch diameter, is at a point where the neutral axis of the belt contacts the sheave groove, thus $d_d = d_p = D$. At pitch diameter, the belt and sheave have identical linear speeds. When the sheave diameter is specified as datum diameter d_d, the belt length is termed datum length L_d. Belt datum length depends on both the datum diameter and centre distance. The principal terminology and dimensions of a V-belt are given in Figure 6.4.

The wedge angle of a V-belt φ is 40°, while the nominal value of corresponding groove angle φ_1 between the two sides of V-groove on a sheave varies from 32° to 38°. This design leads the V-belt to wedge firmly onto the groove during operation so as to increase friction and allow a high torque to be transmitted before slipping occurs.

6.1.3.2 Types of V-Belts

V-belts have been standardized extensively as regards to cross sectional size, length, power ratings and sheave dimensions. There are basically two types, that is, standard or conventional V-belts and narrow V-belts. A narrow V-belt has a greater height h than that of a standard V-belt at the same pitch width b_p. Standard V-belts are designated by an letter of the alphabet for types Y, Z, A, B, C, D and E, with an increased area of cross section; while narrow V-belts are labelled SPZ, SPA, SPB and SPC for different cross section sizes. Each size is suitable for a particular power range.

6.1.3.3 V-Belt Construction

Standard V-belts are fabricated into endless constructions. Since belts are subject to tension, bending and centrifugal forces, suitable belts must be strong in tension, pliable for bending, light in weight and provide sufficient friction against sheaves. V-belts,

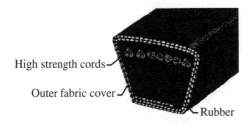

Figure 6.5 Construction of a V-belt.

High strength cords

Outer fabric cover

Rubber

commonly manufactured using leather, fabric, rubber and synthetics, provide a quiet, compact and resilient form of power transmission.

Figure 6.5 shows the construction of a typical V-belt. The outer fabric cover gives a belt good durability. High strength cords, usually made from natural fibres, synthetic strands or steel, are positioned at the pitch line of belt cross section to increase the tensile strength of belts. These cords are embedded in firm rubber compounds that provide flexibility for the belt to pass around sheaves [1]. Different sections are combined so that a belt is constructed to meet specific design requirements.

6.2 Working Condition Analysis

6.2.1 Geometrical Relationships in Belt Drives

A typical open-belt drive with pulley or sheave diameters of D_1 and D_2 and centre distance a is shown in Figure 6.6. From this figure, we have contact angle α_1 on the small pulley as

$$\alpha_1 = \pi - 2\theta$$

where

$$\theta \approx \sin\theta = \frac{D_2 - D_1}{2a}$$

Therefore, the contact angles in degrees are found to be:

$$\alpha_1 = 180^\circ - \frac{D_2 - D_1}{a} \times 57.3^\circ$$

$$\alpha_2 = 180^\circ + \frac{D_2 - D_1}{a} \times 57.3^\circ \tag{6.1}$$

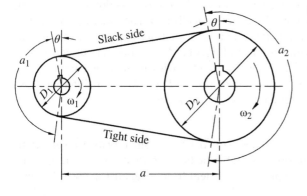

Figure 6.6 Belt drive geometry.

Slack side

a_1

D_1

ω_1

θ

a_2

D_2

ω_2

Tight side

a

The contact angle is one of several factors determining the power transmission capacity of a belt drive. It is equal to 180° only when the speed ratio is one (i.e. no speed change). The contact angle on the small pulley is always less than or equal to 180°, or π rad.

6.2.2 Force Analysis

The mechanics of belt drives can be explained by Figure 6.7a. Initially, the belt is installed by placing it around two pulleys loosely with a reduced centre distance. The two pulleys are then moved apart to a desired centre distance. The belt thus clamps firmly against the pulley surface with an initial tension F_0. This initial tension is critical because it ensures the belt works properly under design loads.

Due to initial tension F_0, normal force N generates at the interface between the belt and pulleys in the radial direction. When the driving pulley rotates at a rotational speed n_1 in the clockwise direction, the belt applies frictional force to the driving pulley in the anticlockwise direction. In return, the driving pulley applies the same amount of frictional force F_f in the opposite direction, that is, in the clockwise direction, to the belt. The belt is thus driven by this frictional force. Similarly, when the belt moves around the driven pulley, the frictional force acting on the driven pulley by the belt is in the clockwise direction. This frictional force acts as tangential force and the driven pulley is thus driven by this force and rotates at a rotational speed n_2. Hence, the belt drives use friction as a useful agent to transmit power.

From startup to operating speed, the tensile force in the bottom strand of belt increases from initial tension F_0 to tight tension F_1; while the tensile force in the upper strand drops from F_0 to a smaller value of F_2, termed slack or loose tension. Correspondingly, the bottom strand of belt is termed the tight side, while the upper strand is the slack side.

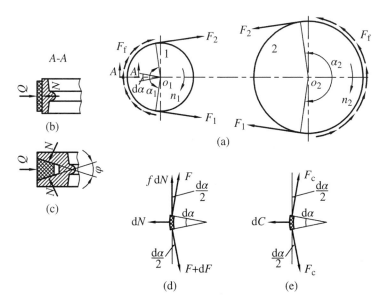

Figure 6.7 Force analysis of a belt drive.

6.2.2.1 Force Analysis of an Element of Belt

Figure 6.7b illustrates an element of flat belt acting on a pulley by a radial force Q. The frictional force on the flat belt working surface is

$$F_f = fN = fQ \tag{6.2}$$

Figure 6.7c illustrates a segment of a V-belt acting on a sheave by the same amount of radial force Q, ignoring friction in the radial direction, the frictional force on the V-belt working surfaces is

$$F_f = 2fN = \frac{f}{\sin \frac{\varphi}{2}} Q = f_v Q \tag{6.3}$$

Consequently, the equivalent coefficient of friction of V-belt f_v is

$$f_v = \frac{f}{\sin \frac{\varphi}{2}} \tag{6.4}$$

Obviously, the equivalent coefficient of friction on a V-belt drive f_v is greater than that on a flat belt drive f. This is called the wedge effect. Because of the increased friction resulting from wedge effect, V-belt drives provide a better overall power transmission capability than a flat belt drive in terms of both low cost and space. The following analysis is on flat belts for simplicity. For V-belts, the coefficient of friction is substituted by f_v.

6.2.2.2 Relations Between Tight Tension F_1, Slack Tension F_2, Initial Tension F_0 and Effective Tension F_e

Assume the total length of belt keeps constant. Therefore, the increase of belt length in the tight side is the same as the decrease of belt length in the slack side; that is, $F_1 - F_0 = F_0 - F_2$. Hence, we have the initial tension as

$$F_0 = \frac{1}{2}(F_1 + F_2) \tag{6.5}$$

Summing torque with respect to the centre of driving pulley in Figure 6.7a, we have

$$\sum T = F_f \frac{D_1}{2} + F_2 \frac{D_1}{2} - F_1 \frac{D_1}{2} = 0$$
$$F_f = F_1 - F_2$$

The effective tension on the belt drive, which is related to the pulley torque, is the sum of distributed frictional forces at the interface between belt and pulley, thus,

$$F_e = F_f = F_1 - F_2 \tag{6.6}$$

Incorporating Eq. (6.5), the tight tension F_1 and slack tension F_2 are given by

$$F_1 = F_0 + \frac{F_e}{2} \tag{6.7}$$

$$F_2 = F_0 - \frac{F_e}{2} \tag{6.8}$$

The power transmission capacity, P, of a belt is

$$P = \frac{F_e v}{1000} \tag{6.9}$$

Therefore, in order to ensure belt to transmit power P at speed v, an effective tension F_e or frictional force F_f must be guaranteed between the belt and pulley. In other words, a minimum initial tension must be maintained on the belt.

In summary, the effective tension F_e is the sum of distributed frictional forces at the interface between the belt and pulley. It also equals the difference between the tensions on the two strands of belt. The greater the difference in belt tension, the greater the power transmitted. Since the frictional force has a critical or maximum value that limits the power a belt drive can transmit, it is therefore necessary to determine the maximum frictional force or the maximum effective tension.

6.2.2.3 Critical or Maximum Effective Tension, F_{ec}

Consider a differential belt element shown in Figure 6.7d: the forces acting on the belt element are tensions F, $F + dF$ and normal force dN. Assume frictional force on the belt is proportional to the normal force, the maximum frictional force at the point of slip is fdN along the contact arc. Neglecting centrifugal tension in the belt for the time being, according to the force equilibrium in both radial and tangential directions we have

$$\begin{cases} dN = F \sin \frac{d\alpha}{2} + (F + dF) \sin \frac{d\alpha}{2} \\ fdN + F \cos \frac{d\alpha}{2} = (F + dF) \cos \frac{d\alpha}{2} \end{cases}$$

For small angles,

$$\begin{cases} \sin \frac{d\alpha}{2} \approx \frac{d\alpha}{2} \\ \cos \frac{d\alpha}{2} \approx 1 \end{cases}$$

We then have

$$\frac{dF}{F} = fd\alpha$$

Integrating this equation over the contact angle α, gives

$$\int_{F_2}^{F_1} \frac{dF}{F} = \int_0^{\alpha} fd\alpha$$

Which leads to

$$\frac{F_1}{F_2} = e^{f\alpha} \tag{6.10}$$

where

f — coefficient of friction between the belt and pulley;

α — contact angle, rad;

e — 2.718, the basis of the natural logarithm.

Equation (6.10) describes the relation between the tight tension F_1 and slack tension F_2 on the verge of slipping. From Eqs. (6.5), (6.6) and (6.10) the maximum effective tension can be expressed as

$$F_{ec} = 2F_0 \frac{1 - 1/e^{f\alpha}}{1 + 1/e^{f\alpha}} \tag{6.11}$$

Equation (6.11) gives a fundamental insight into belt drive forces. Factors affecting the maximum effective tension F_{ec} include initial tension F_0, contact angle α and coefficient of friction f.

The maximum effective tension F_{ec} and the torque are in proportion to the initial tension F_0. With the increase of initial tension F_0, the drive capacity may increase. However, too large an initial tension is not necessary, as increasing F_0 will inevitably increase tight tension F_1 and slack tension F_2, which will in turn reduce belt fatigue life. For a belt drive to transmit power satisfactorily, a proper initial tension must be provided and maintained.

Additionally, increasing contact angle α will increase contact surfaces, and consequently increase total frictional force. Usually, contact angles should be no less than 120°. Similarly, increasing the coefficient of friction f will increase total frictional force and eventually increase the power carrying capacity of a belt drive.

6.2.2.4 Centrifugal Tension, F_c

For a greater power transmitting capacity, most belt drives operate at relatively high speeds. In such cases, as the belt travels around a part of circumference of pulley, centrifugal force acting on the belt creates centrifugal tension F_c. Considering a differential element of belt shown in Figure 6.7e, according to Newton's second law, the differential centrifugal force dC can be expressed as

$$dC = m\frac{v^2}{r} = rd\alpha \cdot q \cdot \frac{v^2}{r} = qv^2 d\alpha$$

where q is the mass of belt per unit length, v for belt speed and r the pulley radius.

From the force equilibrium condition in the radial direction and considering a small angle $d\alpha$, we have

$$\begin{cases} 2F_c \sin\frac{d\alpha}{2} = qv^2 d\alpha \\ \sin\frac{d\alpha}{2} \approx \frac{d\alpha}{2} \end{cases}$$

Therefore, the centrifugal tension F_c is

$$F_c = qv^2 \tag{6.12}$$

When the centrifugal tension is sufficiently large, it should be considered in both F_1 and F_2 in Eq. (6.10) and expressed as [5]

$$\frac{F_1 - qv^2}{F_2 - qv^2} = e^{f\alpha} \tag{6.13}$$

6.2.3 Kinematic Analysis

6.2.3.1 Elastic Creep

When a belt drive is in operation in Figure 6.8, the belt first contacts the driving pulley at point A_1 with tight tension F_1 at speed v, which is the same as the linear speed of driving pulley v_1. The belt then passes through the idle arc of A_1C_1 with no changes in F_1 and v. As the belt moves along the active arc C_1B_1, the tension on a differential belt element close to the tight side is greater than that close to the slack side, resulting in the elastic deformation near the tight side $d\lambda_1$ greater than that on the slack side $d\lambda_2$. The belt contracts backward relative to the driving pulley and elastic creep begins. At the end of the active pulley arc C_1B_1 at point B_1, the belt leaves the pulley with the slack tension F_2 and a reduced speed v. Similarly, for the driven pulley, the belt tension

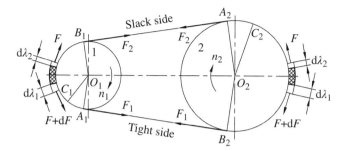

Figure 6.8 Belt elastic creep.

gradually increases from F_2 to F_1, resulting in a gradual increase in elastic deformation of belt along the active arc C_2B_2. The belt elongates forward relative to the driven pulley, leading to a higher belt speed v than the linear speed of driven pulley v_2.

This slight relative sliding between the belt and pulleys due to different tensions on the tight and slack side is called elastic creep. It is the result of minute stretch or contraction of belt as its tension various between F_1 and F_2 while the belt going through contact arcs, causing a loss of driven speed. Since it is essential for different tensions to exist in order to exert a tangential force on the driven pulley, elastic creep is an inevitable physical phenomenon in belting. Except for timing belts, there is always elastic creep in belt drives. Compared with V-belts, flat belts are most affected by this phenomenon. Because of elastic creep, the speed ratio of belt drive is neither constant nor exactly equal to the ratio of two pulley diameters. Therefore, belt drives cannot be used where a constant output speed is required.

6.2.3.2 Slippage of Belts

The contact arc includes the idle arc and active arc. In the driving pulley, when the belt works normally elastic creep only happens within the active arc C_1B_1 where the belt leaves. If the effective tension exceeds the sum of limited frictional force, the length of active arc C_1B_1 will increase. If the active arc extends to the whole contact arc, that is, point C_1 overlaps with point A_1, a total slippage of belt relative to the pulley occurs. This will cause excessive wear and must be avoided.

6.2.3.3 Speed Ratio

Most prime movers rotate at higher speeds than are desirable for driven machines. Theoretically, in the absence of slippage, the linear speeds of driving and driven pulleys are equal to the belt speed, v. Therefore

$$v_1 = \frac{\pi D_1 n_1}{60 \times 1000} \qquad v_2 = \frac{\pi D_2 n_2}{60 \times 1000}$$

Since

$$v = v_1 = v_2$$

The speed ratio, i, is expressed as:

$$i = \frac{n_1}{n_2} = \frac{D_2}{D_1}$$

where

v_1, v_2 — linear speed of driving and driven pulleys, respectively, m s^{-1};

n_1, n_2 — rotational speed of driving and driven pulleys, respectively, rpm;

D_1, D_2 — diameter of driving and driven pulleys, respectively, in mm.

That is, the speed ratio of driving and driven pulleys is inversely proportional to the ratio of their diameters. However, except for timing belts, there is always elastic creep between the belts and pulleys. Creep causes a small reduction in speed, that is, the linear speed of driven pulley is slightly smaller than that of the driving pulley. The difference is measured by a slip ratio, ε, defined as

$$\varepsilon = \frac{v_1 - v_2}{v_1} \times 100\% \tag{6.14}$$

Since slip ratio varies from 1 to 2%, it can therefore be neglected. We then have speed ratio, i, approximately expressed as:

$$i = \frac{n_1}{n_2} = \frac{D_2}{(1 - \varepsilon)D_1} \approx \frac{D_2}{D_1} \tag{6.15}$$

6.2.4 Stress Analysis

As discussed before, a working V-belt is subject to tension, bending and centrifugal forces. Consequently, stresses in a belt include normal tensile stress, bending stress and centrifugal stress.

6.2.4.1 Tensile Stress in Tight Side, σ_1, and Slack Side, σ_2

The tensile stresses caused by belt tensions are normal stresses, with a larger value on the tight side,

$$\text{Tight side} \quad \sigma_1 = \frac{F_1}{A}$$

$$\text{Slack side} \quad \sigma_2 = \frac{F_2}{A} \tag{6.16}$$

6.2.4.2 Centrifugal Stress, σ_c

The centrifugal stress is a normal stress caused by centrifugal tension as belts move around sheaves, expressed as

$$\sigma_c = \frac{F_c}{A} = \frac{qv^2}{A} \tag{6.17}$$

6.2.4.3 Bending Stress, σ_b

In addition to tensile stresses, belts are also subject to cyclic bending stresses when in contact with sheaves. The outer and inner faces of a belt are in tension and compression, respectively. Between these two faces, there is a neutral section that is neither in tension nor in compression. The bending stress is expressed as [6, 7]

$$\sigma_b \approx E\frac{h}{D} \tag{6.18}$$

where h is the belt height and D is the sheave diameter. Since the diameter of driving sheave is normally smaller than that of driven sheave, the bending stress in the driving sheave σ_{b1} is greater than that in the driven sheave σ_{b2}. Too small a dimension of a sheave will drastically increase bending stress and reduce belt life. To avoid excessive bending stress, the minimum sheave diameters for driving sheaves for standard V-belts of types Y, Z, A, B, C, D and E are recommended as 20, 50, 75, 125, 200, 355 and 500 mm, respectively [4].

To sum up, the contributors to the maximum stress σ_{max} in a belt can be expressed as:

$$\sigma_{max} = \sigma_1 + \sigma_{b1} + \sigma_c \qquad (6.19)$$

The maximum stress occurs where the belt enters the small sheave and the bending stress is the major part. A belt experiences a rather complex cycle of stress variation as the belt repeatedly passes around sheaves.

6.2.5 Potential Failure Modes

During operation, a belt is subjected to fluctuating stresses at any cross section, varying from $\sigma_c + \sigma_2$ to $\sigma_1 + \sigma_{b1} + \sigma_c$, as the belt goes through a full revolution. Fatigue, therefore, becomes a potential failure mode in belt drives.

Belt drives are applied where rotational speeds are relatively high, usually at the first stage of speed reduction from an electric motor. If a belt works at a low speed or the belt tension becomes too large, slippage may occur easily, which may cause severe wear on the belt. Both adhesive or abrasive wear may be potential failure modes in belt drives.

Besides, if belt drives operate at an elevated temperature, in a corrosive environment or in adverse loading conditions, degradation in belt material properties, that is, cord breakage and fabric cover cracking, are also potential failure modes [2].

6.3 Power Transmission Capacities

6.3.1 The Maximum Effective Tension

To prevent slippage, there should be a limitation on the maximum effective tension F_{ec}. From Eqs. (6.6) and (6.10), the maximum effective tension at critical slippage state can be expressed as

$$F_{ec} = \left(1 - \frac{1}{e^{f\alpha}}\right) F_1 \qquad (6.20)$$

To prevent fatigue failure and to ensure a belt has sufficient fatigue strength and service life, the maximum stress σ_{max} should not exceed the allowable stress of belt $[\sigma]$, that is,

$$\sigma_{max} = \sigma_1 + \sigma_{b1} + \sigma_c \leq [\sigma]$$

Therefore

$$\sigma_1 \leq [\sigma] - \sigma_{b1} - \sigma_c$$

Since $F_1 = \sigma_1 A$, and using Eq. (6.20), we then have

$$F_{ec} = ([\sigma] - \sigma_{b1} - \sigma_c)A\left(1 - \frac{1}{e^{f\alpha}}\right) \tag{6.21}$$

This is the maximum effective tension within the range of which no slippage and fatigue will occur.

6.3.2 Power Transmission Capacity of a Single V-Belt

6.3.2.1 The Basic Power Rating of a Single Standard V-Belt, P_0

The power transmitted by a belt is the product of the maximum effective tension and belt speed. Substituting f by the equivalent coefficient of friction of a V-belt f_v in Eq. (6.4), we have the basic power rating for a single standard V-belt as

$$P_0 = \frac{F_{ec}v}{1000} = \frac{([\sigma] - \sigma_{b1} - \sigma_c)Av\left(1 - \frac{1}{e^{f_v\alpha}}\right)}{1000} \quad \text{kW} \tag{6.22}$$

The basic power rating a single standard V-belt can transmit is obtained under specific test conditions. The test condition specifies as a moderate length belt running on identical diameter sheaves, transmitting a steady load. Due to limited space, this book only presents data for Type A, B and C V-belts. Detailed data for other belt types can be found in standards or design handbooks [8].

The basic power ratings of single standard Type A, B and C V-belts as the function of pitch diameter of small sheave and selected rotational speeds of 700, 1450 and 2800 rpm are presented in Figure 6.9. The abscissas of the points in the figure give pitch diameters of small sheaves, which can be referred to in Table 6.1. The basic power ratings at other rotational speeds can be obtained by interpolation.

6.3.2.2 The Actual Power Rating of a Single V-Belt, P_r

In practice, deviations from these specific test conditions are acknowledged by introducing correction factors. A given belt can carry a greater power as the speed ratio increases. The increment of transmitted power ΔP_0, as listed in Figure 6.10, is added to the basic power rating to take this factor into account.

Since contact angles affect frictional force and ultimately the transmitted power, a correction factor K_α is introduced for contact angles other than 180°. Corresponding to contact angles of 180°, 150°, 120° and 90°, the correction factor K_α can be selected as 1.0, 0.92, 0.82 or 0.69 [4], respectively. The correction factor K_α for other contact angles can be obtained by interpolation.

Similarly, if a belt length is different from the specified belt length when the basic power rating is obtained, a belt length correction factor K_L is introduced, as illustrated in Figure 6.11. After incorporating these correction factors, the actual power rating of a single V-belt is adjusted as

$$P_r = (P_0 + \Delta P_0)K_\alpha K_L \tag{6.23}$$

where

P_r — actual power rating of a single standard V-belt;
P_0 — basic power rating of a single standard V-belt, see Figure 6.9;
ΔP_0 — increment of the basic power rating of a single standard V-belt, see Figure 6.10;
K_α — correction factor for contact angle;
K_L — correction factor for belt length, see Figure 6.11.

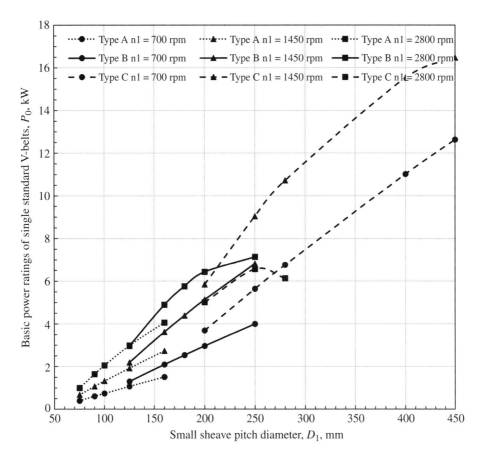

Figure 6.9 Basic power ratings of single standard V-belts P_0, kW [4].

Table 6.1 Datum diameters of V-belt sheaves and standard V-belt datum lengths L_d, mm [4].

Belt type	Datum diameter, D	Standard V-belt datum lengths, L_d
A	75, 80, 85, 90, 95, 100, 106, 112, 118, 125, 132, 140, 150, 160, 180, 200, 224, 250, 280, 315, 355, 400, 450, 500, 560, 630, 710, 800	630, 700, 790, 890, 990, 1 100, 1 250, 1 430, 1 550, 1 640, 1 750, 1 940, 2 050, 2 200, 2 300, 2 480, 2 700
B	125, 132, 140, 150, 160, 170, 180, 200, 224, 250, 280, 315, 355, 400, 450, 500, 560, 600, 630, 710, 750, 800, 900, 1 000, 1 120	930, 1 000, 1 100, 1 210, 1 370, 1 560, 1 760, 1 950, 2 180, 2 300, 2 500, 2 700, 2 870, 3 200, 3 600, 4 050, 4 430, 4 820, 5 370, 6 070
C	200, 212, 224, 236, 250, 265, 280, 300, 315, 335, 355, 400, 450, 500, 560, 600, 630, 710, 750, 800, 900, 1 000, 1 120, 1 250, 1 400, 1 600, 2 000	1 565, 1 760, 1 950, 2 195, 2 420, 2 715, 2 880, 3 080, 3 520, 4 060, 4 600, 5 380, 6 100, 6 815, 7 600, 9 100, 10 700

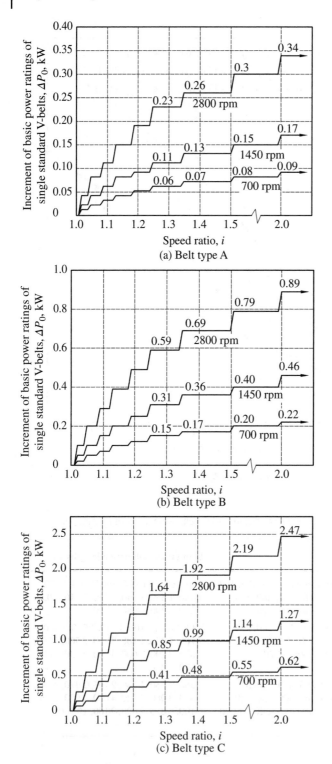

Figure 6.10 Increment of basic power ratings of single standard V-belts ΔP_0, kW [1, 4].

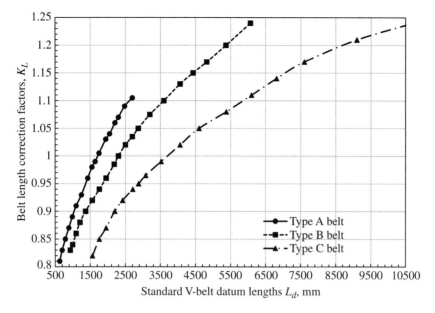

Figure 6.11 Standard V-belt datum lengths L_d versus length correction factors K_L [4].

6.4 Design of Belt Drives

6.4.1 Introduction

Belt drive design involves either the selection of a proper belt to transmit a required power or the determination of a power that can be transmitted by a belt drive. When designing a belt drive, the data provided are the rated power of driving motor or other prime mover, the rotational speed of driving sheave n_1 and driven sheave n_2 (or speed ratio), the type of driver and driven machine and the layout requirement of the belt drive.

The data that need to be determined include belt type, length and the number of belts; centre distance; the size, materials and structure of driving and driven sheaves, as well as the initial tension on the belt. The design procedure and factors involved in the selection of a V-belt and driving and driven sheaves are introduced in this section.

6.4.2 Design Criteria

As discussed in Section 6.2.5, principal failure modes in a belt drive are fatigue and slippage. To prevent these failures, design criteria for a belt drive are to prevent belts from slippage and at the same time to ensure belts have sufficient fatigue strength and service life.

6.4.3 Design Procedure and Guidelines

The design procedure and guidelines for variable selection are given as follows:

6.4.3.1 Compute Design Power, P_{ca}

A belt drive is designed based on the design power P_{ca}, which is obtained by multiplying a service factor with the rated power of a mover, P, expressed as

$$P_{ca} = K_A P \tag{6.24}$$

where K_A is a service factor, depending on the nature of duty of driver and driven machines. Due to the inherent characteristics of belt drives, such as slippage and damping effect of rubble, service factors for belt drives are low compared with those for chain or gear drives. Therefore, the lower range of values from Table 2.1 can be selected as service factors for belt drive design.

6.4.3.2 Specify Suitable Belt Types

The belt type can be tentatively selected from standards or manufacturer's catalogues according to the rotational speed of driving sheave n_1 and design power P_{ca}. When a single V-belt has insufficient capacity, either multiple belts or a lager cross section belt can be selected, determined from an economic point of view. For example, considering the cost and the weight of sheave, a single Type B V-belt may be a better choice than two Type A belts.

6.4.3.3 Determine the Sheave Size

The size of driving sheave should be greater than the minimum diameter recommended for each belt type (see Table 6.1). Although a diameter less than the minimum diameter could lead to a compact design, it will increase bending stress in belts and thereby shorten belt life. The desirable dimension of driven sheave is computed by $D_2 = iD_1$ and rounded to a diameter according to Table 6.1.

Belt speeds should be within a range of 5–$25\,\mathrm{m\,s^{-1}}$. A lower speed may cause an increase of belt tension, the number of belts and the likelihood of slippage, while a higher speed may increase dynamic effects, such as centrifugal force, belt whip and vibration, and eventually affect the working capacity of belt drives.

6.4.3.4 Confirm the Centre Distance, a and Belt Datum Length, L_d

(1) Specify a trial centre distance, a_0

Shorter centre distances are preferred for stability of operation and economy of space. However, too small a centre distance may increase the frequency the belt cycles sheaves, increasing the likelihood of fatigue failure. The minimum centre distance is limited by sheave dimensions and the minimum contact angle. On the other hand, too large a centre distance will increase the size of belt drive and cause vibration if the belt speed is high. The maximum centre distance is limited by available belt datum lengths. The following formula is used to initially estimate a nominal acceptable range of centre distance [6]

$$0.7(D_1 + D_2) < a_0 < 2(D_1 + D_2) \tag{6.25}$$

(2) Compute preliminary belt length, L'_d

Belt length is obtained by summing the contact arc lengths with twice the distance between the beginning and end of contact. As introduced previously, when a sheave diameter is specified as the datum diameter d_d, the belt length is termed datum length L_d. Since the datum diameter d_d is identical to the sheave diameter D, for an open-belt drive in Figure 6.6, the preliminary belt length L'_d is given by

$$L'_d = 2a_0 + \frac{\pi}{2}(D_1 + D_2) + \frac{(D_2 - D_1)^2}{4a_0} \tag{6.26}$$

(3) Select belt datum length, L_d

V-belts are of closed loop and made in standard lengths. The belt datum length L_d of Types A, B and C belt can be selected from Table 6.1, close to the preliminary calculation of L'_d. The corresponding length correction factor K_L can be found in Figure 6.11.

(4) Compute the actual centre distance, a

The actual centre distance a is calculated by

$$a \approx a_0 + \frac{L_d - L'_d}{2} \tag{6.27}$$

Considering the installation adjustment and pretension compensation to take up slack in belts due to stretch and wear, the variation of centre distance is given as

$$a_{min} = a - 0.015L_d$$

$$a_{max} = a + 0.03L_d$$

6.4.3.5 Compute Contact Angle on the Small Sheave, α_1

Since belts transmit power by friction against sheaves, large values of equivalent coefficient of friction f_v and contact angle α favour power transmission capacity. The contact angle should be greater than $120°$ to ensure adequate contact and frictional force, that is, the contact angle on the small sheave α_1 must satisfy

$$\alpha_1 = 180° - \frac{D_2 - D_1}{a} \times 57.3° \geq 120° \tag{6.28}$$

6.4.3.6 Compute the Number of Belts Required to Carry the Design Power

If a single V-belt has insufficient capacity, multiple belts may be used. The number of belts required is calculated from

$$z = \frac{P_{ca}}{P_r} = \frac{P_{ca}}{(P_0 + \Delta P_0)K_a K_L} \leq 10 \tag{6.29}$$

The number of belts required is normally less than 10 to ensure each belt carries a similar load. When one belt needs replacement, a complete new set should be installed.

6.4.3.7 Decide Initial Tension, F_0

Belts must be installed with an initial tension recommended by manufacturers. From Eqs. (6.9) and (6.11), and considering the centrifugal force and the effect of contact angle, the initial tension can be estimated by [6]

$$F_0 = 500\frac{P_{ca}}{zv}\left(\frac{2.5}{K_a} - 1\right) + qv^2 \tag{6.30}$$

6.4.3.8 Compute the Force Acting on the Sheave Shaft, F_Q

For proper design and installation of shafts and bearings, it is desirable to determine the force acting on the shaft mounting the sheave. From Figure 6.7, the force acting on the sheave shaft is calculated by:

$$F_Q = 2zF_0 \sin \frac{a_1}{2} \tag{6.31}$$

6.4.4 Design of V-Belt Sheaves

V-belt sheaves should have uniformly distributed mass, a highly furnished groove surface and be lightweight. For a high-speed sheave, the inertia force must be well balanced.

Materials used in a sheave include cast iron, cast steel, aluminium, Ultra High Molecular Weight (UHMW) polyethylene and so on. Cast iron is used for sheaves with a rotating speed less than $25\,\mathrm{m\,s^{-1}}$, while cast steel is recommended for higher speed and aluminium and UHMW for light-duty drives.

Figure 6.12 shows various structures of V-belt sheaves. The selection of sheave structure largely depends on the size, indicated by datum diameter d_d, which is slightly smaller than the outside diameter of sheave. Figure 6.12a shows a solid sheave whose datum diameter is less than 2.5–3 times the diameter of the shaft the sheave is mounted on. Figure 6.12b,c are sheaves with thinned web and hole plate, respectively, whose datum diameters are less than 300 mm. Figure 6.12d is a spoke sheave whose datum diameter is more than 300 mm. Detailed dimensions can be referred to in design handbooks [8].

6.4.5 Design Cases

Example Problem 6.1

Design a V-belt drive for a conveyor to be driven by a 7.5 kW motor running at 1450 rpm. The belt drive has a speed ratio of 3.5 and works less than 12 hours daily. Select Type A belt, and the mass per unit length of Type A belt is $q = 0.105\,\mathrm{kg\,m^{-1}}$.

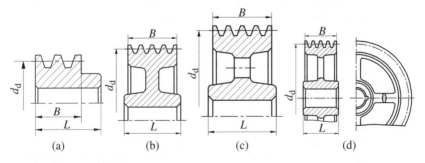

Figure 6.12 Sheave structures of V-belt drives.

Solution

Steps	Computation	Results	Units
1. Compute the design power P_{ca}	Select service factor K_A from Table 2.1 as $K_A = 1.1$, therefore the design power is $P_{ca} = K_A P = 1.1 \times 7.5 = 8.25$ kW	$P_{ca} = 8.25$	kW
2. Determine the diameters of the sheaves	Select the diameter of the driving sheave as $D_1 = 100$ mm according to Table 6.1. Select slip ratio $\varepsilon = 1.5\%$, the diameter of the driven sheave can be calculated by Eq. (6.15) $D_2 = i(1 - \varepsilon)D_1 = 3.5 \times (1 - 1.5\%) \times 100 = 344.8$ mm Select $D_2 = 355$ mm according to Table 6.1. Check the speed of the belt by $v = \dfrac{\pi D_1 n_1}{60 \times 1000} = \dfrac{\pi \times 100 \times 1450}{60 \times 1000} = 7.59 < 25$ m/s The speed of the belt is acceptable. Check the error of speed ratio $\Delta i = \left\lvert \dfrac{D_2/D_1 - 3.5}{3.5} \right\rvert \times 100\% = \left\lvert \dfrac{355/100 - 3.5}{3.5} \right\rvert = 1.43\% < 5\%$ The speed ratio is acceptable.	$D_1 = 100$ $D_2 = 355$ $v = 7.59$	mm mm $\mathrm{m\,s^{-1}}$
3. Confirm centre distance a and belt datum length L_d	Specify a trial centre distance according to $0.7(D_1 + D_2) < a_0 < 2(D_1 + D_2)$ The initial centre distance is selected as $a_0 = 500$ mm. The preliminary belt length is calculated by Eq. (6.26) $L'_d = 2a_0 + \dfrac{\pi}{2}(D_1 + D_2) + \dfrac{(D_2 - D_1)^2}{4a_0}$ $= \left[2 \times 500 + \dfrac{\pi}{2} \times (100 + 355) + \dfrac{(355 - 100)^2}{4 \times 500} \right] = 1747$mm Select the belt datum length from Table 6.1 as $L_d = 1750$ mm. The actual centre distance a is calculated from Eq. (6.27) $a \approx a_0 + \dfrac{L_d - L'_d}{2} = \left(500 + \dfrac{1750 - 1747}{2} \right) \approx 502$ mm $a_{min} = a - 0.015L_d = 502 - 0.015 \times 1750 = 476$ mm $a_{max} = a + 0.03L_d = 502 + 0.03 \times 1750 = 555$ mm	$L_d = 1750$ $a = 502$ $a_{min} = 476$ $a_{max} = 555$	mm mm mm mm
4. Compute the contact angle on the driving sheave	From Eq. (6.28) we have $\alpha_1 = 180^\circ - \dfrac{D_2 - D_1}{a} \times 57.3^\circ$ $= 180^\circ - \dfrac{355 - 100}{502} \times 57.3^\circ = 150.9^\circ > 120^\circ$ So the contact angle on the driving sheave is acceptable.	$\alpha = 150.9$	$^\circ$
5. Compute the number of belts required to carry the design power	From Eq. (6.29) $z = \dfrac{P_{ca}}{(P_0 + \Delta P_0)K_a K_L}$ Since $n_1 = 1450$ rpm, $D_1 = 100$ mm, $i = 3.5$, from Figures 6.9 and 6.10a, we have $P_0 = 1.32$ kW, $\Delta P_0 = 0.17$ kW Select $K_a = 0.92$. From Figure 6.11, select $K_L = 1.00$. $z = \dfrac{8.25}{(1.32 + 0.17) \times 0.92 \times 1.00} = 6.02$ Select $z = 7$	$z = 7$	

Steps	Computation	Results	Units
6. Confirm initial tension, F_0	From Eq. (6.30) $$F_0 = 500\frac{P_{ca}}{zv}\left(\frac{2.5}{K_a} - 1\right) + qv^2$$ $$= 500 \times \frac{8.25}{7 \times 7.59}\left(\frac{2.5}{0.92} - 1\right) + 0.105 \times 7.59^2 = 139\,\text{N}$$	$F_0 = 139$	N
7. Compute the force acting on the sheave shaft, F_Q	From Eq. (6.31) $$F_Q = 2zF_0 \sin\frac{\alpha_1}{2} = 2 \times 7 \times 139 \times \sin\frac{150.9^\circ}{2} = 1884\,\text{N}$$	$F_Q = 1884$	N
8. Structural design of V-belt sheaves	$D_1 = 100\,\text{mm}$ $D_2 = 355\,\text{mm}$	Select solid sheave for driving sheave and spoke sheave for driven sheave	

6.5 Installation and Maintenance

At installation, the two sheaves are first moved closer to facilitate belt assembly. The centre distance is then increased to a designed value to generate initial tension in the belt. A proper amount of initial tension must be well maintained. An excessive initial tension shortens belt life and overloads bearings and shafts, while insufficient initial tension may cause slippage and generate heat and wear, which may also reduce belt life. Besides, belts stretch permanently and loose initial tension after a period of service. Therefore, a belt drive must have provisions to allow for belt adjustment and replacement.

When the centre distance of a belt drive is fixed, a grooved idler is preferably used on the inside of slack strand close to the large sheave to take up slack (Figure 6.13). If the centre distance is adjustable, belt tension can be maintained by regular manual tensioning or by the weight of pivoted, overhung motor [9]. Manufacturers' catalogues and design handbooks give details for proper belt-tensioning procedures.

Figure 6.13 Belt tensioning by an idler pulley.

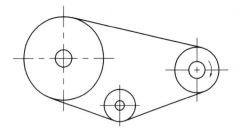

References

1 Mott, R.L. (2003). *Machine Elements in Mechanical Design*, 4e. Prentice Hall.
2 Collins, J.A. (2002). *Mechanical Design of Machine Elements and Machines: A Failure Prevention Perspective*, 1e. New York: Wiley.
3 Hindhede, U., Zimmerman, J.R., Hopkins, R.B. et al. (1983). *Machine Design Fundamentals: A Practical Approach*. New York: Wiley.
4 GB/T13575.1-2008 (1995). *Classical and Narrow V-Belt Drives-Part 1: System Based on Datum Width*. Beijing: Standardization Administration of the People's Republic of China.
5 Qiu, X.H. (1997). *Mechanical Design*, 4e. Beijing: Higher Education Press.
6 Pu, L.G. and Ji, M.G. (2006). *Mechanical Design*, 8e. Beijing: Higher Education Press.
7 Gere, J.M. and Timoshenko, S.P. (1996). *Mechanics of Materials*, 4e. CL Engineering.
8 Wen, B.C. (2010). *Machine Design Handbook*, 5e, vol. 2. Beijing: China Machine Press.
9 Juvinall, R.C. and Marshek, K.M. (2011). *Fundamentals of Machine Component Design*, 5e. New York: Wiley.

Problems

Review Questions

1 Why are belt drives are usually arranged at a high-speed stage instead of low speed stage in a power transmission?

2 What causes a V-belt to have a greater load carrying capacity than a flat belt?

3 Which measure could be used to increase power transmission capacity of a belt drive?

4 While designing a V-belt drive, how should one solve the following problems?
   (1) the speed of the belt is either too fast or too slow;
   (2) the contact angle on the driving sheave is too small;
   (3) too many number of stress cycles;
   (4) too many number of belts.

5 During belt drive design, why do we need to set a limitation on belt speeds?

Objective Questions

1   The variable speed ratio in a belt drive is because _____.
    (a) the contact angles of the driving sheave and driven sheave are different
    (b) of belt slippage
    (c) the coefficient of friction is unstable
    (d) of the elastic creep of the belt

2   In V-belt drive design, the purpose of limiting the minimum diameter of small sheave is _____.
    (a) to limit bending stress
    (b) to make structure compact
    (c) to ensure enough friction on the contact surface between the belt and sheave
    (d) to limit contact angle on the small sheave

3   The difference between tensions on the tight side and slack side of a belt drive is 3000 N. If the belt speed is $15\,\mathrm{m\,s^{-1}}$, the transmitted power is _____ kW.
    (a) 90
    (b) 22.5
    (c) 100
    (d) 45

4   The selection of a belt type is according to _____.
    (a) the diameter of driving sheave
    (b) the rotational speed of a belt drive
    (c) the calculated power and the rotational speed of the driving sheave
    (d) the transmitted power

5   The reason of elastic creep in a belt drive is because _____.
    (a) the belt is not absolutely flexible
    (b) the coefficient of friction is small
    (c) the centrifugal force appears
    (d) of the difference between tight side tension and slack side tension

Calculation Questions

1   A flat belt transmits power between two 600 mm diameter pulleys. The coefficient of friction between the belt and pulley is 0.3. The force acting on the sheave shaft should not exceed 1200 N. Find the limiting values of the tight side tension F_1 and slack side tension F_2. Disregard service factors.

2   For a flat belt drive with diameters of $D_1 = 125\,\mathrm{mm}$ and $D_2 = 400\,\mathrm{mm}$, if the input shaft rotates at 1440 rpm, calculate
    (1) the rotational speed of the driven shaft if no elastic creep is considered.
    (2) the rotational speed of the driven shaft if a slip ratio is 2%.
    (3) how to compensate for the elastic creep.

3   A belt drive can transmit a maximum power of $P = 3\,\text{kW}$. The driving sheave has a diameter of $d_1 = 150\,\text{mm}$, rotating at $n_1 = 1420\,\text{r}\,\text{min}^{-1}$. The contact angle on the driving sheave is $\alpha_1 = 160°$. The coefficient of friction between the belt and belt sheave is 0.3. Decide the maximum effective force F_{ec} and tight side tension F_1.

4   A single belt drive with an initial tension of $F_0 = 300\,\text{N}$ has a contact angle of $\alpha_1 = 150°$ on the small drive. The belt speed is $v = 8\,\text{m}\,\text{s}^{-1}$. If the equivalent coefficient of friction between the belt and belt drive is $f = 0.5$, neglecting centrifugal force, decide the maximum power P the belt drive can transmit.

5   A belt drive has contact angles of $180°$ on the driving and driven sheaves with an initial tension of $F_0 = 150\,\text{N}$. The equivalent coefficient of friction is 0.5. Neglecting centrifugal force, if the transmitted effective force is $F_e = 180\,\text{N}$, decide whether the belt will slip and explain why.

Design Problems

1   Design a V-belt drive for a 5.5 kW motor running at 1440 rpm driving an air compressor at 520 rpm. Ensure it has the shortest possible centre distance.

2   A single V-belt is selected to deliver 3 kW engine power to a tractor. Approximately 70% of this power is transmitted to the belt. The maximum engine speed is 3000 rpm. The driving sheave has a diameter of 150 mm, the driven, 300 mm. The belt selected should be as close to a 2300 mm datum length as possible. Select a satisfactory belt.

3   A 1.5 kW electric motor running at 1460 rpm is to drive a blower at a speed of 240 rpm. Select a V-belt drive for this application and specify standard V-belts, sheave sizes and centre distance.

4   The compressor of a small air-conditioning unit requires a 1.5 kW, 1200 rpm motor. The speed of compressor is 300 rpm. The centre distance should be minimized to provide a compact design. Design a suitable V-belt drive.

Structure Design Problems

1   On a V-belt drive in Figure P6.1, the angle between the two working surfaces of the V-belt is $\phi = 40°$, then the corresponding angle on the sheave groove ϕ_1 is _____ $40°$.
    (a) >
    (b) <
    (c) =

2   In Figure P6.2, which one best depicts the position of a V-belt in a sheave groove? _____.

3   Figure P6.3 is a layout of power transmission. Find the errors and correct them.

4   For the tensioning scheme shown in Figure P6.4, part (a) is for a flat belt, while (b) is for a V-belt. Make corrections where you see the errors occur.

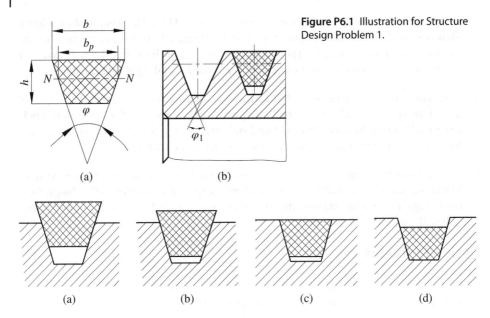

Figure P6.1 Illustration for Structure Design Problem 1.

Figure P6.2 Illustration for Structure Design Problem 2.

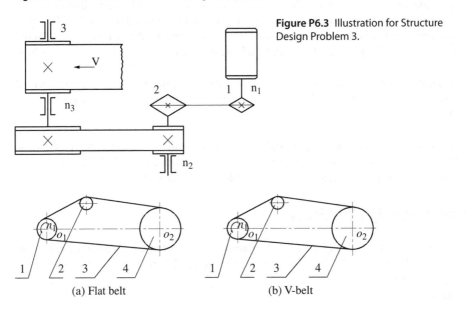

Figure P6.3 Illustration for Structure Design Problem 3.

(a) Flat belt   (b) V-belt

Figure P6.4 Illustration for Structure Design Problem 4.

5   The power from an electric motor is used to drive a conveyor running at a low speed. Propose the layout and connect components to realize the power transmission. Components are shown in Figure P6.5. At least two drives need to be used in your design.

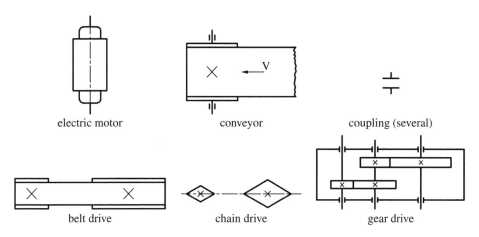

Figure P6.5 Illustration for Structure Design Problem 5.

CAD Problems

1 Write a flow chart for belt drive design process to complete the Example Problem 6.1.

2 Develop a program to implement a user interface similar to Figure P6.6 and complete the Example Problem 6.1. Select different trial centre distance of a_0 and compare the results.

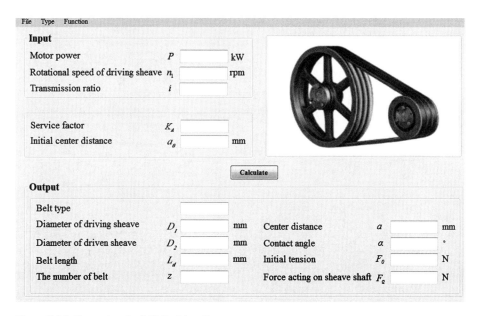

Figure P6.6 Illustration for CAD Problem 2.

7

Chain Drives

Nomenclature

a	centre distance, mm	F_d	dynamic force, N
a_c	acceleration, m s^{-2}	F_e	effective peripheral force, N
b_1	width between inner link plate, mm	F_f	sagging force, N
		F_Q	force acting on the shaft, N
D	pitch diameter of a sprocket, mm	i	speed ratio
d	chordal diameter of a sprocket, mm	K_A	service factor
		K_p	multiple-strand correction factor
d_1	roller diameter, mm	K_Q	factor
d_2	pin diameter, mm	K_z	tooth correction factor for a sprocket
F_1	tight side tension, N	L_p	chain length in pitches, the number of links
F_2	slack side tension, N	m	mass of tight side chain, kg
F_c	tension induced by centrifugal force, N	N	number of strand
		n	rotational speed of a sprocket, rpm

Analysis and Design of Machine Elements, First Edition. Wei Jiang.
© 2019 John Wiley & Sons Singapore Pte. Ltd. Published 2019 by John Wiley & Sons Singapore Pte. Ltd.
Companion website: www.wiley.com/go/Jiang/analysis_of_machine_elements

P transmitted power, kW
P_0 basic power rating of a chain drive, kW
P_r actual power rating of a chain drive, kW
p pitch, mm
p_t strand spacing, mm
q mass of chain per unit length, kg m^{-1}
Q limiting tensile load of single-strand chain, N
R pitch radius of a sprocket, mm
r chordal radius of a sprocket, mm
S_{ca} safety factor

t link plate thickness, mm
v_{avg} average speed of a chain, m s^{-1}
v instantaneous velocity along the chain, m s^{-1}
v' instantaneous velocity vertical to the chain, m s^{-1}
z number of teeth of a sprocket
α contact angle, °or rad
φ pitch angle, °
ω angular velocity, rad s^{-1}

Subscripts
1   driving sprocket
2   driven sprocket

7.1 Introduction

7.1.1 Applications, Characteristics and Structures

A chain drive is another major type of flexible drive used to transmit power over comparatively long centre distances. They are commonly used in conveyor systems, automobiles, motorcycles, bicycles and many other similar applications. Contrary to belt drives, chain drives are desirable at low to moderate speeds, high torque applications and usually at lower speed stage of a power transmission system. Figure 6.1 shows a chain drive together with a gear reducer and a belt drive in a power transmission system.

Chain drives combine some features of belt drives and gear drives. Compared with belt drives, chain drives have high transmission efficiency, great power transmitting capacity and long life, as no slippage or creep is involved between the chain and sprocket teeth. Besides, chain drives can operate in hostile environments, such as high ambient temperature, high moisture, oily, dusty or dirty situations. Furthermore, chain drives can be applied to an arbitrary centre distance compared with gear drives. The main disadvantage of a chain drive is the impact and noise due to variable angular velocity ratio and velocity. Besides, chains may elongate and even jump off sprockets due to wear [1–4].

A chain drive consists of a driving sprocket 1, a driven sprocket 3 and a chain loop 2, as illustrated in Figure 7.1. A chain is a flexible power transmission element composed of a series of pin-connected links. When transmitting power, the chain engages toothed wheels, called sprockets. The driving sprocket imparts constant tension to the chain, forcing the driven sprocket rotates. As a joint enters and leaves a sprocket, adjacent links rotate relatively to each other. Chains obtain flexibility from pin-connected links that articulate at each joint during operation [5].

7.1.2 Types of Chains

Chains include three types according to their applications, that is, power transmission chains, drag chains and lifting chains. Drag chains and lifting chains are used mostly in

Figure 7.1 A chain drive.

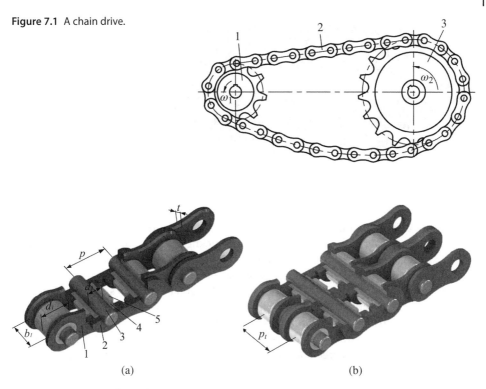

Figure 7.2 Structure of a single-strand and a double-strand roller chain.

conveying machinery and materials handling machinery, respectively [2]. Power transmission chains are mainly for transmitting mechanical energy.

Power transmission chains can be further categorized as roller chains, silent or inverted tooth chains [3], with the former being most widely used. The rollers on each pin rotate when contacting sprocket teeth, providing low friction between the chain and sprockets, leading to a high transmission efficiency around 96%. The power transmission capacity of commonly used roller chains is below 100 kW and the chain speed is less than $15 \, \mathrm{m \, s^{-1}}$ [2], although higher capacity and speed chains are also available [4]. This chapter only discusses the analysis and design of roller chains.

Figure 7.2a shows the structure of a single-strand roller chain. It is composed of inner link plate 1, outer link plate 2, pin 3, bushing 4 and roller 5. In a roller chain, the pin and outer link plate, bushing and inner link plate are press fitted, respectively. The roller and bushing, bushing and pin are slip fitted, respectively. Therefore, pins pivot inside bushings and rollers encircle each bushing to provide rolling engagement and contact with sprocket teeth.

When a higher power is to be transmitted, a multiple-strand roller chain is required. A multiple-strand roller chain consists of two or more parallel strands assembled on common pins. Chains can be manufactured in single, double, triple and quadruple strands. The number of strand is usually less than four to avoid uneven loading. Figure 7.2b shows the structure of a double strand chain.

Figure 7.3 Chain joint construction.

The variables in a roller chain are pitch p, roller diameter d_1, pin diameter d_2, width between inner link plates b_1, link plate thickness t, the number of links L_p and strand spacing p_t for a multiple-strand chain, as illustrated in Figure 7.2. Pitch is the centre distance between corresponding hinges of adjacent links, which identifies the dimension of a roller chain. The larger the pitch, the larger the chain size and the greater the power transmission capacity of a chain. The width between inner link plates b_1 is an internal dimension corresponding to the width of sprocket tooth. Usually, an even number of links with straight links is preferred, as an odd number of links needs an offset link that may introduce additional bending stress when forming a loop chain, as illustrated in Figure 7.3.

Roller chains are standardized products, with the pitch as a base for all other dimensions. Standardization organizations worldwide have issued standards, for example, ANSI/ASME B29.1-2011 [6], ISO 606:2015 [7], GB/T 1243-2006 [8] and so on, which give standard dimensions for chains and sprockets to ensure interchangeability. An identification code is adopted to facilitate specifying and ordering. In the ANSI/ASME standard, the digits (other than the final zero) indicate the pitch of chain in eighths of an inch [1]. For example, the number 100 chain has a pitch of 10/8 in. In GB/T 1243-2006 standard in China, the designation is indicated by chain number, number of strands and number of links plus the Standard No. The chain number indicates the pitch of chain in sixteenths of an inch. Chain number is usually followed by a serial number A or B. Serial A is used in America and many other countries, while serial B is mainly used in European countries. Both serials are available in China. For example, 08A-1-86 GB/T 1243-2006 indicates a serial A single-strand chain with a pitch of 12.7 mm and 86 links under Standard GB/T 1243-2006. More detailed specifications can be found in relevant references [6–8].

Silent chains are named because they operate more smoothly and are quieter than conventional roller chains. A silent chain is made from a series of toothed link plates pinned together across the width of chain to permit articulation. The teeth of links may have straight-sided or involute profiles. Power is transmitted through engagement of chain teeth with sprocket teeth. Compared with roller chains, silent chains can operate at higher speeds and carry greater loads, yet they are more expensive and heavier.

7.2 Working Condition Analysis

7.2.1 Geometrical Relationships in Chain Drives

When a roller chain sequentially meshes with sprocket teeth, the sprocket can be regarded as a polygon. The length of the side of polygon is pitch p, and the number of sides of the polygon is the number of sprocket teeth z, as shown in Figure 7.4.

From trigonometry, the relation between pitch p and pitch diameter of sprocket D can be expressed as

$$p = D \sin \frac{\varphi}{2} \tag{7.1}$$

Since pitch angle $\varphi = 360°/z$, Eq. (7.1) can be written as

$$p = D \sin \frac{180°}{z} \tag{7.2}$$

The magnitude of pitch angle φ is closely related to the number of sprocket teeth z. Since pitch angle φ may affect wear in chain joints, it is important to reduce this angle or increase the number of sprocket teeth.

The contact angle in a chain drive reflects the number of sprocket teeth in meshing or carrying the load. The calculation of contact angle on a sprocket is similar to that of a belt drive, expressed as

$$\alpha_1 = 180° - \frac{z_2 - z_1}{\pi a} p \times 57.3°$$
$$\alpha_2 = 180° + \frac{z_2 - z_1}{\pi a} p \times 57.3° \tag{7.3}$$

7.2.2 Kinematic Analysis

7.2.2.1 Speed Ratio

Figure 7.5 shows a small sprocket rotates at a constant speed in the clockwise direction, driving the large sprocket through a chain. Since the sprocket can be regarded as

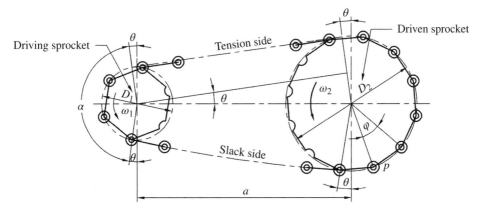

Figure 7.4 Chain drive geometry.

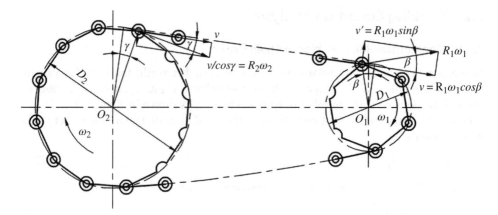

Figure 7.5 Velocity analysis of a chain drive.

a polygon, with the length of side as pitch p, and the number of sides as the number of teeth z, the circumference of the sprocket is zp. If the rotational speeds of the small and large sprockets are n_1 and n_2, respectively, the calculation of average speed of a chain drive is similar to that of a belt drive, expressed as

$$V_{avg} = \frac{z_1 n_1 p}{60 \times 1000} = \frac{z_2 n_2 p}{60 \times 1000} \tag{7.4}$$

Thus, the speed ratio is

$$i = \frac{n_1}{n_2} = \frac{z_2}{z_1} \tag{7.5}$$

7.2.2.2 Angular Velocity Ratio

From Figure 7.5, the velocity component along the chain at angle β is

$$v = R_1 \omega_1 \cos \beta \tag{7.6}$$

The velocity component vertical to the chain at angle β is

$$v' = R_1 \omega_1 \sin \beta \tag{7.7}$$

Since angle β varies from $-180°/z_1$ to $180°/z_1$, the instantaneous velocity of chain varies periodically, even if the driving sprocket rotates at a constant speed. The variation of velocity component v causes an uneven chain velocity. Larger pitch p or fewer teeth z in a sprocket leads to a greater velocity variation. At the same time, the periodical variation of velocity component v' causes the rise and fall of chain. At any instant,

$$v = R_1 \omega_1 \cos \beta = R_2 \omega_2 \cos \gamma \tag{7.8}$$

The angular velocity ratio is thus

$$\frac{\omega_1}{\omega_2} = \frac{R_2 \cos \gamma}{R_1 \cos \beta} \tag{7.9}$$

Therefore, even if a driving sprocket rotates at a constant angular velocity, the driven sprocket experiences fluctuation in angular velocity. The angular velocity ratio also varies. Only when the numbers of teeth of driving sprocket and driven sprocket are identical and the centre distance is an integer multiple of pitch p will the angular velocity ratio be constant [2].

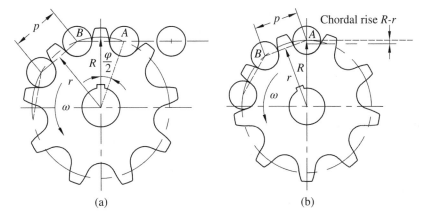

Figure 7.6 Chordal action of roller chain. Source: Adapted from Juvinall & Marshek 2001, Figure 19.7, p. 790. Reproduced with permission of John Wiley & Sons, Inc.

7.2.2.3 Chordal Action

As introduced before, when a chain sequentially meshes with sprocket teeth, the sprocket resembles a polygon. Figure 7.6a,b illustrates two positions where the centre-lines of chain are at chordal radius r and sprocket pitch radius R, respectively. As the sprocket rotates, the amount of chain rise and the fall of the chain centreline is

$$\Delta r = R - r = R \left(1 - \cos \frac{180^\circ}{z} \right) \tag{7.10}$$

Similarly, the velocity reaches the minimum and maximum value at these two positions. The minimum velocity occurs at a chordal radius r is

$$v_{min} = \frac{2\pi r n}{60 \times 1000} = \frac{\pi n}{60 \times 1000} \times \frac{p}{\sin \frac{180^\circ}{z}} \times \cos \frac{180^\circ}{z} \tag{7.11}$$

The maximum velocity occurs at pitch radius R, expressed as

$$v_{max} = \frac{2\pi R n}{60 \times 1000} = \frac{\pi n}{60 \times 1000} \times \frac{p}{\sin \frac{180^\circ}{z}} \tag{7.12}$$

Employing Eq. (7.4), we have the chordal speed variation as

$$\frac{\Delta v}{v_{avg}} = \frac{v_{max} - v_{min}}{v_{avg}} = \frac{\pi}{z} \left[\frac{1}{\sin \frac{180^\circ}{z}} - \frac{1}{\tan \frac{180^\circ}{z}} \right] \tag{7.13}$$

The rise and fall of chain becomes harmful when resonance occurs, which is known as chain whip [5]. The rise and fall of chain, as well as the variation of instantaneous velocity, are caused by the cyclic fluctuation between the sprocket pitch radius R and chordal radius r, or by the polygon, as the chain engages the sprocket. This is called chordal action, or polygonal action. Chordal action is a kinematic consequence of the polygon due to the pitch length in chains.

Chordal action affects operating smoothness of a roller chain drive, particularly in high speed applications. Both chordal speed variation and chordal action decrease as the number of sprocket teeth is increased. If the number of sprocket teeth is sufficiently large, that is, more than 21, the chordal action may be hardly noticeable [5]. Therefore, a large number of sprocket teeth or a chain with a small pitch give a uniform chain speed and smooth operation.

7.2.3 Force Analysis

7.2.3.1 Tension in Tight Side

In a chain drive, the upper side is in tension, called the tight side, and the lower side is called the slack side. Power is transmitted by the tight tension side. Figure 7.4 illustrates a chain drive transmitting power. Initially, the load is applied to a chain roller by the driving sprocket teeth in contact. From the roller, the load is transmitted, in turn, to a bushing, pin and a pair of link plates (see Figure 7.2). Along the tight side of chain, the load is further transmitted to the driven sprocket by successive link plates, pins, bushings and rollers. The driven sprocket is then forced to rotate under constant tension imparted to the chain from the driving sprocket [3, 5]. At high speeds, centrifugal force may add significantly to the chain.

The tight side tension force F_1, therefore, consists of effective peripheral force F_e from the transmitted power, the tension induced by centrifugal force F_c and sagging force F_f due to the weight of chain, expressed as

$$F_1 = F_e + F_c + F_f \tag{7.14}$$

where the effective peripheral force F_e is determined by the transmitted power P and the speed of chain v, expressed as

$$F_e = \frac{1000P}{v} \tag{7.15}$$

The tension induced by centrifugal force F_c is analogous to that of belts as

$$F_c = qv^2 \tag{7.16}$$

where q is mass of chain per unit length and can be found in design handbooks or manufacturers' catalogues. Chain drives are thus usually used in a low speed stage in a power train where the speed is lower and the torque and force larger.

The sagging force F_f relates to the weight of chain, the centre distance and layout of chain drive. Detailed calculations can be found in reference [9].

7.2.3.2 Tension in Slack Side

The slack side tension force includes only centrifugal force F_c and sagging force F_f, expressed as

$$F_2 = F_c + F_f \tag{7.17}$$

7.2.3.3 Dynamic Forces

The periodical variation of both chain speed and angular velocity of driven sprocket, as well as the impact during meshing between the chain and sprocket, inevitably induce dynamic forces. The dynamic force caused by periodical variation of chain speed is computed by

$$F_d = ma_c \tag{7.18}$$

where the acceleration is

$$a_c = \frac{dv}{dt} = \frac{d}{dt}(R_1\omega_1 \cos\beta) = -R_1\omega_1^2 \sin\beta \tag{7.19}$$

when $\beta = \pm\frac{180^\circ}{z_1}$, incorporating Eq. (7.2), the maximum acceleration is

$$a_{cmax} = \mp R_1\omega_1^2 \sin\frac{180^\circ}{z_1} = \mp\frac{\omega_1^2 p}{2} \tag{7.20}$$

Therefore, low angular velocity, small pitch or large number of teeth, as well as light mass and small angular acceleration favour the reduction of dynamic forces.

7.2.4 Potential Failure Modes

As a chain drive in operation, the tensile force in the chain fluctuates from the tight side tension to the slack side tension, and back, for each cycle. Fatigue, therefore, becomes a primary potential failure mode for roller chains. Fatigue may occur on link plates, rollers and bushings or sprocket tooth surfaces. Fatigue fracture may disrupt power transmission totally, often with catastrophic consequences. Fatigue strength is thus the basis for chain ratings and essential for chain design.

Since rollers and bushings, bushings and pins are slip fitted, respectively, each time a link articulates over a sprocket tooth, wear may occur in the chain and on the sprockets. Abrasive wear, adhesive wear and fretting wear may be potential failure modes. As wear progresses slowly, the pitch of each pin link increases gradually. When the wear on pins, bushings, rollers and teeth produces sufficient dimensional change, rollers may climb high on the sprocket teeth or even skip off from them.

When a roller chain comes into meshing with a sprocket, impact occurs between the roller and sprocket teeth. Hertz contact stress also generates at the interface between the chain roller and sprocket tooth. At heavy loads and high speeds, the intensity of impact and contact stress increases dramatically, resulting in local welding at contact surfaces. As the chain drive rotates continuously, the welded surfaces are forced to separate, resulting in excessive wear or galling.

When a low speed chain drive ($v < 0.6\ \text{m s}^{-1}$) carries heavy static loads, links are likely deform plastically or even break due to insufficient tensile strength.

7.3 Power Transmission Capacities

7.3.1 Limiting Power Curves

The power performance of a chain drive is limited by potential failure modes. Normally, three modes of failure are considered in determining the power rating of a chain drive for its power transmission capacity, as shown in Figure 7.7. At low to moderate speeds, the transmitted power is limited by fatigue failure of link plates. At higher speeds, roller and bushing fatigue governs power transmission capacity of a chain drive. At even higher speeds and heavier loads, lubrication breakdown may cause instantaneous local welding and galling between pins and bushings, resulting in excessive wear and thus limiting the transmitted power and the maximum speed of a chain drive [10].

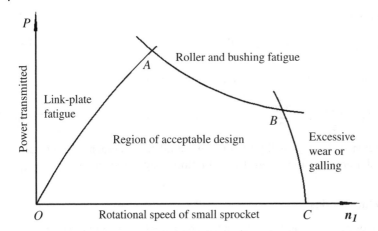

P

Power transmitted

Link-plate fatigue

Roller and bushing fatigue

A

Region of acceptable design

B

Excessive wear or galling

O

Rotational speed of small sprocket

C

n_1

Figure 7.7 Design-acceptable region bounded by limiting failure curves for roller chains. Source: Adapted from Collins 2002, Figure 17.11, p. 720. Reproduced with permission of John Wiley & Sons, Inc.

Precise curves of Figure 7.7 for different chain specifications are obtained by similar experimental conditions, that is, a horizontal chain drive with a speed ratio from one-third to three carries a steady load with an expected service life of approximately 15 000 hours. The small sprocket has 19 teeth. The single-strand standard roller chain has 120 pitches in length. The power ratings are obtained under recommended lubrication, with relative elongation of pitch due to wear less than 3% [11]. They are the rated transmitted power curves of a standard size roller chain as a function of the small sprocket rotational speed, and are provided by standards or manufacturer's catalogues.

7.3.2 Actually Transmitted Power

Under the specified experiment conditions, the important variables that determine the basic power rating of a roller chain are the pitch of chain and the rotational speed of small sprocket [11]. In actual operating conditions, the number of small sprocket teeth and the number of strands are most likely not the same as those specified in the experiment. Correction factors are introduced to account for these differences and the power that a selected chain actually transmits is given by

$$P_r = \frac{K_p P_0}{K_z} \tag{7.21}$$

where
P_r – actual power rating of a chain drive, kW;
P_0 – basic power rating, kW, which is obtained by experiment and can be found in standards or design handbooks;
K_z – tooth correction factor for the number of teeth of a small sprocket other than 19 teeth. K_z is calculated by [11]

$$K_z = \left(\frac{19}{z_1}\right)^{1.08} \tag{7.22}$$

K_p – multiple-strand correction factor.

The ratings in design standards and handbooks are for single-strand chains. If multiple-strand chains are used, the single-strand rating is multiplied by 1.7, 2.5 and 3.3 for double-, triple- and quadruple-strand chains, respectively [9, 12], to account for the uneven load sharing among parallel strands.

7.4 Design of Chain Drives

7.4.1 Introduction

The design of a chain drive requires the selection of an appropriate roller chain and sprockets to transmit a given power to meet specified design requirements. The information provided includes the transmitted power P, the rotational speed of driving sprocket n_1 and driven sprocket n_2, or speed ratio i, the allowable speed fluctuation, layout requirements, dimensional limitations or space constraints and operating conditions.

After analysis and design, decisions must be made about primary design variables for a chain drive, including the designation of chain, the number of links L_p and strands; the number of teeth of driving and driven sprockets, their materials, structure and dimensions; the centre distance a between sprockets; drive arrangements, forces acting on the shafts; lubrication, mounting and tensioning methods and other details. Finally, drawings of sprockets are provided.

7.4.2 Materials

The life of a roller chain depends to a great extent on materials, heat treatments and manufacturing precision. Based on the failure modes discussed in Section 7.2.4, candidate materials for roller chains should have high strength and good resistance to fatigue, wear and corrosion.

Low capacity chains are usually made of cast iron or carbon steel, while high capacity chains require high-strength alloy steels. All parts, including bushings, pins, rollers and link plates are heat treated by either carburizing, case hardening or through hardening, as application requires, to improve strength, wear resistance and impact resistance [5, 10].

The materials for sprockets should also have sufficient strength and wear resistance. Low or medium carbon steels are selected with the surface hardened to 59–63 HRC. Materials and surface treatments for small sprockets are usually better than large sprockets, as small sprockets experience more cycles during operation.

7.4.3 Design Criteria

For a medium- or high-speed chain drive, that is, $v \geq 0.6$ m s^{-1}, fatigue is a primary failure mode candidate. According to power criteria, the selected chain should have sufficient capacity to transmit the required power, that is

$$\frac{K_p P_0}{K_z} \geq K_A P \qquad (7.23)$$

Therefore, the required rating of a chain is

$$P_0 \geq \frac{K_A K_z}{K_p} P \qquad (7.24)$$

where K_A is service factor, considering the inherent characteristics of power sources and driven machines, operating conditions and duration of service. Service factors slightly higher than the lowest values in the recommended range in Table 2.1 can be selected for a given application.

For low speed, that is, $v < 0.6\,\mathrm{m\,s^{-1}}$, or heavily loaded chain drives, the failure mode may be static breakage. The design is then followed strength criteria, that is,

$$S_{ca} = \frac{QN}{K_A F_1} \geq 4 \sim 8 \qquad (7.25)$$

where

N – the number of strands;

Q – limiting tensile load of a single-strand chain, N, which can be found in design handbooks or manufacturer's catalogues;

F_1 – tensile force on the tight side of chain, N.

7.4.4 Design Procedure and Guidelines

This section introduces a systematic step-by-step procedure and suggestions for the design of medium- to high-speed chain drives. The design of low speed chain can refer to relevant design handbooks [9] following relevant design criteria.

7.4.4.1 Tentatively Select the Number of Sprocket Teeth *z* and Speed Ratio *i*

The number of teeth of sprockets, *z*, should be within a proper range. When the size of the driving sprocket is constant, small sprocket teeth number z_1 leads to a large pitch *p* and consequently increases the undesirable effect of chordal action, noise and dynamic force. It will also increase pitch angle φ, resulting in the increase of wear.

On the other hand, increasing small sprocket teeth number z_1 will consequently increase large sprocket teeth number z_2, resulting in a large chain drive and will make a chain more likely to shift outwards on the sprocket teeth, as illustrated in Figure 7.8.

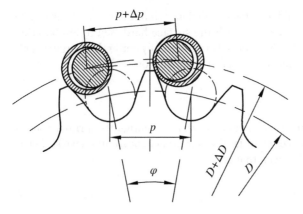

Figure 7.8 Chain engagement before and after wear.

Figure 7.8 shows a chain engaging a sprocket before and after wear. From the geometrical relationship in the figure,

$$D + \Delta D = \frac{p + \Delta p}{\sin \dfrac{180^{\circ}}{z}}$$

Incorporating Eq. (7.2), we have

$$\Delta D = \frac{\Delta p}{\sin \dfrac{180^{\circ}}{z}} \tag{7.26}$$

When wear happens, pitch increases. If the pitch increment, Δp, that is the amount of wear, keeps constant, increasing the number of teeth z will consequently increase the variation of diameter, ΔD. This implies that the chain will ride high to the top of sprocket teeth or, more likely, even jump off the sprocket. Normally, when wear-induced chain elongation reaches 3%, the chain should be replaced [3].

When the height of sprocket teeth stays constant, the allowable variation of diameter ΔD is limited. Increasing the number of teeth z will reduce pitch elongation, thus limiting the allowable amount of wear and eventually reducing the life of the chain.

In order for a chain to operate smoothly at moderate and high speeds, driving sprockets should have more than 17 teeth and driven sprockets should normally have no more than 120 teeth. It is preferable to have an odd number or, better yet, prime number of teeth on sprockets and an even number of links in the chain [4]. Such an arrangement will allow each sprocket tooth to mesh with all links instead of meshing with the same link continually. Thus, wear is expected to be evenly distributed and chain life is eventually prolonged. The recommended numbers of teeth for sprockets are 17, 19, 21, 23, 25, 38, 57, 76, 95 and 114 [2].

The speed ratio of a chain drive is limited to eight, usually within the range of 2–3.5 to ensure a sufficient contact angle. A large speed ratio reduces the contact angle and the number of teeth in engagement and, consequently, increases the load carried by each tooth and wear on the sprocket teeth. The contact angle on a small sprocket should be no less than 120°.

7.4.4.2   Determine the Required Power Rating of a Single-Strand Chain, P_0

Considering the operating condition, the number of sprocket teeth and the number of strands, the required power rating of the selected chain P_0 can be modified by

$$P_0 = \frac{K_A K_z}{K_p} P \tag{7.27}$$

7.4.4.3   Select Types of Chain and Pitch, p

Select the type of chain and pitch p according to the required power rating of a single-strand chain P_0 and the speed of a small sprocket. Pitch p is the characteristic parameter of a chain drive. It indicates the relative size of a chain and sprockets. The larger the pitch, the larger the chain size and the higher the load carrying capacity. However, chordal action and resultant vibration, impact and noise are more severe with a large pitch.

Therefore, a small pitch chain is preferred for a light load and high-speed chain drive, whereas a large pitch chain is better for a heavy load and low-speed chain drive. Generally, to transmit large power at a high speed, a multiply stranded chain with a small pitch is recommended. From the aspect of cost, a small pitch, multiply stranded chain is suitable for small centre distances and large speed ratio transmission while a large pitch, single-stranded chain is for large centre distance and small speed ratio transmission [2]. A multiply stranded chain needs less radial but more axial space than a single-stranded chain does.

7.4.4.4 Determine the Centre Distance Between the Sprocket Shafts, a and Chain Length, L_p

A short centre distance provides a compact design and allows for a shorter, less expensive chain. However, a shorter chain also means more articulation and thus more wear and short life. On the other hand, large centre distance easily causes whip and surge in the slack side [5]. Therefore, the initially selected centre distance should be within 30–50 pitches, that is, $a_0 = (30-50)p$, and the maximum value is no greater than $80p$, that is, $a_{max} = 80p$.

The length of chain is generally measured in pitches, expressed by the number of links L_p. The initial value L_{p0} can be calculated by [2]

$$L_{p0} = \frac{2a_0}{p} + \frac{z_1 + z_2}{2} + \left(\frac{z_2 - z_1}{2\pi}\right)^2 \frac{p}{a_0} \tag{7.28}$$

Since the length of chain must be an integer multiple of pitch and an even number of links is recommended to avoid an offset link, the initial value L_{p0} should be then rounded to a whole number L_p.

The theoretical centre distance for a given chain length, expressed in pitches is calculated by

$$a = \frac{p}{4}\left[\left(L_p - \frac{z_1 + z_2}{2}\right) + \sqrt{\left(L_p - \frac{z_1 + z_2}{2}\right)^2 - 8\left(\frac{z_2 - z_1}{2\pi}\right)^2}\right] \tag{7.29}$$

The theoretical centre distance assumes no sag on either the tight or slack side of the chain. However, the actual centre distance should be adjustable to accommodate variation in chain length due to sag and wear. The allowance is usually selected as $\Delta a = (0.002-0.004)a$. Therefore, the actual centre distance is

$$a' = a - \Delta a \tag{7.30}$$

7.4.4.5 Select an Appropriate Lubrication According to the Speed of Chain

Lubrication method is selected from Table 7.1 according to the speed of chain calculated by Eq. (7.4).

7.4.4.6 Forces Acting on the Shaft

The force acting on the shaft F_Q is calculated approximately by

$$F_Q \approx K_Q F_e \tag{7.31}$$

Table 7.1 Lubrication methods [1, 5].

Lubrication methods	Figures	Description
Manual		For chain velocity between 0.86 and 3.3 m s^{-1}; Oil is supplied by a brush or an oil can regularly.
Drip		For chain velocity between 0.86 and 3.3 m s^{-1}; Oil is fed directly onto the link plates of each chain strand by drip cup.
Oil bath		For chain velocity between 3.3 and 7.6 m s^{-1}; The lower strand of chain runs through a sump of oil in the chain housing.
Oil disc		For chain velocity between 3.3 and 7.6 m s^{-1}; An oil disc attached to the shaft picks up lubricant from the sump and throws it on to the chain.
Oil stream		For chain velocity greater than 7.6 m s^{-1}; An oil pump sprays a continuous stream of oil on the inside of chain loop at the slack side, evenly across the chain width.

where

F_e – effective peripheral force, N;

K_Q – factor, for horizontal transmission, select $K_Q = 1.15$; for vertical transmission select $K_Q = 1.05$.

7.4.5   Design Cases

Example Problem 7.1

Design a chain drive for a heavily loaded conveyor to be driven by a gasoline engine with a mechanical drive. The input speed is 900 rpm and the desired output speed is from 230 to 240 rpm. The nominal power to be transmitted is estimated as 10 kW. Select chain 16 A with a pitch of $p = 25.4$ mm.

Solution

Steps	Computation	Results	Units
1. Specify the number of teeth of sprockets	Use the average of output speed to compute speed ratio, i.e. $$i = \frac{n_1}{n_2} = \frac{900}{235} = 3.83$$	$i = 3.83$	
	Select the number of teeth of small sprocket as $z_1 = 21$	$z_1 = 21$	
	The number of teeth of the large sprocket is $$z_2 = iz_1 = 3.83 \times 21 = 80.4$$ Select an odd integer $z_2 = 81$	$z_2 = 81$	
2. Compute the design power P_{ca} to be transmitted by the chain	Select service factor K_A from Table 2.1 as $K_A = 1.5$. From Eq. (7.22), we have $$K_z = \left(\frac{19}{z_1}\right)^{1.08} = \left(\frac{19}{21}\right)^{1.08} = 0.9$$	Single strand	kW
	For a single-strand chain, $K_p = 1.0$. From Eq. (7.24), the required power rating of the selected chain P_0 is $$P_0 = \frac{K_A K_z P}{K_p} = \frac{1.5 \times 0.9 \times 10}{1.0} = 13.5$$	$P_0 = 13.5$	
3. Select type of chain, and pitch p	Tentatively select chain 16A with pitch $p = 25.4$ mm to start design	16A $p = 25.4$	mm
4. Determine initial centre distance a_0 and chain length L_p	Initially select centre distance $a_0 = 40p$. The initial value of the required chain length in pitches is calculated from Eq. (7.28) as $$L_{p0} = \frac{2a_0}{p} + \frac{z_1 + z_2}{2} + \left(\frac{z_2 - z_1}{2\pi}\right)^2 \frac{p}{a_0}$$ $$L_{p0} = \frac{2 \times 40p}{p} + \frac{21 + 81}{2} + \left(\frac{81 - 21}{2\pi}\right)^2 \frac{p}{40p} = 133.3$$ Specify an even integral number of pitches for the chain length as $L_p = 134$.	$L_p = 134$	pitches
5. The theoretical centre distance a	From Eq. (7.29) $$a = \frac{p}{4}\left[\left(L_p - \frac{z_1 + z_2}{2}\right) + \sqrt{\left(L_p - \frac{z_1 + z_2}{2}\right)^2 - 8\left(\frac{z_2 - z_1}{2\pi}\right)^2}\right]$$ $$a = \frac{p}{4}\left[\left(134 - \frac{21 + 81}{2}\right) + \sqrt{\left(134 - \frac{21 + 81}{2}\right)^2 - 8\left(\frac{81 - 21}{2\pi}\right)^2}\right] = 1025.4mm$$	$a = 1025.4$	mm
6. The actual centre distance a'	From Eq. (7.30), the actual centre distance is $$a' = a - \Delta a = 1025.4 - 0.004 \times 1025.4 \approx 1021mm$$	$a' = 1021$	mm

Steps	Computation	Results	Units
7. Calculate chain speed and decide lubrication method	From Eq. (7.4) $$v_{avg} = \frac{z_1 n_1 p}{60 \times 1000} = \frac{21 \times 900 \times 25.4}{60 \times 1000} = 8 \text{ ms}^{-1}$$ From Table 7.1, we can select spray lubrication	$v_{avg} = 8$	m s^{-1}
8. Forces acting on the shaft	The effective force is calculated from Eq. (7.15) as $$F_e = \frac{1000P}{v} = \frac{1000 \times 10}{8} = 1250N$$ For horizontal transmission select $K_Q = 1.15$. From Eq. (7.31), the force acting on the shaft is $$F_Q \approx K_Q F_e = 1.15 \times 1250 = 1438 \text{ N}$$	$F_Q = 1438$	N
9. Contact angle on the small sprocket	From Eq. (7.3), the contact angle on the small sprocket is $$\alpha_1 = 180° - \frac{z_2 - z_1}{\pi a} p \times 57.3°$$ $$\alpha_1 = 180° - \frac{81 - 21}{\pi \times 1021} \times 25.4 \times 57.3° = 152.8°$$ Because this is greater than 120°, it is acceptable.	$\alpha = 152.8°$	
10. Sprocket structure and dimension	Omitted		

7.5 Drive Layout, Tension and Lubrication

7.5.1 Drive Layout

Due to gravity, the preferred arrangement for a chain drive is with the centreline of sprockets horizontal and with the tight side on the top to prevent chain pinching between sprockets (Figure 7.9a). Vertical drives should be avoided. If unavoidable, the centreline should be inclined (Figure 7.9b). The rotation of sprocket should let accumulated slack away from the tight strand. Excessive sag on the slack side should be avoided, especially on drives that are not horizontal. Otherwise, a idler should be used.

7.5.2 Tensioning

Periodic maintenance for adequate tension is important for proper operation and long life of chain drives. Since wear is inevitable after long service, wear-induced elongation must be compensated to avoid improper mesh or whip in the slack strand.

Tensioning can be realized either by increasing the centre distance between sprockets or by using idlers. Adjusting centre distance is often a recommended approach. However, for a fixed centre distance, chain idlers or a small adjustable idler sprocket should be provided to remove excessive slack due to wear. Figure 7.10 illustrates the arrangement of idlers for a horizontal and vertical chain drive. The idlers should be located on the outside of chain on the slack side, close to the driving or driven sprockets to increase contact angle [9].

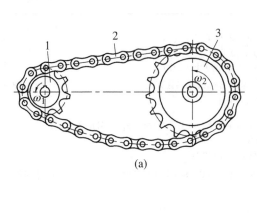

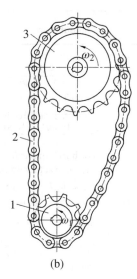

(a)

(b)

Figure 7.9 Chain drive arrangements.

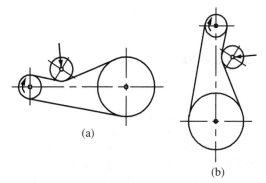

(a)

(b)

Figure 7.10 Location of chain idlers.

7.5.3 Lubrication

Since a roller chain contains several relatively moving parts, such as pins, bushings and rollers, a constant supply of clean, medium or light mineral oil to ensure adequate lubrication is essential for smooth operation and satisfactory life of a chain drive. Proper lubrication could reduce wear, protect against dust, corrosion and heat, and also prevent seizing of pins and bushings.

To lubricate properly, lubricant oil must penetrate into small clearances between the relative moving surfaces, like the bushing interfaces with the roller on the outside and the pin on the inside. Lubricant oil is preferably to be applied to the slack side for maximum penetration and on the inside of the strand so that centrifugal force and gravity tend to move oil into links [5].

The selection of lubrication method depends on the speed of operation and the power being transmitted. Commonly used methods are illustrated in Table 7.1. Where the chain would be exposed to considerable dirt, dust or moisture, a protective casing should be used.

References

1 Mott, R.L. (2003). *Machine Elements in Mechanical Design*, 4e. Prentice Hall.
2 Pu, L.G. and Ji, M.G. (2006). *Mechanical Design*, 8e. Beijing: Higher Education Press.
3 Juvinall, R.C. and Marshek, K.M. (2011). *Fundamentals of Machine Component Design*, 5e. New York: Wiley.
4 Budynas, R.G. and Nisbett, J.K. (2011). *Shigley's Mechanical Engineering Design*, 9e. New York: McGraw-Hill.
5 Hindhede, U., Zimmerman, J.R., Hopkins, R.B. et al. (1983). *Machine Design Fundamentals: A Practical Approach*. New York: Wiley.
6 ANSI/ASME B29.1-2011 (2011). *Precision Power Transmission Roller Chains, Attachments, and Sprocket*. Three Park Avenue, New York, NY: The American Society of Mechanical Engineers.
7 ISO 606:2015 (2015). *Short-Pitch Transmission Precision Roller and Bush Chains, Attachments and Associated Chain Sprockets*. Geneva, Switzerland: International Organization for Standards.
8 GB/T 1243-2006 (2006). *Short-Pitch Transmission Precision Roller and Bush Chains, Attachments and Associated Chain Sprockets*. Beijing: Standardization Administration of the People's Republic of China.
9 Wen, B.C. (2015). *Machine Design Handbook*, 5e, vol. 2. Beijing: China Machine Press.
10 Collins, J.A. (2002). *Mechanical Design of Machine Elements and Machines: A Failure Prevention Perspective*, 1e. New York: Wiley.
11 ISO 10823:2004 (2004). *Guidelines for the Selection of Roller Chain Drives*. Geneva, Switzerland: International Organization for Standards.
12 American Chain Association (1982). *Chains for Power Transmission and Material Handling*. New York: Marcel Dekker.

Problems

Review Questions

1 List three typical failure modes of roller chains.

2 Is it possible to express the speed ratio of a chain drive as $i = \frac{n_1}{n_2} = \frac{z_2}{z_1} = \frac{d_2}{d_1}$? Explain why.

3 What factors affect the uniformity of chain speed transmission?

4 What lubrication methods are available for chain drives? How does one select a suitable method?

5 How could one improve operating conditions if wear occurs on a roller chain?

Objective Questions

1 A power transmission includes a belt drive, a chain drive and a gear drive. Which of the following is the best arrangement between a motor and a working machine?
(a) belt drive   gear drive   chain drive
(b) chain drive   gear drive   belt drive
(c) belt drive   chain drive   gear drive
(d) chain drive   belt drive   gear drive

2 In a chain drive, reducing the number of teeth of driving sprocket will cause_____.
(a) the severity of speed unevenness of chain drive
(b) the chain to jump off the sprocket more easily after wear
(c) poor lubrication that may lead to gluing
(d) the increase of rotational speed of driving sprocket

3 At a constant rotational speed, if we want to reduce speed unevenness and dynamics load of a chain drive, we can _____.
(a) increase p and z_1
(b) increase p and reduce z_1
(c) reduce p and increase z_1
(d) reduce p and z_1
(where z_1 is the number of teeth of driving sprocket, p is pitch)

4 The number of rows of a multiple-strand chain is generally no more than four to ___.
(a) facilitate installation
(b) limit axial width of chain drive
(c) ensure even loading on each row
(d) reduce the weight of chain

5 The purpose of limiting the maximum number of sprocket teeth is _____.
(a) to ensure the strength of chain
(b) to ensure the transmission stability of chain drive
(c) to ensure the speed ratio selection of chain drive
(d) to prevent the chain from jumping off the sprocket

Calculation Questions

1   A chain drive with pitch $p = 25.4$ mm rotates at $n_1 = 750$ rpm. The number of teeth of driving sprocket is $z_1 = 23$. Determine the average speed v, the maximum speed v_{max} and the minimum speed v_{min}.

2   A single-strand roller chain is used to transmit power between a 23-tooth driving sprocket and a 69-tooth driven sprocket. The rotational speed of driving sprocket is 960 rpm. If the rotational speed of driven sprocket is reduced to 210 rpm:
   (a) What is the number of teeth of driving sprocket if the teeth number of the driven sprocket is kept unchanged? What will be the power transmitted?
   (b) What is the number of teeth of a driven sprocket if the teeth number of the driving sprocket is kept unchanged? What will be the power transmitted?
   (c) Compare the two approaches: which one is better?

3   A single-strand 08A chain operates on a 20-tooth sprocket at 750 rpm. The small sprocket is applied to the shaft of an electric motor and the output sprocket to a conveyor. Determine the power rating for the single-strand chain. What would be the rating for four strands?

4   A four strand 08A roller chain is proposed to transmit power from a 21-tooth driving sprocket that rotates at 1200 rpm. The driven sprocket is to rotate at one-third the speed of the driving sprocket.
   (a) Determine the power that can be transmitted by the chain drive;
   (b) Estimate the centre distance if the chain length is 82 pitches;
   (c) Describe the preferred method of lubrication.

Design Problems

1   Design a roller chain drive to transmit 10 kW. The rotational speed of motor is $n = 970$ rpm. The speed ratio of the chain drive is $i = 3$, uniform power transmission and centre distance no less than 550 mm.

2   A 5 kW 750 rpm electric motor is to drive an agitator at 325 rpm. The motor shaft will be in approximately the same horizontal plane as the agitator shaft. Specify the chain size, the sizes and number of teeth of sprockets, the number of chain pitches and the centre distance.

3   A heavy conveyor is to be driven at 250 rpm by a 30 kW six-cylinder engine with an input speed of 500 rpm. The centre distance is approximately 1000 mm. Moderate shock loading is expected. Select a suitable roller chain and associated sprockets for this application.

4 A rock crusher is driven by a 75 kW hydraulic drive at a speed of 625 rpm. The rock crusher is to operate at 225 rpm, subject to heavy service. A centre distance of from 1000 to 1500 mm will be acceptable. Design a roller chain drive.

Structure Design Problems

1 In the design of chain sprocket, the number of teeth of driving and driven sprockets are limited by $z_{min} = 9$ and $z_{max} = 120$, respectively. Explain the reason.

2 In chain design, why the speed ratio has a limitation of less than 8?

3 In Figure P7.1, the large sprocket is the driven sprocket. Decide on the rotational direction of the small sprocket.

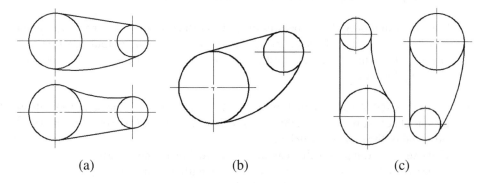

(a)   (b)   (c)

Figure P7.1 Illustration for Structure Design Problem 3.

4 In a chain drive design, two choices are available, that is, (1) $p = 12.7$ mm, $z_1 = 38$; (2) $p = 25.4$ mm, $z_1 = 19$. Compare the two selections in respective of operating smoothness and dynamic loads.

5 An electric motor at a speed of 1500 rpm is to drive a device at 15 rpm. A belt drive, gear drive and chain drive can be used. Propose at least two layouts to transmit the motion and compare them.

CAD Problems

1 Write a flow chart for the design process of a chain drive.

2 Develop a program to implement a user interface similar to Figure P7.2 and complete the Example Problem 7.1.

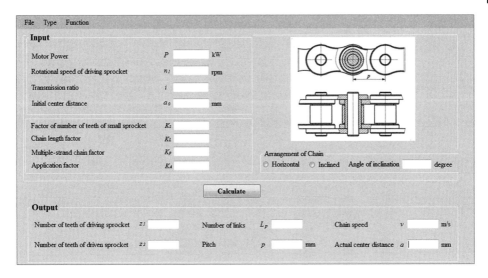

Figure P7.2 Illustration for CAD Problem 2.

8

Gear Drives

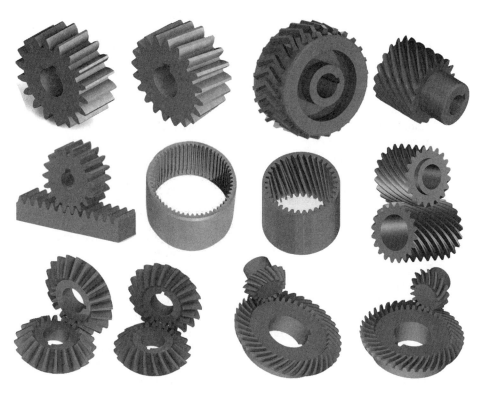

Nomenclature

a	centre distance, mm	d_{1t}	trial diameter of pitch circle, mm
b	face width, mm	d_v	pitch diameter of virtual spur gear, mm
c	clearance, mm	E	elastic modulus, MPa
d	diameter of reference or pitch circle, mm	F_a	axial force, or thrust force, N
		F_{ca}	design load, or actual load, N
d_a	diameter of addendum circle, mm	F_n	normal force, N
d_b	diameter of base circle, mm	F_r	radial force, or separating force, N
d_f	diameter of dedendum circle, mm	F_t	tangential force, or transmitted force, N
d_m	mean diameter, mm	h	tooth height, mm

Analysis and Design of Machine Elements, First Edition. Wei Jiang.
© 2019 John Wiley & Sons Singapore Pte. Ltd. Published 2019 by John Wiley & Sons Singapore Pte. Ltd.
Companion website: www.wiley.com/go/Jiang/analysis_of_machine_elements

h_a	addendum, mm	v	pitch line velocity, m s^{-1}
h_F	bending moment arm, mm	x	profile shift coefficient
h_f	dedendum, mm	Y_{Fa}	tooth form factor
i	Speed or velocity ratio, transmission ratio	Y_{Sa}	stress correction factor
		Y_{ST}	stress correction factor for testing gear
j	number of gear teeth meshing per revolution	Y_β	helix angle factor for bending stress
K	load factor		
K_A	application factor	Y_ε	contact ratio factor for bending stress
K_{FN}	life factor for bending stress		
K_{HN}	life factor for contact stress	Z_E	elastic coefficient, MPa$^{1/2}$, (N mm^{-2})$^{1/2}$
K_t	trial load factor		
K_v	dynamic factor	Z_H	zone factor
K_α	transverse load factor	Z_β	helix angle factor for contact stress
K_β	face load factor		
L	contact length, mm	Z_ε	contact ratio factor for contact stress
L_h	design life in hours, h		
l	lead, mm	z	number of teeth
m	module, mm	z_v	virtual number of teeth
m_m	mean module , mm	α	pressure angle, °
m_n	normal module, mm	α_n	normal pressure angle, °
m_{nt}	trial normal module, mm	α_t	transverse pressure angle, °
m_t	transverse module, mm	β	helix angle, °
N	number of load cycles	β_b	helix angle at base circle, °
n	rotational speed, rpm	γ	pressure angle at the tip of tooth, °
P	transmitted power, kW	δ	pitch cone angle in a bevel gear, °
p_a	axial pitch, mm	ε	contact ratio
p_b	base pitch, mm	ε_a	transverse contact ratio
p_c	circular pitch, mm	ε_{av}	transverse contact ratio of a virtual gear
p_n	normal circular pitch, mm		
p_t	transverse circular pitch, mm	ε_β	face contact ratio, overlap ratio
Q	gear accuracy grade in ISO/GB standard	μ	Poisson's ratio
		ρ	radius of curvature, mm
Q_{AGMA}	gear quality number in AGMA standard	ρ_n	normal radius of curvature, mm
		ρ_t	transverse radius of curvature, mm
R	cone distance, mm		
r	radius of reference or pitch circle, mm	σ_F	bending stress, MPa
		σ_{Flim}	endurance limit for bending stress, MPa
r_a	radius of addendum circle, mm		
r_b	radius of base circle, mm	$[\sigma_F]$	allowable bending stress, MPa
r_{bv}	developed back cone radius, mm	σ_H	contact stress, MPa
S_F	safety factor for bending strength	σ_{Hlim}	endurance limit for contact stress, MPa
S_H	safety factor for contact strength		
s_F	root thickness of a tooth, mm	$[\sigma_H]$	allowable contact stress, MPa
T	torque, N mm	φ_d	face width factor
u	teeth ratio, gear ratio	φ_R	face width factor of a bevel gear
u_v	virtual spur gear teeth ratio	ω	angular velocity, rad/ s^{-1}

Subscripts

		H	contact stress
		n	normal
1	pinion	r	radial
2	gear	t	transverse
F	bending stress		

8.1 Introduction

8.1.1 Applications, Characteristics and Structures

Gears are toothed wheels used for transmitting motion and power from one rotating shaft to another. A gear drive, or gearing, is an assembly of two or more gears. Gear drives are the most widely used and most important form of mechanical drives. They are used to transmit torque and power in a wide variety of applications, from delicate instruments to the heaviest and most powerful machinery. The power transmitted various from negligibly small values in watches to megawatts in wind turbines. The gear diameters can vary from a fraction of a millimetre to 10 and more metres [1–5].

Compared with other mechanical power transmission, such as belt drives and chain drives, gear drives have advantages of high efficiency and reliability, compact design, long service life, constant transmission ratio and a wide range of speed and speed ratio transmission capabilities. However, as precision machine elements, gears must be designed, manufactured and installed with great care to ensure proper function. Otherwise, noise, vibration and dynamic loads will generate during operation. As such, the manufacturing and assembly cost of gears is high. Besides, they are not suitable for long distance power transmission.

A typical gear drive composes of a driving gear, usually the small gear called pinion, and a driven gear, usually the larger one, called a gear. Both of them are installed on shafts supported by bearings. When a single pair of gears cannot achieve desired output speed because of speed ratio limitation, a combination of interconnected gears, called the gear train, can be used. A gear reducer is a kind of gear train. They are desirable because they can accomplish a large speed reduction in a rather small package, having the possibility to be attached to driven machines with a wide range of torques, speeds and speed ratios.

This chapter discusses the fundamentals, kinematics, strength analysis and design of several types of gears, with emphasis on spur gears, as many fundamental principles of these are applicable to other gear types.

8.1.2 Types of Gear Drives

Gear drives can be classified into three types according to relative shaft positions. Three shaft orientations, that is, parallel, intersect and neither parallel nor intersect, account for the three basic types of gear drive, namely parallel-shaft gearings, intersecting-shaft gearings and crossed gearings, as presented in Table 8.1. Each of these can be studied by observing a single pair of gears.

Parallel-shaft gearings are the simplest and most popular type of drives. They use spur gears, helical gears, herringbone gears, pinion and rack and internal gears, to connect parallel shafts and transmit large power with high efficiency.

Table 8.1 Types of gear drives.

Gear drives	Parallel-shaft gearings	Spur gear drives
		Helical gear drives
		Herringbone gear drives
		Pinion and rack drives
		Internal gear drives
	Intersecting-shaft gearings	Straight bevel gear drives
		Spiral bevel gear drives
	Crossed gearings	Crossed helical gear drives
		Hypoid gear drives
		Worm and wormgear drives

Spur gears are the most widely used type because of simplicity, low fabrication cost and great precision. They have straight involute teeth parallel to the axis of the shaft that carries them. Spur gears impose only radial loads on the supporting bearings. They are usually limited to pitch line velocities around 20 m s^{-1} to avoid high frequency vibration and unacceptable noise [6].

Helical gears are superior to spur gears in terms of tooth strength, operation speed and power transmission capacity because of the helix angle they possess. The gradual engagement of helical gear due to angled teeth leads to a smoother and quieter operation than spur gears. On the other hand, the angled teeth impose both radial and thrust loads on the supporting bearings. To eliminate thrust loads on the supporting bearings when transmitting high power, a herringbone gear can be used.

The rack and pinion drive is a special case of parallel-shaft gearing, as the rack can be regarded as a cylindrical gear with an infinite pitch diameter. It may have either straight spur gear teeth or helical gear teeth. Most gears are external gears. Internal gears are used to achieve a short centre distance and are a necessity in epicyclic gear trains.

Intersecting-shaft gearings use straight bevel gears and spiral bevel gears to transmit motion between intersecting shafts. Bevel gears have teeth formed on conical surfaces. The straight bevel gear teeth appear similar to spur gear teeth, except they are tapped in both tooth thickness and height. Bevel gears impose both radial and thrust loads on the supporting bearings. They must be accurately mounted at an accurate axial distance from the pitch cone apex for proper meshing. When bevel gears are made with teeth that form a helix angle similar to that in helical gears, they are called spiral bevel gears. Spiral bevel gears provide the advantage of gradual engagement along the tooth face. Therefore, they operate more smoothly than straight bevel gears and can be made smaller for a given power transmission capacity.

Shafts carrying a pair of helical gears are typically arranged parallel to each other. However, when a pair of helical gears mesh with each other with nonintersecting shaft axes, they are called crossed helical gear drives. In crossed helical gear drives, teeth initially mesh at a point and, later, a line. Therefore, the load carrying capacity gradually increases after a wear-in period. Hypoid gear drives resemble spiral bevel gear drives, except that the shafts have a small offset and nonintersecting. For larger offsets, the pinion begins to resemble a tapered worm and the set is then called a spiroid gearing.

Another example of crossed gearings whose shaft axes are neither parallel nor intersecting is the worm and wormgear drives. A worm and its mating wormgear operate on shafts that are perpendicular to each other. A worm resembles a screw, in meshing with a wormgear, whose teeth are similar to those of helical gear, except that they are contoured to envelop the worm. Both radial and thrust loads are imposed on the supporting bearings of worm and wormgear shafts. Worm drives accomplish a large speed reduction ratio compared with other types of gear drives. The details about worm gearing will be discussed in Chapter 9.

According to the hardness of gear tooth surface, we have soft tooth surface gears, whose hardness is less than 350HBW and hard tooth surface gears, whose hardness is greater than 350HBW.

Gears can also be classified by their working conditions. If gears and shafts are enclosed in a housing, they are enclosed gearings; otherwise, they are open. Enclosed gearings predominate in engineering practice, as in such arrangement, gears are in a sealed environment that provides protection against contamination.

8.1.3 Geometry and Terminology

While transmitting power and motion between rotating shafts, the teeth of the driving pinion mesh accurately in the spaces between the teeth on the driven gear. To make further discussion more meaningful, the assigned terminology and defined geometric variables of spur gears are shown in Figure 8.1 and introduced next.

1) *Reference circle and reference diameter, d.* Reference circles are circles that have a standard module and pressure angle.
2) *Pitch circle and pitch diameter.* Pitch circles are two imaginary circles at a tangent to pitch point *P*. The diameter of a pitch circle may be different from the reference diameter if meshing gears are required to respond to centre distance variations. This book is limited to discussion on standard installation where the pitch circle and reference circle coincide with each other. Therefore, pitch diameter equals reference diameter.
3) *Base circle and base diameter, d_b.* The base circle is the circle from which an involute tooth curve is developed. The base diameter never changes. From Figure 8.1b, we show the relationship between pitch diameter *d* and base diameter d_b as

$$d_b = d \cos \alpha \tag{8.1}$$

4) *Addendum circle, addendum circle diameter, d_a, and addendum, h_a.* Addendum h_a is the radial distance between the top land and pitch circle. Addendum circle is obtained by adding the addendum to the pitch radius *r*.
5) *Dedendum circle, dedendum circle diameter, d_f and dedendum, h_f.* Dedendum h_f is the radial distance from the bottom land to the pitch circle. The dedendum circle is obtained by subtracting the dedendum from the pitch radius *r*. The involute tooth profile extends only as far as the base circle. The portion of the tooth inside the base circle cannot participate in conjugate action and therefore must be shaped to provide tip clearance for the mating teeth. The sum of addendum and dedendum is tooth height *h*.
6) *Clearance circle and clearance, c.* The clearance circle is the circle tangent to the addendum circle of the mating gear. The dedendum is typically made slightly larger than the addendum so as to provide clearance *c* between the tooth tip of one gear and the bottom land of the mating gear. A root fillet is used to combine the tooth flank and the bottom land.

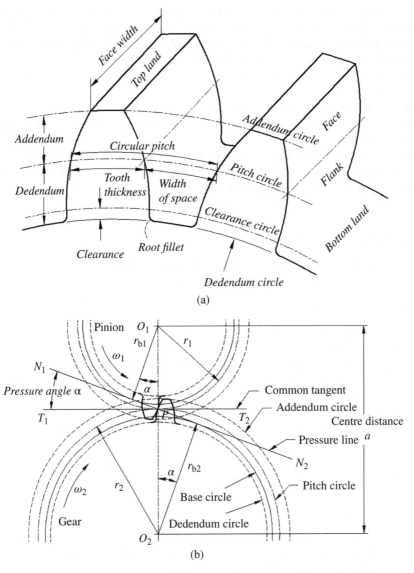

Figure 8.1 Geometry and terminology of spur gears. Source: Budynas and Nisbett 2011, Figure 13.5, p. 676. Reproduced with permission of McGraw-Hill.

7) *Centre distance, a.* Centre distance is the distance between the centres of the pitch circles of the mating pinion and gear. For standard installation, the centre distance is

$$a = \frac{1}{2}(d_1 + d_2) \tag{8.2}$$

8) *Circular pitch, p_c.* Circular pitch is the distance along the pitch circle between corresponding points of adjacent teeth, and is equal to the sum of space width and tooth

thickness. It is the arc length equal to the circumference of pitch circle divided by the number of teeth, expressed as

$$p_c = \frac{\pi d}{z} \tag{8.3}$$

9) *Base pitch, p_b.* Base pitch is the arc length equal to the circumference of base circle divided by the number of teeth

$$p_b = \frac{\pi d_b}{z} \tag{8.4}$$

Combined with Eq. (8.1), the relation between base pitch p_b and circular pitch p_c is expressed as

$$p_b = p_c \cos \alpha \tag{8.5}$$

Mating teeth must have the same base pitch and circular pitch.

10) *Module, m.* Module is the amount of pitch diameter per tooth. The module is the index of tooth size in SI units. A higher module number denotes a larger tooth and vice versa.

$$m = \frac{d}{z} \tag{8.6}$$

Combined with Eq. (8.3), circular pitch, p_c, is

$$p_c = \pi m \tag{8.7}$$

11) *The diametral pitch.* The diametral pitch is the ratio of the number of teeth to the pitch diameter, which is the reciprocal of module.

12) *Pressure angle, α.* The pressure angle is the angle between the common tangent to the pitch circles and the pressure line. The preferred pressure angle is 20°. For the proper meshing of a pair of gears, they must have same module and same pressure angle.

13) *Circular tooth thickness and width of tooth space.* The circular tooth thickness and the width of tooth space are measured along the pitch circle, each being nominally equal to $p_c/2$. The tooth space is usually made slightly larger than the tooth thickness in order to provide a small amount of backlash for smooth operation.

14) *Profile shift coefficient, x.* To avoid undercutting, or to adapt to required centre distance, a gear cutter needs to move a distance xm away from standard assembly position and x is the profile shift coefficient. Positive profile shift coefficient results in a thicker and stronger tooth.

15) *Backlash.* Backlash is the difference between tooth space and tooth thickness. It tends to eliminate tooth jamming, compensate for machining errors and heat expansion and promote proper lubrication, but may also cause impact, vibration and noise. Backlash is obtained by reducing the tooth thickness or by increasing the centre distance between mating gears.

16) *Face width, b.* Face width is the length of tooth in the axial direction.

17) *Top land and bottom land.* The top land is the top surface of gear, while the bottom land is the surface of gear between the flanks of adjacent teeth.

18) *Tooth face, tooth flank and tooth surface.* The tooth face is the surface between the pitch line and the top of tooth, while tooth flank is the surface between the pitch line and the bottom land. Therefore, tooth surface is tooth face and tooth flank combined.

19) *Root fillet.* This is the portion of tooth flank joined to the bottom land.

8.2 Working Condition Analysis

8.2.1 Kinematic Analysis

8.2.1.1 Speed Ratio and Pitch Line Velocity

Gears are employed to produce rotational speed change in the driven gear relative to the driving gear. This change is measured by speed ratio, defined as the ratio of the rotational speed of driving gear to the driven gear. For proper rotational speed transmission, speed ratio should be maintained at a constant value continuously through meshing.

To satisfy this kinematic requirement, it is required that, from Figure 8.1b, as gears rotate, the common normal N_1N_2 to the tooth surfaces at contact point must intersect the line of centres O_1O_2 at pitch point P, where the pitch line velocities of the pinion and gear are identical and each may be calculated by

$$v = \frac{\pi d_1 n_1}{60 \times 1000} = \frac{\pi d_2 n_2}{60 \times 1000} \tag{8.8}$$

The speed ratio, therefore, is

$$i = \frac{n_1}{n_2} = \frac{d_2}{d_1} = \frac{z_2}{z_1} = u \tag{8.9}$$

Equation (8.9) is equally valid for spur, helical and bevel gear drives.

8.2.1.2 Contact Ratio

Figure 8.2 shows a pinion centred at O_1, rotating clockwise at an angular velocity of ω_1, drives a gear centred at O_2, rotating counterclockwise at an angular velocity of ω_2. As gears rotate, smooth and continuous motion transfer from one pair of meshing teeth to the succeeding pair is especially important. This is achieved by keeping the first pair in contact until the following pair establishes initial contact. Or in other words, a degree of contact overlapping must be maintained.

Contact ratio is an effective means for measuring this overlapping tooth contact. It is defined as contact length divided by base pitch p_b. During tooth meshing, contact points move along the line of action N_1N_2 or the pressure line. The contact length is the distance on the line of action located between the two addendum circles. From Figure 8.2, we have [2]

$$\varepsilon = \frac{\overline{B_1 B_2}}{p_b} = \frac{\sqrt{r_{a1}^2 - r_{b1}^2} + \sqrt{r_{a2}^2 - r_{b2}^2} - a \sin \alpha}{p_c \cos \alpha} \tag{8.10}$$

To ensure a smooth and continuous meshing, the contact ratio must be greater than one. Contact ratio can be regarded as the average number of teeth in contact or, on the time basis, the number of pairs of teeth simultaneously engaged. For example, a contact

Figure 8.2 Contact ratio calculation.

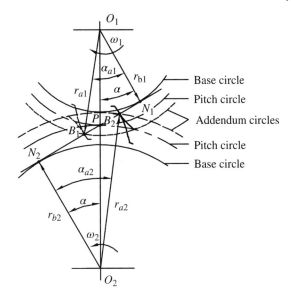

ratio of 1.35 means that one pair of teeth is in contact at all times and two pairs of teeth are in contact 35% of the time.

During gear life time, manufacturing, mounting and operation inevitably induce tooth spacing errors and deflection. These inaccuracies may reduce contact ratios and increase the possibility of impact and noise. Therefore, contact ratios are usually designed within a range of 1.4–1.8 to compensate for contact delays and to ensure a smooth motion transmission [6]. Also, a large contact ratio usually implies load sharing among teeth, which benefits the increase of load carrying capacities of gears.

8.2.2 Design Loads

In practice, gears in mesh never operate under a smooth, continuous load. Various manufacturing, assembly, geometric, loading and material variabilities influence the actual load on the gear teeth. To account for the influence of these factors, an extensive list of empirical adjustment factors is introduced in the calculation of design load F_{ca}, or actual load, by

$$F_{ca} = KF_n \tag{8.11}$$

where the load factor K has four components, expressed as

$$K = K_A K_v K_\alpha K_\beta \tag{8.12}$$

The application factor K_A considers load increment due to shock or impact characteristic of prime movers and driven machines, which may result in peak loads greater than nominal loads. A relatively high value within the recommended range from Table 2.1 can be selected.

The dynamic factor K_v is used to evaluate the effect of dynamic loads caused by the deflection of gear teeth, and the inaccuracy of base pitch and tooth profile, which are usually generated by manufacturing and operation. The value of dynamic factor K_v

depends on the accuracy of tooth profile and base pitch, teeth elastic properties and pitch line velocities, and can be estimated by empirical formula as [3, 7]

$$K_v = \left(1 + \frac{\sqrt{200v}}{A}\right)^B \tag{8.13}$$

and

$$A = 50 + 56(1.0 - B) \tag{8.14}$$

$$B = 0.25(12 - Q_{AGMA})^{0.667} \tag{8.15}$$

where Q_{AGMA} is the gear quality number in the American Gear Manufacturers Association (AGMA) standard, ranging from 5 to 11 with increasing precision. Contrary to the definition in AGMA 2008 standard, gear accuracy grades in ISO, DIN and GB standards have small numbers to indicate high precision (see Table 8.5). The sum of the quality number from AGMA 2008 and the corresponding accuracy classification number from ISO 1328 or AGMA 2015 is always 17 [3]. Therefore, Eq. (8.15) can be rewritten as

$$B = 0.25(Q - 5)^{0.667} \tag{8.16}$$

where Q is gear accuracy grade in ISO and GB standards, ranging from 6 to 12.

The transverse load factor K_α reflects nonuniform load distribution among two or more pairs of simultaneously meshing teeth. It is caused by manufacturing inaccuracy and teeth deflections under loads, and is determined by many factors, such as base pitch deviations, contact ratio and gear stiffness. Table 8.2 shows transverse load factors for contact stress $K_{H\alpha}$ and transverse load factors for bending stress $K_{F\alpha}$. When the pinion and gear are of different accuracy grades, select the transverse load factor according to the lower gear accuracy grade.

The face load factor K_β reflects the nonuniform load distribution over the face width. Although it is usually assumed that nominal load is uniformly distributed along the face width of teeth, manufacturing and assembly inaccuracy, bearing clearances and deflections of gear teeth, shafts, bearings and housing all affect face load factor. Table 8.3 lists

Table 8.2 Transverse load factors for contact stress $K_{H\alpha}$ and bending stress $K_{F\alpha}$.

Unit load $K_A F_t/b$			$\geq 100\,\text{N mm}^{-1}$							$<100\,\text{N mm}^{-1}$
Accuracy grade by ISO/GB			5	6	7	8	9	10	11	<5
Hard tooth surface	Spur gears	$K_{H\alpha}$	1.0		1.1	1.2				$1/Z_\varepsilon^2 (\geq 1.2)$
		$K_{F\alpha}$								$1/Y_\varepsilon (\geq 1.2)$
	Helical gears	$K_{H\alpha}$	1.0	1.1	1.2	1.4				$\varepsilon_\alpha / \cos^2 \beta_b (\geq 1.4)$
		$K_{F\alpha}$								
Soft tooth surface	Spur gears	$K_{H\alpha}$	1.0			1.1	1.2			$1/Z_\varepsilon^2 (\geq 1.2)$
		$K_{F\alpha}$								$1/Y_\varepsilon (\geq 1.2)$
	Helical gears	$K_{H\alpha}$	1.0		1.1	1.2	1.4			$\varepsilon_\alpha / \cos^2 \beta_b (\geq 1.4)$
		$K_{F\alpha}$								

Source: Adapted from Jelaska 2012, Table 3.7, p176. Reproduced with permission of John Wiley & Sons Inc.

Table 8.3 Face load factor for contact stress calculation, $K_{H\beta}$.

$$K_{H\beta} = a_1 + a_2(1 + a_3\varphi_d^2)\varphi_d^2 + a_4 b$$

Quality level		a_1	a_2	a_3 (Supporting methods)			a_4
				Symmetric	Asymmetric	Cantilever	
Soft tooth surface	5	1.10	0.18	0	0.6	6.7	1.2×10^{-4}
	6	1.11	0.18	0	0.6	6.7	1.5×10^{-4}
	7	1.12	0.18	0	0.6	6.7	2.3×10^{-4}
	8	1.15	0.18	0	0.6	6.7	3.1×10^{-4}
Hard tooth surface	5	$K_{H\beta} \leq 1.34$: 1.05	0.26	0	0.6	6.7	1.0×10^{-4}
		$K_{H\beta} > 1.34$: 0.99	0.31	0	0.6	6.7	1.2×10^{-4}
	6	$K_{H\beta} \leq 1.34$: 1.05	0.26	0	0.6	6.7	1.6×10^{-4}
		$K_{H\beta} > 1.34$: 1.00	0.31	0	0.6	6.7	1.9×10^{-4}

Source: Adapted from Wen 2015.

the face load factors for contact stress calculation $K_{H\beta}$. Face load factor for bending stress calculation $K_{F\beta}$ depends on $K_{H\beta}$ and the ratio of face width to tooth height, b/h and can be determined by [8]

$$K_{F\beta} = (K_{H\beta})^{N_0} \tag{8.17}$$

where

$$N_0 = \frac{(b/h)^2}{1 + (b/h) + (b/h)^2} \tag{8.18}$$

8.2.3 Potential Failure Modes

Gears may be vulnerable to failure by several of the following potential failure modes, depending upon operating loads and speeds, materials and lubrication, as well as manufacturing and assembly.

1) Fracture of gear teeth
   Gear tooth fracture is a catastrophic failure that may cause break down of a machine. It may be a static brittle or ductile fracture caused by an unexpected overload or, more frequently, a bending fatigue fracture. Whenever a pair of teeth meshing with each other during operation, the root of gear tooth is subjected to cyclic bending stresses. Besides, the root fillet is also a site of stress concentration. Microscopic cracks may initiate in the tooth fillet on the tensile side. These microscopic cracks may propagate across the root cross section after millions of accumulated cycles and ultimately lead to the breakage of teeth from gears.
2) Pitting of gear tooth surface
   Pitting is a surface fatigue failure in which small metal particles are dislodged from tooth surfaces. As gears mesh repeatedly, cyclic Hertz contact stresses generate on the curved tooth surfaces and subsurface. The cyclic Hertzian stresses may initiate minute cracks below the surface. After a large number of stress cycles, these minute

cracks may grow and join, and small bits of metal are eventually separated leaving surface pits (see Figure 2.9). Pitting is the result of slow surface destruction, leading to the roughness of tooth surface, vibration and noise.

3) Wear

As gears mesh, both rolling and sliding occur. Pitting is primarily a consequence of rolling contact while wear is the result of sliding contact [6]. In enclosed gearings when lubrication is insufficient, or in open gearings the presence of foreign particles, together with sliding motion between meshing surfaces, will cause abrasive wear on tooth surfaces. Wear may cause alteration of tooth profile and loss of conjugate motion, resulting in impact or intolerable levels of vibration or noise.

4) Scuffing

Scuffing is a severe form of destructive adhesive wear. When a pair of gears carries heavy loads at high speeds, high sliding motion between meshing teeth surface generates frictional heat, leading to a high temperature, reducing the viscosity of lubricant. Combined with the high pressure due to heavy loads on the mating surfaces, gears confront inadequate lubrication or even lubrication failure, causing direct metal-to-metal contact on tooth surfaces. As the gears rotate continuously, local micro-welding and surface dragging and scoring happen eventually, especially on the soft tooth surface. Scuffing significantly rough tooth profiles, resulting improper meshing, erratic output angular velocity, dynamic loads and noise.

5) Plastic deformation

When soft teeth are subjected to heavy loads and particularly impact loads, the surface material may flow plastically. Materials on the pinion tooth surface flow away from the pitch line; while materials on the gear tooth surface flow towards the pitch line. The deformed tooth profiles also lead to improper meshing, vibration or noise.

The failure modes mentioned before, although are described separately, two or more of them may occur simultaneously in practice. In fact, sometimes one type of failure may promote or even aggravate the other. It is designers' responsibility to prevent these failures at early design stage. This chapter mainly introduces the analysis and design process for the prevention of tooth breakage and tooth surface pitting. The design process for preventing other failures can be found in relevant design handbooks [9, 10].

8.3   Strength Analysis for Spur Gears

Tooth surface pitting and tooth breakage are the two most prevailing failure modes in gear drives. For proper power transmission, gears must be capable of running for an expected life without significant pitting and gear teeth must be safe from breakage. These are guaranteed by tooth surface contact strength analysis and tooth bending strength analysis, respectively, starting with force analysis.

8.3.1   Forces on Spur Gear Teeth

When a pair of gears is in operation, the teeth of driving pinion push on the driven teeth, exerting force on the gear. Thus, a torque is transmitted and, because the gear rotates, power is also transmitted. Power transmission actually involves the application of a torque during rotation at a given speed. The relationship between the nominal torque and the transmitted power is governed by

$$T_1 = 9.55 \times 10^6 \frac{P_1}{n_1} \tag{8.19}$$

If friction is neglected, the power transmitted by the pinion and gear are exactly the same. Therefore, a reduction in the rotational speed of a gear will proportionally increase the torque transmitted to the gear shaft.

Figure 8.3 shows a pinion rotates counterclockwise, transmitting a torque T_1. Assume the force acts at the midpoint of face width. At pitch point P, a normal force F_n from the driven gear exerts normally on the involute tooth profile of the pinion. Neglecting friction, this force can be resolved into two components, that is, a tangential force F_t in the tangential direction and a radial force F_r in the radial direction. The tangential force F_t is directly related to power transmission; therefore, this force is often called transmitted force. By moment equilibrium about the axis of rotation, the magnitude of tangential force F_t can be found from:

$$F_t = \frac{2T_1}{d_1} \tag{8.20}$$

The radial force F_r, which tends to separate the driving pinion and driven gear, is also called separating force. It is calculated by

$$F_r = F_t \tan \alpha \tag{8.21}$$

And the normal force F_n is

$$F_n = \frac{F_t}{\cos \alpha} = \frac{2T_1}{d_1 \cos \alpha} \tag{8.22}$$

The magnitude of forces on the pinion and gear are the same, except that they act in the opposite directions. Therefore, Eqs. (8.20)–(8.22) can be used to compute forces on either the pinion or the gear by appropriate substitution. Each tooth experiences repeated loading during operation.

Figure 8.3 Force analysis of a spur gear.

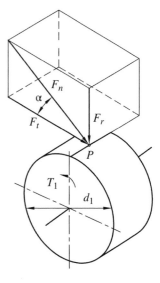

8.3.2 Tooth Surface Fatigue Strength Analysis

Pitting is a subsurface fatigue phenomenon caused by excessive cyclic contact stresses, resulting in loss of surface materials. Pitting frequently concentrates on a band along the pitch line near dedendum, where high contact stresses occur. Small pinions usually pit first as their teeth experience larger number of stress cycles than teeth on large gears.

To resist pitting failure, tooth surface fatigue strength analysis or tooth surface durability analysis is required. The contact stress on the meshing teeth surface is calculated by idealized Hertz contact stress equation, modified by a list of adjustment factors to account for the influence of manufacturing, assembly, geometric, loading and material variabilities. To control pitting, contact stresses should not exceed the allowable contact stress of tooth materials.

8.3.2.1 Hertz Formula

When curved tooth surfaces of meshing teeth contact with each other, they resemble a pair of contact cylinders on parallel axes, as shown in Figures 2.8 and 8.4. Therefore, the Hertz cylindrical contact stress model can be adopted for the calculation of contact stress for the assessment of surface durability of cylindrical gears. The Hertz formula introduced in Chapter 2 is rewritten here as

$$\sigma_{H\max} = \sqrt{\frac{F_n}{L} \cdot \frac{\left(\frac{1}{\rho_1} \pm \frac{1}{\rho_2}\right)}{\pi\left(\frac{1-\mu_1^2}{E_1} + \frac{1-\mu_2^2}{E_2}\right)}} \tag{2.51}$$

where F_n is the applied normal force, L is the contact length, μ_1, μ_2, E_1, E_2 are the Poisson's ratio and elastic modulus of two cylinders and ρ_1, ρ_2 are the radius of curvature at contact point. In accordance with convention, the algebraic 'plus' sign refers to external gearing and the minus refers to internal gearing.

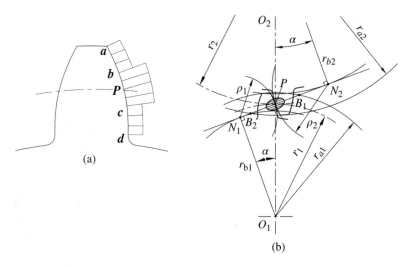

Figure 8.4 Contact stress analysis on a tooth surface.

8.3.2.2 Contact Stress Calculation

During gear mesh, contact points between the engaged teeth move from the tip towards the baseline along tooth profiles. For smooth operation, contact ratio is normally greater than one so that the load is carried by either a single pair or more than one pair of teeth. Therefore, the maximum stress occurs when only a single pair of teeth carries the total load, that is, at points where another pair of teeth is about to pick up its share of load, most probably near the middle of tooth profile, or near pitch point P.

Figure 8.4a shows schematically the approximate contact stress distribution along the tooth profile. Since the exact position of point b or c where a single pair of teeth carries the load is difficult to locate precisely; the stress calculation is too complicated at these two points and, also, the actual value of contact stresses at points b, c and P do not vary much; combined with the fact that pitting actually always happens near the pitch line, pitch point P is selected to evaluate contact strength to simplify calculation.

When Hertz formula is applied to the contact stress analysis in gears, the radii of cylinders are taken to be the curvature radii of the involute tooth profiles of mating teeth at contact points, as shown in Figure 8.4b. For the mating teeth, the radii of curvature on the pinion and gear tooth profiles at pitch point P are

$$\rho_1 = \frac{d_1}{2}\sin\alpha \qquad \rho_2 = \frac{d_2}{2}\sin\alpha \tag{8.23}$$

Therefore

$$\frac{\rho_2}{\rho_1} = \frac{d_2}{d_1} = \frac{z_2}{z_1} = u$$

and

$$\frac{1}{\rho_1} \pm \frac{1}{\rho_2} = \frac{2}{d_1\sin\alpha} \pm \frac{2}{d_2\sin\alpha} = \frac{2(d_2 \pm d_1)}{d_1 d_2 \sin\alpha} = \frac{2}{\sin\alpha}\frac{\frac{d_2}{d_1} \pm 1}{d_1 \frac{d_2}{d_1}} = \frac{2}{d_1\sin\alpha}\frac{u \pm 1}{u}$$

The total normal load on the teeth is the design load, found from Eqs. (8.11) and (8.22) as

$$KF_n = \frac{2KT_1}{d_1\cos\alpha}$$

The contact length is face width b. Introducing face width factor φ_d, defined as $\varphi_d = \frac{b}{d_1}$. Rearranging Hertz formula Eq. (2.51) using the notations used in gearing, the Hertzian contact stress on the gear tooth surface can be written as

$$\sigma_H = \sqrt{\frac{KF_t}{b\cos\alpha} \cdot \frac{2}{d_1\sin\alpha} \cdot \frac{u \pm 1}{u} \cdot \frac{1}{\pi\left[\frac{1-\mu_1^2}{E_1} + \frac{1-\mu_2^2}{E_2}\right]}}$$

$$= \sqrt{\frac{KF_t}{bd_1} \cdot \frac{u \pm 1}{u} \cdot \frac{2}{\sin\alpha\cos\alpha} \cdot \frac{1}{\pi\left[\frac{1-\mu_1^2}{E_1} + \frac{1-\mu_2^2}{E_2}\right]}}$$

$$= \sqrt{\frac{K}{\varphi_d d_1 \cdot d_1} \cdot \frac{2T_1}{d_1} \cdot \frac{u \pm 1}{u}} \times \sqrt{\frac{2}{\sin\alpha\cos\alpha}} \times \sqrt{\frac{1}{\pi\left[\frac{1-\mu_1^2}{E_1} + \frac{1-\mu_2^2}{E_2}\right]}}$$

Let zone factor as $Z_H = \sqrt{\dfrac{2}{\sin\alpha\cos\alpha}}$, which considers the effect of profile curvature at pitch point on the contact stress. For a standard spur gear, the pressure angle is 20°, so $Z_H = 2.5$.

The elastic coefficient $Z_E = \sqrt{\dfrac{1}{\pi\left[\frac{1-\mu_1^2}{E_1}+\frac{1-\mu_2^2}{E_2}\right]}}$ considers the mechanical properties of pinion and gear. The elastic modulus values of steels, cast steel, nodular cast iron and grey cast iron are 206, 202, 173 and 118 GPa [9], respectively. Poisson's ratio is usually assumed 0.3. The elastic coefficient of matching materials can thus be calculated. For the common match of steel pinion and steel gear, the elastic coefficient is 189.8 MPa$^{1/2}$.

With these simplifications, the contact stress can be expressed as

$$\sigma_H = Z_H Z_E \sqrt{\frac{2KT_1}{\varphi_d d_1^3}\cdot\frac{u\pm1}{u}} = Z_H Z_E \sqrt{\frac{KF_t}{bd_1}\cdot\frac{u\pm1}{u}} \tag{8.24}$$

8.3.2.3 Contact Strength Analysis

Excessive contact stress is the real reason for surface fatigue pitting. To prevent pitting, contact stresses on tooth surface should not exceed the allowable values, that is

$$\sigma_H = Z_H Z_E \sqrt{\frac{KF_t}{bd_1}\cdot\frac{u\pm1}{u}} \leq [\sigma_H] \tag{8.25}$$

The design formula can be derived from Eq. (8.25) for the determination of pitch diameter of the pinion of a spur gear for surface durability, expressed as

$$d_1 \geq \sqrt[3]{\frac{2KT_1}{\varphi_d}\cdot\frac{u\pm1}{u}\cdot\left(\frac{Z_H Z_E}{[\sigma_H]}\right)^2} \tag{8.26}$$

In Eq. (8.26), the unit for torque is N·mm, the unit for face width and diameter is mm and the unit for allowable contact stress $[\sigma_H]$ is MPa. The determination of allowable contact stresses will be introduced in Section 8.6.4.

Contact stresses, σ_H, are the same for both meshing teeth of pinion and gear, that is, $\sigma_{H1} = \sigma_{H2}$. However, the ratio of $[\sigma_H]/\sigma_H$ is different for the meshing gears. This is because the pinion and gear are usually of different materials, heat treatments and hardness, resulting in different allowable stresses $[\sigma_H]$. When using Eq. (8.26) in design, the smaller value of allowable contact stress should be used.

It is observed from Eq. (8.24) that contact stress σ_H is affected by pinion diameter d_1. The contact stress σ_H has no relation with module m. It can be interpreted that, with the increase of pinion diameter d_1, the diameter of base circle also increases, leading to a decreased curvature of involuted tooth profile, thus reducing the contact stress.

8.3.3 Tooth Bending Strength Analysis

Tooth breakage is a bending fatigue phenomenon caused by the bending action of a load. Tooth breakage usually happens near the root fillet in a spur gear where high cyclic bending stresses occur. To prevent tooth breakage, tooth bending strength analysis is required. The bending stress on the meshing teeth near the root fillet is calculated by the Lewis formula, modified by a list of adjustment factors to account for the influence of manufacturing, assembly, geometric, loading and material variabilities. To prevent tooth breakage, bending stresses should not exceed the allowable bending stress of gear materials.

8.3.3.1 Bending Stress Calculation

During gear mesh, the load on the gear sweeps downward from the tip towards the baseline. To simplify calculations and also for safety, the load is assumed to be carried by a single pair of teeth and acts at the tip of tooth. Such assumption implies a unit contact ratio and a loading position that produces a maximum stress at the tooth root, which is conservative for tooth design. Resolve the normal force into tangential components $F_{ca}\cos\gamma$ that bends the tooth and radial component $F_{ca}\sin\gamma$ that compresses the tooth, as shown in Figure 8.5.

The compressive stress generated by $F_{ca}\sin\gamma$ is usually negligible as they are less serious than the bending stress generated by $F_{ca}\cos\gamma$. The angle at tooth tip γ is somewhat greater than pressure angle α. When calculating bending stress, the gear tooth is idealized as a cantilever beam. Drawing 30° tangents to the root fillets, the critical section is the connection of point A and B at tooth root in the zone of maximum stress concentration, as shown in Figure 8.5.

In practice, fatigue cracks and failure begin on the tensile side of teeth, strength is then determined for this side. The nominal bending stress at the tensile side of root fillet designated as point A is [6, 11]

$$\sigma_{F0} = \frac{M}{W} = \frac{F_{ca}\cos\gamma \cdot h_F}{\frac{bs_F^2}{6}} = \frac{KF_t}{b} \cdot \frac{6\cos\gamma \cdot h_F}{\cos\alpha \cdot s_F^2} = \frac{KF_t}{bm} \cdot \frac{6\cos\gamma \cdot \left(\frac{h_F}{m}\right)}{\cos\alpha \cdot \left(\frac{s_F}{m}\right)^2} = \frac{KF_t}{bm}Y_{Fa}$$

where Y_{Fa} is the tooth form factor, defined as $Y_{Fa} = \dfrac{6\cos\gamma \cdot \left(\frac{h_F}{m}\right)}{\cos\alpha \cdot \left(\frac{s_F}{m}\right)^2}$. It considers the effect of tooth form, as well as tooth tip loading assumption on the tooth root bending stress.

To consider the effect of other factors, including stress concentration at tooth root fillet, on the tooth root bending stress [12], stress correction factor Y_{Sa} is introduced. Thus, the bending stress at root fillet is

$$\sigma_F = \frac{KF_t}{bm}Y_{Fa}Y_{Sa} \tag{8.27}$$

Figure 8.5 Bending stress analysis.

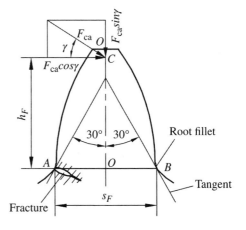

Table 8.4 Tooth form factor Y_{Fa} and stress correction factor Y_{Sa} [9].

Profile shift coefficient	$z(z_v)$	17	18	19	20	25	30	40	60	80	100	150	200	∞
$x = -0.5$	Y_{Fa}					3.50	3.25	2.93	2.60	2.46	2.37	2.26	2.22	2.063
	Y_{Sa}					1.37	1.41	1.46	1.54	1.60	1.64	1.73	1.79	1.966
$x = 0$	Y_{Fa}	2.97	2.91	2.85	2.80	2.62	2.52	2.40	2.28	2.22	2.18	2.14	2.12	2.063
	Y_{Sa}	1.52	1.53	1.54	1.55	1.59	1.63	1.67	1.73	1.77	1.79	1.83	1.87	1.966
$x = 0.5$	Y_{Fa}	2.22	2.20	2.18	2.17	2.14	2.12	2.10	2.08	2.075	2.07	2.068	2.065	2.063
	Y_{Sa}	1.76	1.77	1.78	1.80	1.82	1.84	1.86	1.89	1.91	1.92	1.94	1.95	1.966

Both tooth form factor Y_{Fa} and stress correction factor Y_{Sa} vary with the number of teeth and profile shift coefficient x. For helical gears, the virtual number of teeth z_v should be used. Table 8.4 lists tooth form factor Y_{Fa} and stress correction factor, Y_{Sa} for standard gears and gears with ± 0.5 profile shift coefficient. More data for gears with other profile shift coefficients can be found in the standards or design handbooks [9].

8.3.3.2 Bending Strength Analysis

A check of calculated stress against the allowable stress would disclose the margin of safety. Substituting $F_t = 2T_1/d_1$, $m = d_1/z_1$, $\varphi_d = b/d_1$ into Eq. (8.27), the bending strength can be calculated and checked by

$$\sigma_F = \frac{KF_t}{bm} Y_{Fa} Y_{Sa} = \frac{2KT_1}{\varphi_d m^3 z_1^2} Y_{Fa} Y_{Sa} \le [\sigma_F] \tag{8.28}$$

The design equation is then deduced as

$$m \ge \sqrt[3]{\frac{2KT_1}{\varphi_d z_1^2} \cdot \frac{Y_{Fa} Y_{Sa}}{[\sigma_F]}} \tag{8.29}$$

While using these equations, the unit of torque is N mm, the unit of face width and module is mm and the unit of allowable bending stress $[\sigma_F]$ is MPa. The determination of allowable bending stresses will be introduced in Section 8.6.4.

Since $Y_{Fa1} Y_{Sa1} \ne Y_{Fa2} Y_{Sa2}$ and $[\sigma_{F1}] \ne [\sigma_{F2}]$, the actual and allowable bending stress in the pinion and gear are different and the bending strength of both meshing gears should be checked. While using Eq. (8.29) for design, greater value of $Y_{Fa} Y_{Sa}/[\sigma_F]$ should be used and the result should be rounded to a standard module. For opening gearing, the allowable stress $[\sigma_F]_{open}$ is selected as 70–80% of material allowable bending stress to take into account the effect of wear.

It is noticeable in Eq. (8.28) that bending stress σ_F closely relates to module m, which affects the size of gear teeth. When the material and load have been determined, the larger the module, the thicker the gear teeth and, therefore, the smaller the bending stress.

8.4 Strength Analysis for Helical Gears

Helical gears share many attributes of spur gears when used to transmit power or motion between parallel shafts. The distinguishing geometrical difference is the orientation of their teeth. Spur gears have straight teeth aligned with the axis of the gear, while helical gears have inclined teeth at a helix angle β with respect to the axis of the gear. A spur gear may be regarded as a special helical gear with a zero helix angle.

The teeth of a helical gear may be either right- or left-hand. When a helical gear is lying on a flat surface, if the teeth lean to the right then it is a right-hand helical gear and if the teeth lean to the left it is a left-hand helical gear. Two externally mating helical gears on parallel shafts must have the same helix angle but with the opposite hand. An internal helical gear and its pinion must be of same hands.

8.4.1 Geometry and Terminology

8.4.1.1 The Geometry of a Helical Gear

The basic terminology of helical gears is substantially the same as that of spur gears. Additional variables are introduced to account for different aspects due to helix angle, as shown in Figure 8.6.

The description of the geometry of a helical gear is related to three primary planes; namely, the transverse plane, the normal plane and the tangential plane. The transverse plane or rotation plane is perpendicular to the axis of rotation. The transverse circular pitch p_t and transverse pressure angle α_t are measured in the transverse plane. The transverse plane contains the involute feature and is mainly used for the calculation of geometrical dimensions. The normal plane is perpendicular to the curved surface of teeth. The normal circular pitch p_n and normal pressure angle α_n are measured in the normal plane. The normal plane is used for strength analysis and standard values of module and pressure angle are defined in the normal plane; for example, normal pressure angle $\alpha_n = 20°$. The normal plane contains the geometry needed for cutting teeth,

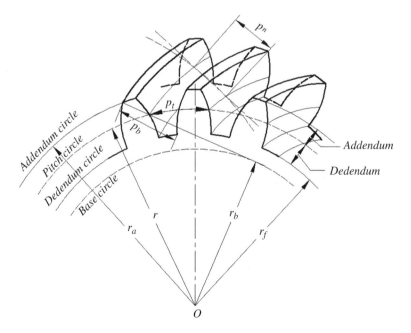

Figure 8.6 Geometry and terminology of a helical gear.

and helical gear teeth can be cut by the same tools as spur gears by moving or feeding the tool perpendicularly to the normal plane. The tangential plane or axial plane is tangential to the pitch surface of the gear. The axial pitch p_a is measured in the axial plane, which is the distance between corresponding points of adjacent teeth measured parallel to the gear axis. The angle between the normal plane and transverse plane is the helix angle β.

By definition, the transverse circular pitch is the same as the circular pitch defined for spur gears. Thus

$$p_t = p_c = \frac{\pi d}{z} \tag{8.30}$$

From Figure 8.7, we can have the relations between normal circular pitch and transverse circular pitch

$$p_n = p_t \cos \beta \tag{8.31}$$

Therefore, the relation between the transverse module and normal module is

$$m_t = \frac{m_n}{\cos \beta} \tag{8.32}$$

Also, from Figure 8.7, we have axial pitch p_a expressed as

$$p_a = \frac{p_t}{\tan \beta} = \frac{p_n}{\sin \beta} \tag{8.33}$$

Since the pitch diameter of a helical gear is calculated at transverse plane as

$$d = m_t z = \frac{m_n z}{\cos \beta} \tag{8.34}$$

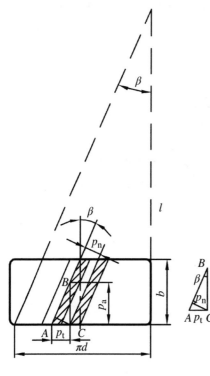

Figure 8.7 Geometrical relationship of transverse, normal and axial pitches.

Centre distance is

$$a = \frac{1}{2}(d_1 + d_2) = \frac{m_n}{2\cos\beta}(z_1 + z_2) \tag{8.35}$$

The relationship between transverse and normal pressure angle is [13]

$$\tan\alpha_t = \frac{\tan\alpha_n}{\cos\beta} \tag{8.36}$$

From Figure 8.7, the helix angle β is

$$\tan\beta = \frac{\pi d}{l}$$

Similar, the base circle helix angle β_b is

$$\tan\beta_b = \frac{\pi d_b}{l}$$

Therefore, the relationship between helix angle β and base circle helix angle β_b is [13]

$$\tan\beta_b = \tan\beta\cos\alpha_t \tag{8.37}$$

For a pair of helical gears mounted on parallel shafts to mesh properly, they must have the same module and pressure angle, as well as the same yet opposite helical angle.

8.4.1.2 Contact Ratio

When a pair of spur gears engage, the line of contact is initially near the tip of driven gear tooth across the face width. It then moves smoothly along the involute profile to the root of driven gear tooth where the mating teeth separate as the gears continue to rotate. The contact length actually changes from one face width to two face widths and back. Comparatively, when a pair of helical gears engage, the contact line is incline due to helix angle and the contact length changes smoothly from minimum to maximum as the gears rotate, leading to a gradual and smooth engagement. The contact ratio of a helical gear drive is normally much higher than that of a spur gear drive.

The contact ratio of a helical gear drive is a measure of overall load sharing among the teeth in contact. It is the sum of two contributions, transverse contact ratio ε_a and face contact ratio, or overlap ratio ε_β, expressed as

$$\varepsilon = \varepsilon_a + \varepsilon_\beta \tag{8.38}$$

Since tooth profiles are involute in the transverse plane, the transverse contact ratio is calculated in the transverse plane by means similar to that of a spur gear drive. Transverse contact ratio reflects the load sharing among multiple teeth simultaneously in contact, which can be obtained by Eq. (8.10) or estimated by [12]

$$\varepsilon_a = \left[1.88 - 3.2\left(\frac{1}{z_1} \pm \frac{1}{z_2}\right)\right]\cos\beta \tag{8.39}$$

The positive sign is used for external gears and the negative for internal gears.

The face contact ratio is the contact ratio in the axial or face direction. It corresponds to distribution of tooth loading along the contact length. The face contact ratio closely relates to face width and helix angle, from Figure 8.7, we have

$$\varepsilon_\beta = \frac{b}{p_a} = \frac{b\sin\beta}{\pi m_n} = 0.318\varphi_d z_1 \tan\beta \tag{8.40}$$

8.4.1.3 Virtual Number of Teeth

Figure 8.8 illustrates the concept of the virtual spur gear. The intersection of transverse cutting plane $T-T$ with the pitch cylinder of helical gear is a circle with diameter d, while the intersection of normal cutting plane $N-N$ with the pitch cylinder is an ellipse with a curvature radius ρ at pitch point P. The curvature radius ρ at pitch point P is calculated by [13]

$$\rho = \frac{d}{2\cos^2\beta} \tag{8.41}$$

In the elliptical intersection at pitch point P, the tooth profile of the helical gear is approximately the same as the tooth profile of a spur gear with a pitch radius ρ, or a virtual spur gear. The curvature of the ellipse at point P and curvature of the virtual pitch circle are matched. The virtual number of teeth z_v is defined as the quotient of the circumference of the virtual pitch circle with radius ρ and the normal circular pitch p_n, that is,

$$z_v = \frac{2\pi\rho}{p_n}$$

Incorporating Eqs. (8.30), (8.31) and (8.41), we have

$$z_v = \frac{z}{\cos^3\beta} \tag{8.42}$$

Equation (8.42) reveals the relations between the virtual number of teeth and the physical number of teeth of a helical gear. The virtual gear is equivalent to a spur gear with a virtual number of teeth z_v, which gives stronger teeth in both bending and surface fatigue strength than a spur gear with the same physical number of teeth as the helical gear. The larger number of virtual teeth also reduces undercutting tendency. Helical gears thus could have a smaller minimum number of teeth than spur gears. The strength analysis of a helical gear is performed on the virtual teeth in the normal plane.

8.4.2 Forces on Helical Gear Teeth

Figure 8.9 represents the forces acting on the teeth of a driving helical gear. The normal force F_n acts perpendicular to the face of helical tooth in the normal plane. It is

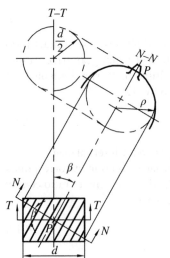

Figure 8.8 Virtual spur gear.

resolved into three orthogonal components, tangential force F_t, radial force F_r and axial force F_a.

The tangential force F_t acts in the transverse plane and is at a tangent to the pitch circle of the helical gear. It associates with power transmission and transmits torque from the driving pinion to the driven gear. The tangential force can be derived from the nominal torque as

$$F_t = \frac{2T_1}{d_1} \tag{8.43}$$

The radial force F_r acts towards the centre of gears. It tends to separate the driving pinion and driven gear, and contributes to the shaft bending and bearing loads. It gives

$$F_r = \frac{F_t \tan \alpha_n}{\cos \beta} \tag{8.44}$$

The axial force F_a acts parallel to the axis of gear and causes an axial or thrust load that must be resisted by bearings. Its direction can be determined by the Right- or Left-Hand Rule. With the tangential force already known, the magnitude of axial force is computed from

$$F_a = F_t \tan \beta \tag{8.45}$$

Since helical gears impose both radial and axial loads on supporting bearings, when two or more helical gears are mounted on the same shaft, the hand of gears should be properly selected so that the thrust loads produced by helical gears can counteract each other.

From Figure 8.9, the normal force can be obtained from

$$F_n = \frac{F_t}{\cos \beta \cos \alpha_n} = \frac{F_t}{\cos \alpha_t \cos \beta_b} \tag{8.46}$$

Figure 8.9 Force analysis of a helical gear.

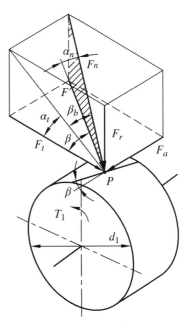

8.4.3 Tooth Surface Fatigue Strength Analysis

A helical gearing has an additional margin for increasing load carrying capacity. In a helical gear drive, the inclined contact line due to helix angle progress from the tip of teeth across the pitch line to the lower flank of tooth. The gradually changed contact length leads to a smooth engagement. Moreover, a large contact ratio in a helical gear drive implies more teeth are engaged simultaneously and share the applied loads. It is the gradual teeth engagement and lower average load per tooth that gives helical gears the ability to transmit heavy loads at high speeds.

8.4.3.1 Contact Stress Calculation

Pitting resistance for helical gear teeth is evaluated by the same approach as discussed for spur gears, with minor adjustment to account for geometrical differences due to helix angle. The Hertz formula is employed in contact stress calculations.

Similar to spur gears, incorporating Eqs. (8.11) and (8.46), the design load is calculated by

$$F_{ca} = KF_n = \frac{KF_t}{\cos \alpha_t \cos \beta_b} \tag{8.47}$$

When helical gears mesh, engagement is gradual and load is propagated diagonally across the tooth surface. The load is shared by all the teeth in engagement. The total length of inclined contact lines is affected by transverse and face contact ratio, and contact ratio factor Z_ε is introduced. The total length of contact line is calculated by [5]

$$L = \frac{b}{Z_\varepsilon^2 \cos \beta_b} \tag{8.48}$$

The relationship between normal radius of curvature ρ_n and transverse radius of curvature ρ_t in a helical gear can be expressed as [12]

$$\rho_n = \frac{\rho_t}{\cos \beta_b} \tag{8.49}$$

Similar to a spur gear, the transverse radius of curvature ρ_t is calculated in the transverse plane by

$$\rho_t = \frac{1}{2} d_1 \sin \alpha_t \tag{8.50}$$

Since strength analysis is performed in the normal plane, therefore,

$$\frac{1}{\rho_{n1}} \pm \frac{1}{\rho_{n2}} = \frac{2 \cos \beta_b}{d_1 \sin \alpha_t} \pm \frac{2 \cos \beta_b}{u d_1 \sin \alpha_t} = \frac{2 \cos \beta_b}{d_1 \sin \alpha_t} \cdot \frac{u \pm 1}{u} \tag{8.51}$$

Incorporating Eqs. (8.47), (8.48) and (8.51) and Hertz formula Eq. (2.51), we have

$$\sigma_H = \sqrt{\frac{F_n}{L} \cdot \frac{\left(\frac{1}{\rho_1} \pm \frac{1}{\rho_2}\right)}{\pi \left(\frac{1-\mu_1^2}{E_1} + \frac{1-\mu_2^2}{E_2}\right)}} = \sqrt{\frac{KF_t}{b \cos \alpha_t} \cdot \frac{2 \cos \beta_b}{d_1 \sin \alpha_t} \cdot \frac{u \pm 1}{u} \cdot Z_E Z_\varepsilon}$$

$$= \sqrt{\frac{KF_t}{bd_1} \cdot \frac{u \pm 1}{u} \cdot \sqrt{\frac{2 \cos \beta_b}{\sin \alpha_t \cos \alpha_t}} \cdot Z_E Z_\varepsilon}$$

Let zone factor $Z_H = \sqrt{\frac{2 \cos \beta_b}{\sin \alpha_t \cos \alpha_t}}$, which reflects the effect of profile curvature at pitch point on the contact stress. The transverse pressure angle α_t can be obtained from Eq. (8.36) and the helix angle at base circle β_b from Eq. (8.37).

Contact ratio factor Z_ε affects the effective length of contact line and consequently the unit face width load. When $\varepsilon_\beta < 1$, contact ratio factor is calculated by [14]

$$Z_\varepsilon = \sqrt{\frac{4 - \varepsilon_\alpha}{3}(1 - \varepsilon_\beta) + \frac{\varepsilon_\beta}{\varepsilon_\alpha}} \qquad (8.52)$$

When $\varepsilon_\beta \geq 1$, use $\varepsilon_\beta = 1$ to calculate Z_ε.

In a helical gear, the inclined contact line due to helix angle will also affect contact stress on the tooth surface. The helix angle factor Z_β is introduced to account for this effect. The helix angle factor Z_β is calculated by [14]

$$Z_\beta = \sqrt{\cos \beta} \qquad (8.53)$$

We then have contact stress in a helical gear expressed as

$$\sigma_H = Z_H Z_E Z_\varepsilon Z_\beta \sqrt{\frac{KF_t}{bd_1} \cdot \frac{u \pm 1}{u}} \qquad (8.54)$$

8.4.3.2 Contact Strength Analysis

For a design to be acceptable based on contact strength of helical gears, it must ensure that

$$\sigma_H = Z_H Z_E Z_\varepsilon Z_\beta \sqrt{\frac{KF_t}{bd_1} \cdot \frac{u \pm 1}{u}} \leq [\sigma_H] \qquad (8.55)$$

Let face width factor $\varphi_d = b/d_1$; Eq. (8.55) can be converted to the design formula as

$$d_1 \geq \sqrt[3]{\frac{2KT_1}{\varphi_d} \cdot \frac{u \pm 1}{u} \left(\frac{Z_H Z_E Z_\varepsilon Z_\beta}{[\sigma_H]} \right)^2} \qquad (8.56)$$

8.4.4 Tooth Bending Strength Analysis

Contrary to a spur gear whose teeth generally break near tooth root fillets, tooth breakage in a helical gear frequently happens along an inclined line where high bending stresses occur. The bending strength analysis is similar to that of a spur gear, with minor modifications to account for geometrical differences between helical and spur gears.

8.4.4.1 Bending Stress Calculation

The approaches to the design evaluation of tooth bending strength of spur gears are applicable to helical gears, yet a normal virtual spur gear is used. Besides, the effect of contact ratio and helix angle are considered by introducing contact ratio factor Y_ε and helix angle factor Y_β. Thus, the maximum local bending stress in a helical gear tooth is

$$\sigma_F = \frac{KF_t}{bm_n} Y_{Fa} Y_{Sa} Y_\varepsilon Y_\beta \qquad (8.57)$$

where tooth form factor Y_{Fa} and stress correction factor Y_{sa} can be selected from Table 8.4 by the virtual number of teeth z_v.

Contact ratio factor Y_ε for bending stress calculation is calculated by [9]

$$Y_\varepsilon = 0.25 + \frac{0.75}{\varepsilon_{av}} \tag{8.58}$$

where ε_{av} is transverse contact ratio of virtual spur gear, calculated by [9, 15]

$$\varepsilon_{av} = \frac{\varepsilon_\alpha}{\cos^2 \beta_b} \tag{8.59}$$

and the base circle helix angle β_b can be obtained from [15]

$$\cos \beta_b = \sqrt{1 - (\sin \beta \cos \alpha_n)^2} \tag{8.60}$$

Y_β is helix angle factor for bending stress calculation. It is introduced to consider the influence of oblique lines of contact in a helical gear on the tooth root bending stress and can be calculated by [15]

$$Y_\beta = 1 - \varepsilon_\beta \frac{\beta}{120°} \tag{8.61}$$

where β is the helix angle. If β is greater than $30°$, select $\beta = 30°$. ε_β is face contact ratio, which is calculated by Eq. (8.40) and the value 1.0 is substituted for ε_β if $\varepsilon_\beta > 1.0$.

8.4.4.2 Bending Strength Analysis

For safe operation of a helical gear, the bending stress should not exceed the allowable stress, that is

$$\sigma_F = \frac{KF_t}{bm_n} Y_{Fa} Y_{Sa} Y_\varepsilon Y_\beta \leq [\sigma_F] \tag{8.62}$$

If tooth strength is the main criterion determining load carrying capacity, let $F_t = 2T/d_1$, $\varphi_d = b/d_1$ and $d_1 = m_t z_1 = m_n z_1/\cos\beta$, the normal module of a helical gear can be derived from Eq. (8.62) as,

$$m_n \geq \sqrt[3]{\frac{2KT_1\cos^2\beta}{\varphi_d z_1^2} \cdot \frac{Y_{Fa} Y_{Sa} Y_\varepsilon Y_\beta}{[\sigma_F]}} \tag{8.63}$$

8.5 Strength Analysis for Bevel Gears

8.5.1 Geometry and Terminology

Bevel gears provide the most efficient methods of transmitting power between intersecting shafts, usually at right angles. Notwithstanding the fact that bevel gears are more complex than spur and helical gears in manufacturing and assembly, they find wide applications for power transmission between intersecting shafts.

Bevel gears may have straight, spiral or other curvilinear teeth. The differences between spiral and straight bevel gears is similar to those between helical and spur gears. The spiral teeth have a gradual pitch line contact and a larger number of teeth in contact. This make it possible to obtain smoother engagement and higher load carrying capacities than straight bevel gears. Straight bevel gears are widely used in low-speed operation; while spiral bevel gears are preferable for higher speed and heavy load applications. Hypoid gears are similar to spiral bevel gears, but have a relatively small shaft offset. Their pitch surfaces are hyperboloids of revolution [1]. For larger offsets, the pinion begins to resemble a tapered worm and the set is then called a spiroid gearing, which will be discussed in Chapter 9.

Figure 8.10 shows the sectional view of a pair of bevel gears meshing when the shafts are perpendicular to each other. AOP and BOP are pinion pitch cone and gear pitch cone, respectively. AO_1P and BO_2P are pinion back cone and gear back cone, respectively. The pinion back cone AO_1P and gear back cone BO_2P can be developed into two sector gears with pitch radii as the back cone distance of the bevel pinion and gear, respectively. Supplement teeth to the sector gears form whole imaginary virtual gears and the number of teeth increases to the virtual number of teeth z_{v1} and z_{v2}, respectively. Thus, the tooth profiles of a bevel pinion or gear resemble the tooth profiles of imaginary virtual spur gears with pitch radii equal to the developed pinion back cone radius r_{bv1} and the developed gear back cone radius r_{bv2} [6]. This similarity is also applicable to the midpoint of bevel gear face width where the developed virtual spur gear will be used for force and stress analysis.

When a pair of bevel gears are in operation, the surface of pinion pitch cone AOP and gear pitch cone BOP roll together without slipping. They share a common apex at the shaft axis intersection O. Since the pitch cones of mating gears have coincident apexes, the mounting of bevel gears is critical for achieving satisfactory performance. Proper mounting is realized by adjusting mounting distance, which is the distance from the end surface of the pinion hub to the apex of pitch cone O.

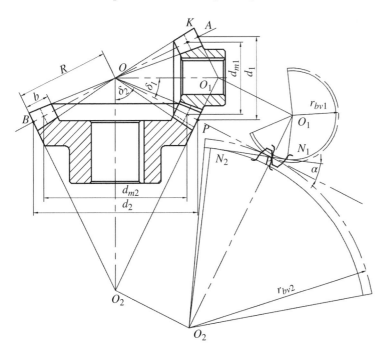

Figure 8.10 Geometry and terminology of bevel gears.

Bevel gear teeth are tapped, both in tooth thickness and height, from the large tooth profile at one end to the small tooth profile at the other end. Customarily, tooth dimensions are specified either at the large end or at the midpoint of face width. Dimensions at the large end are generally used for geometrical calculation; while dimensions at face width midpoint are used for strength analysis and design. The following discussion is limited to straight bevel gears mounted on perpendicular shafts with a pressure angle of 20°.

1) Pitch diameter, d

When a standard modulus is specified at the large end, the pitch diameter d at the large end is calculated in the same way as for spur gears, that is,

$$d = mz \tag{8.64}$$

2) Gear ratio, u

The gear ratio for bevel gears are determined from the number of teeth, the pitch diameters or the pitch cone angles, as

$$u = \frac{n_1}{n_2} = \frac{z_2}{z_1} = \frac{d_2}{d_1} = \frac{\sin \delta_2}{\sin \delta_1} = \cot \delta_1 = \tan \delta_2 \tag{8.65}$$

Therefore, the pitch cone angles for the pinion δ_1 and gear δ_2 are determined by the ratio of the number of teeth. A pair of bevel gears having a unit ratio is used simply for the purpose of rotational direction change.

3) Cone distance, R

Figure 8.10, the cone distance is calculated by

$$R = \sqrt{\left(\frac{d_1}{2}\right)^2 + \left(\frac{d_2}{2}\right)^2} = \frac{m}{2}\sqrt{z_1^2 + z_2^2} = \frac{d_1}{2}\sqrt{u^2 + 1} \tag{8.66}$$

4) The relationship between the diameter of virtual spur gear d_v, mean diameter d_m and pitch diameter d. Similar to obtaining a virtual spur gear on the developed back cone on the large ends [6, 13], we can have a virtual spur gear at the midpoint of face width for strength analysis. The relationship between the diameter of virtual spur gear at the midpoint of face width d_v (not shown in Figure 8.10) and mean diameter d_m is [12]

$$d_v = \frac{d_m}{\cos \delta} \tag{8.67}$$

and

$$\frac{d_{m1}}{d_1} = \frac{d_{m2}}{d_2} = \frac{R - 0.5b}{R} = 1 - 0.5\frac{b}{R} = 1 - 0.5\varphi_R \tag{8.68}$$

The face width b is bounded by large and small ends. Let the face width factor of a bevel gear be $\varphi_R = b/R$, which is normally within the range of 0.25~0.35. The most commonly used value is 1/3. So

$$d_{m1} = (1 - 0.5\varphi_R)d_1 \qquad d_{m2} = (1 - 0.5\varphi_R)d_2 \tag{8.69}$$

5) The relationship between mean module m_m and module m

Since $\frac{d_m}{d} = 1 - 0.5\varphi_R$, we have $\frac{m_m}{m} = 1 - 0.5\varphi_R$, so

$$m_m = (1 - 0.5\varphi_R)m \tag{8.70}$$

6) The relationship between the number of virtual spur gear teeth z_v and the number of straight bevel gear teeth z
From $d_v = m_m z_v$ and Eq. (8.67), we have

$$z_v = \frac{z}{\cos \delta} \tag{8.71}$$

7) The virtual spur gear ratio u_v

$$u_v = \frac{z_{v2}}{z_{v1}} = \frac{z_2 \cos \delta_1}{z_1 \cos \delta_2} = u^2 \tag{8.72}$$

8.5.2 Forces on Straight Bevel Gear Teeth

Since bevel gear teeth are tapped in both tooth thickness and height, the load applied on the bevel gear is nonuniform, proportionately greater at the large end. To simplify calculation, the normal force F_n is usually assumed to be acting at the midpoint of face width of a bevel gear teeth on the pitch cone, as shown in Figure 8.11. Although the actual acting point is a little displaced from the midpoint, no serious error will result.

The normal force, F_n, on the pinion can be resolved into three mutually perpendicular components: a tangential force, F_t, a radial force, F_r, and an axial force, F_a. The tangential force F_t acts tangentially to the pitch cone and is the force that produces the torque on the pinion and gear. It is calculated from

$$F_{t1} = \frac{2T_1}{d_{m1}} = F_{t2} \tag{8.73}$$

The radial force F_r acts towards the centre of the pinion, perpendicular to its axis. Thus,

$$F_{r1} = F_{t1} \tan \alpha \cos \delta_1 = F_{a2} \tag{8.74}$$

The axial force F_a acts parallel to the axis of the pinion, tending to push it away from the mating gears. It generates a thrust load on the shaft bearings and also produces a bending moment on the shaft. Thus,

$$F_{a1} = F_{t1} \tan \alpha \sin \delta_1 = F_{r2} \tag{8.75}$$

Figure 8.11 Force analysis of a bevel gear.

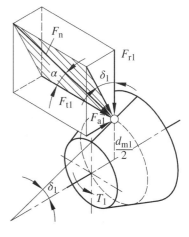

The normal force F_n can be obtained from

$$F_n = \frac{F_{t1}}{\cos \alpha} \tag{8.76}$$

8.5.3 Tooth Surface Fatigue Strength Analysis

The calculation of surface contact stress essentially follows the same approach previously presented for spur and helical gears, except for some minor corrections associated with bevel gear geometry. The contact stress for a bevel gear is calculated on the virtual spur gear developed at the midpoint of the face width, employing the Hertz formula.

8.5.3.1 Contact Stress Calculation

For a straight bevel gear tooth, the variables that will be used in the Hertz formula in the virtual gear are

$$L = b \tag{a}$$

$$KF_n = \frac{KF_{t1}}{\cos \alpha} = \frac{2KT_1}{d_{m1} \cos \alpha} \tag{b}$$

$$\rho_{v1} = \frac{d_{v1}}{2} \sin \alpha = \frac{d_{m1} \sin \alpha}{2 \cos \delta_1} \tag{c}$$

$$\rho_{v2} = \frac{d_{v2}}{2} \sin \alpha = \frac{u_v d_{v1} \sin \alpha}{2} \tag{d}$$

$$\frac{1}{\rho_{v1}} + \frac{1}{\rho_{v2}} = \frac{2 \cos \delta_1}{d_{m1} \sin \alpha} \left(1 + \frac{1}{u_v}\right) \tag{e}$$

$$u_v = u^2 \tag{f}$$

$$b = \varphi_R R = \frac{d_1}{2} \varphi_R \sqrt{1 + u^2} \tag{g}$$

$$d_{m1} = (1 - 0.5\varphi_R)d_1 \tag{h}$$

so

$$\cos \delta_1 = \sqrt{1 - \sin^2 \delta_1} = \sqrt{1 - \frac{1}{1 + \cot^2 \delta_1}} = \sqrt{1 - \frac{1}{1 + u^2}} = \frac{u}{\sqrt{1 + u^2}} \tag{i}$$

Integrating Eqs. (a)–(i) into the Hertz formula, we have

$$\sigma_H = \sqrt{\frac{F_n}{L} \cdot \frac{\left(\frac{1}{\rho_1} \pm \frac{1}{\rho_2}\right)}{\pi \left(\frac{1-\mu_1^2}{E_1} + \frac{1-\mu_2^2}{E_2}\right)}} = Z_E \sqrt{\frac{2KT_1}{bd_{m1} \cos \alpha} \cdot \frac{2 \cos \delta_1 \left(1 + \frac{1}{u_v}\right)}{d_{m1} \sin \alpha}}$$

$$= Z_E \sqrt{\frac{2}{\sin \alpha \cos \alpha} \cdot \frac{2KT_1 \cos \delta_1 \left(1 + \frac{1}{u_v}\right)}{bd_{m1}^2}} = Z_E Z_H \sqrt{\frac{2KT_1 \frac{u}{\sqrt{1+u^2}} \frac{1+u^2}{u^2}}{\frac{\varphi_R}{2} \sqrt{1 + u^2}(1 - 0.5\varphi_R)^2 d_1^3}}$$

Therefore, the contact stress on a bevel gear surface is calculated by

$$\sigma_H = Z_E Z_H \sqrt{\frac{4KT_1}{\varphi_R(1 - 0.5\varphi_R)^2 d_1^3 u}} \tag{8.77}$$

8.5.3.2 Contact Strength Analysis

For a pair of bevel gears to work safely, the contact stress should not exceed the allowable stress, that is,

$$\sigma_H = Z_E Z_H \sqrt{\frac{4KT_1}{\varphi_R(1 - 0.5\varphi_R)^2 d_1^3 u}} \leq [\sigma_H] \tag{8.78}$$

The derived equation from Eq. (8.78) can be used for the design of bevel gear teeth for surface durability, thus the design equation is

$$d_1 \geq \sqrt[3]{\left(\frac{Z_E Z_H}{[\sigma_H]}\right)^2 \frac{4KT_1}{\varphi_R(1 - 0.5\varphi_R)^2 u}} \tag{8.79}$$

8.5.4 Tooth Bending Strength Analysis

Likewise, bending stress analysis for bevel gear teeth is also similar to that already presented for spur and helical gear teeth, with minor alterations according to bevel gear geometry. The maximum bending stress is calculated at the tooth root fillet of the midpoint virtual spur gear.

8.5.4.1 Bending Stress Calculation

From the Lewis formula Eq. (8.27), the maximum bending stress occurs at the root of the tooth is computed from

$$\sigma_F = \frac{KF_t}{bm_m} Y_{Fa} Y_{Sa}$$

The tangential force is computed at the midpoint of the tooth, thus,

$$F_{t1} = \frac{2T_1}{d_{m1}} = \frac{2T_1}{m_m z_1} = \frac{2T_1}{m(1 - 0.5\varphi_R)z_1}$$

Face width, b, by definition is

$$b = \varphi_R R = \frac{d_1}{2}\varphi_R \sqrt{1 + u^2} = \frac{m z_1}{2}\varphi_R \sqrt{1 + u^2}$$

The mean module m_m is

$$m_m = (1 - 0.5\varphi_R)m$$

Integrating these equations into Lewis formula, therefore, the bending stress can be calculated by

$$\sigma_F = \frac{4KT_1}{m^3 z_1^2(1 - 0.5\varphi_R)^2 \varphi_R \sqrt{1 + u^2}} Y_{Fa} Y_{Sa} \tag{8.80}$$

8.5.4.2 Bending Strength Analysis

To prevent tooth breakage, bending stress must satisfy

$$\sigma_F = \frac{4KT_1}{m^3 z_1^2 (1 - 0.5\varphi_R)^2 \varphi_R \sqrt{1 + u^2}} Y_{Fa} Y_{Sa} \leq [\sigma_F] \tag{8.81}$$

Equation (8.81) can be used for the design of bevel gear teeth, as

$$m \geq \sqrt[3]{\frac{4KT_1}{\varphi_R (1 - 0.5\varphi_R)^2 z_1^2 \sqrt{1 + u^2}} \cdot \frac{Y_{Fa} Y_{Sa}}{[\sigma_F]}} \tag{8.82}$$

The factors Y_{Fa} and Y_{sa} can be found by the virtual number of teeth z_v from Table 8.4.

8.6 Design of Gear Drives

8.6.1 Introduction

In gear drive design, the provided design specifications include the transmitted power, the rotational speeds of pinion and gear, or gear ratio, the operating conditions and environment. The types of driving device and driven machine are also required for the selection of application factors.

The design starts by deciding on the gear types and accuracy grades according to the application. It also involves the selection of tentative materials and heat treatments, as well as initial variables, like the number of teeth, helix angle and so on. Design must satisfy two design criteria; that is, bending strength of gear teeth and pitting resistance of tooth surface.

Gears can be designed in either order as long as both design criteria are satisfied. Different design approaches are demonstrated in Section 8.6.7. Since information in the problem statement is insufficient for solving the unknowns directly, trial and iteration in the gear design process are unavoidable. There are usually a number of acceptable designs and computer-aided design techniques, for example, iteratively running developed design programs by a computer, are thus preferred to find an optimized design.

The objective of gear design is to determine design variables, such as the module, the number of teeth, face width, pitch diameters and so on to achieve smooth and quiet operation with a compact, long life, low cost, good manufacturability gear drive. As gears are in a power transmission system, they must be compatible with surrounding elements, such as bearings, shafts, housing, driver and driven machines.

8.6.2 Materials and Heat Treatments

Gears can be made from a wide variety of materials to achieve properties required for the application. Considering the general failure modes discussed in Section 8.2.3, candidate gear materials should have good strength, especially fatigue strength, good resistance to wear, surface fatigue and corrosion, high surface hardness and internal ductility, good machinability and reasonable cost. Compromise is often necessary when selecting materials to satisfy some of incompatible requirements.

8.6.2.1 Commonly Used Gear Materials

Common gear materials include ferrous, nonferrous metals and nonmetallic materials. Steels combine superior characteristics of high strength and moderate cost. Their properties, as well as surface hardness, can be improved via proper heat treatments. Most of moderate to heavy-duty speed reducers use medium carbon steel and alloy steels for gear materials. Wrought steels and cast steels can be used for gear blanks. For large sized gears, cast steels are preferred.

Both grey and nodular cast iron are used for large gears in low speed open gearings. The inexpensive cast iron gears have good wear resistance and high vibration damping capacity. They are usually paired with steel pinions to obtain reasonable strength with quiet operation.

Other commonly used materials are nonferrous metals, like bronze. The lower modulus of elasticity of copper alloys provides greater tooth deflection and improves load sharing among teeth. Since bronze and steel run well together, the combination of a steel pinion and a bronze gear is often used. Brass is an inexpensive material usually for clock gears.

Nonmetallic gears, such as plastic gears, can operate quietly under light load without lubrication. They have low weight, good wear and corrosion resistance, Plastic gears can be moulded into final form without subsequent machining, which greatly reduce production cost. Nonmetallic gearings are increasingly used in the area where low cost and quiet operation take precedence over strength requirements.

8.6.2.2 Heat Treatments

Gear blanks made of medium carbon steels or alloy steels are usually first normalized and tempered before being machined. Such treatment produces soft tooth surfaces with hardness less than 350HBW. These gears are limited to general applications. If both meshing gears have soft tooth surfaces, the pinion tooth surfaces are treated 30–50HBW higher than those of the gear. This is because pinions have to experience more stress cycles with relatively thin root thicknesses compared with mating gears.

The gears with hard tooth surfaces (hardness $\geq$38–65HRC) are firstly produced by milling, shaping or hobbing, followed by case hardening processes, such as carburizing or nitriding, to improve surface hardness. Carburizing produces surface hardness within a range of 55–65 HRC, resulting in almost the highest strength for gears. Nitriding produces a hard but thin case and should be avoided if overload or shock will be experienced, because this case is not sufficiently strong to resist such loads [3]. Local surface heating by induction hardening or oxyacetylene flame hardening followed by rapid quenching are used to produce a high-hardness, wear-resistant surface layer of gear teeth. Finally, gear teeth are finished by grinding, shaving or honing to achieve high precision.

8.6.3 Gear Manufacturing and Quality

Gears are made by a variety of manufacturing methods to achieve various qualities. The selection of the proper method involves a balanced consideration of design requirements, the capability and limitation of manufacturing methods and costs. Gears may be initially cut by hobbing, shaping, milling and so on, followed by a finishing process such as shaving, grinding, lapping, honing or burnishing to obtain high precision gear teeth.

Gear teeth can be obtained by forming methods and generating methods. When form cutting, the tooth space takes the exact shape of cutter. When generating, a generating tool is moved relative to the gear blank like two gears in mesh for the purpose of gear tooth profile cutting [1]. A standardized tooth system has a specified pressure angle and modulus to facilitate interchangeability and availability.

Hobbing is employed to make gears with a high degree of profile accuracy and excellent surface furnishing at a high production rate. A hob takes a shape similar to a worm with periodic relief grooves and is sharpened to provide cutting edges. While hobbing, the hob is fed across the face width of gear blank as both the gear blank and the hob rotate at a proper speed ratio. Shaping utilizes either a rack cutter or a pinion cutter to cut gear teeth. The cutter is reciprocated back-and-forth across the face of gear blank as the gear blank rotates. Milling uses a milling cutter to cut tooth spaces sequentially by rotating the gear blank one circular pitch each time until a complete gear forms. Hobbing, shaping and milling produce spur and helical gear teeth with good accuracy for low speed, light load applications.

When gears operate at high speed or under heavy loads, gears may be subjected to additional dynamic loads if errors exist in tooth profiles. Tooth surface quality must, therefore, be enhanced by a finishing process following gear cutting operation. Gear finishing, like shaving, grinding and honing, uses highly accurate tools to remove minute amounts of material off teeth surfaces.

Shaving is a corrective process of scarping away tiny material to improve profile accuracy and surface finish. Grinding can be realized by form grinders or generating grinders, similar to the cutting method just discussed. Both utilize abrasive wheels to grind both sides of the space between two teeth. Honing tends to average surface irregularities, removing nicks and burrs by a hone, a fine-grit abrasive gear-shaped tool [6]. Gear finishing can improve dimensional accuracy, increase surface hardness, enhance load sharing among gears and reduce vibration and noise. Gears in high-speed, high-load and high-precision applications are often honed after grinding in order to produce a very smooth surface finish.

Different manufacturing methods produce diverse accuracy grade gears. Gear accuracy grade defines the tolerances of various size gears manufactured to a specified accuracy. An experience-based guide to accuracy levels achieved by different manufacturing methods for various applications are provided in Table 8.5. In ISO, GB and DIN standards, small number represents high accuracy, while in AGMA 2008 standard, large number specifies high precision.

Gears are more expensive than belts and chains. Heavy loads, high speeds, long lives and quiet operation require high quality gear manufacture; when loads and speeds are less demanding, accuracy requirements may be relaxed. As expected, the manufacturing cost increases sharply with increased precision.

8.6.4 Allowable Stresses

Tooth surface pitting and tooth breakage are two forms of failure in gears due to fatigue. The strength and pitting resistance against failure can be evaluated by endurance limits of gear materials. Endurance limits are important properties obtained from fatigue tests, using a pair of accuracy grades 4~6 ISO/GB standard spur gears, with module $m = 3$~5 mm, centre distance $a = 100$ mm, pressure angle $\alpha = 20°$, face

Table 8.5 Gear accuracy grades and various gearing applications [6, 9].

Accuracy levels	Manufacturing methods	Gear accuracy grade ISO, GB, DIN	Gear quality number AGMA 2008	Applications
Ultra-high accuracy	Lapping, honing, or burnishing	2–3	14–15	Master gears, high speed, high load and high reliability gears
High accuracy	Grinding or shaving with first-rate machine tools	4–5	12–13	High speed and precision gears in turbine and aerospace gearings
Relatively high accuracy	Grinding, shaving, hobbing with best machine tools	6–7	10–11	Medium speed vehicle gears and industrial gears
Medium accuracy	Hobbing, shaping, milling with best machine tools	8–9	8–9	Low speed vehicle gears and industrial gears
Low accuracy	Hobbing, shaping, milling with ordinary machine tools	10–11	6–7	Low speed gears
Very low accuracy	Casting, moulding	12	4–5	Low speed and light load gears

width $b = 10{\sim}50$ mm, linear velocity $v = 10$ m s^{-1} and reliability of 99% for a life of 5×10^7 cycles under repeated, unidirectional stable loads [16]. Although other factors, such as gear dimensions, lubrication, temperature and so on are considered in design handbooks and standards [9, 14], they are deemed to have minor effects on endurance limits and therefore neglected for simplicity in this book. Contact endurance limits, σ_{Hlim}, and bending endurance limits, σ_{Flim}, have a linear relationship with surface hardness HBW for a soft tooth surface or HV for a hard tooth surface [16]. They are also closely related to material quality grade. Table 8.6 lists the contact endurance

Table 8.6 Contact endurance limits, σ_{Hlim}, and bending endurance limits, σ_{Flim}, of selected gear materials [16].

Material	Type	Hardness HBW/HV	Contact endurance limits, σ_{Hlim}, MPa	Bending endurance limits, σ_{Flim}, MPa
Normalized cast steels	Carbon steels	140–210 HBW	269–338	106–128
Cast iron materials	Nodular cast iron	175–300 HBW	462–641	180–224
	Grey cast iron	150–240 HBW	287–380	46–69
Through hardened wrought steels	Carbon steels	115–215 HV	484–554	191–215
	Alloy steels	200–360 HV	636–846	272–340
Through hardened cast steels	Carbon steels	130–215 HV	408–479	146–165
	Alloy steels	200–360 HV	553–757	234–292
Flame or induction	Wrought and cast steels	500–615 HV	1153–1215	359–369
Nitriding steels/through hardening steels nitrided	Nitriding steels	650–900 HV	1250	420
	Through hardening steels	450–650 HV	998	363

limits, σ_{Hlim} and bending endurance limits, σ_{Flim} of commonly used gear materials for a medium material quality grade. General design practice would use the data below the average, that is, between MQ (medium quality) and ML (low quality). The bending endurance limits of a completely reverse stress are 70% of the unidirectional value [16].

When design lives required are different from those in the tests, endurance limits obtained from the tests may be modified by multiplying life factors K_{HN} for contact stress and K_{FN} for bending stress The number of load cycles is calculated by the same equation for both contact and bending stress, as

$$N = 60njL_h \tag{8.84}$$

where N is the number of load cycles; L_h is design life in hours; n is rotational speed of gears in rpm and j is the number of gear teeth meshing per revolution.

Thus, the allowable contact stress corresponding to the design life and safety factor is calculated by

$$[\sigma_H] = \frac{K_{HN}\sigma_{H\lim}}{S_H} \tag{8.85}$$

where

K_{HN} —life factor for contact stress. It accounts for the actual number of contact stress cycles different from 5×10^7 cycles when the data were produced for the contact endurance limits. The life factor for contact stress K_{HN} varies from 0.85 to 1.6, depending on the actual number of contact stress cycles, materials and heat treatments.

S_H —safety factor for contact strength. Since gears can still operate even after surface failure gives a noise warning, safety factors are selected within a range of 0.85–1.6 corresponding to reliabilities from low to high. The usual selection is $S_H = 1.0$.

The allowable bending stress corresponding to the design life and safety factor is calculated in a way similar to that of allowable contact stress, as

$$[\sigma_F] = \frac{K_{FN}\sigma_{F\lim}}{S_F}Y_{ST} \tag{8.86}$$

where

K_{FN} – life factor for bending stress, which is used to adjust bending endurance limit to a life other than 3×10^6 cycles when the data were produced for the bending endurance limits. The life factor for bending stress varies from 0.85 to 2.5, depending on the actual number of bending stress cycles, materials and heat treatments.

Y_{ST} – stress correction factor for testing gears, usually select $Y_{ST} = 2$.

S_F – safety factor for bending strength. Since tooth breakage is sudden and catastrophic, resulting in the end service of a transmission system, safety factors are usually selected in a range from 1.25 to 1.5; larger than the safety factor against pitting.

Life factors for contact or bending stress are selected according to the actual number of stress cycles, materials and heat treatments. The smaller the actual number of stress cycles and the harder the materials, the greater the life factor that can be used. For a higher number of cycles or for critical applications where the tolerance for pitting and tooth wear is minimal, the average or below the average value is used for most cases to ensure a conservative design. Detailed data for life factors can be found in references [9, 14, 15].

8.6.5   Design Criteria

As mentioned before, the two most likely potential failure modes governing a typical gear design are bending fatigue failure at tooth root fillets and surface fatigue failure at the tooth surface. To ensure the proposed design configuration is acceptable, both bending stress and contact stress must be less than corresponding allowable stresses.

For enclosed soft tooth surface gears, pitting is the most common failure mode. To provide adequate performance for pitting resistance, contact stress σ_H should not exceed the allowable contact stress $[\sigma_H]$. Thus the gear should first be designed by $\sigma_H \leq [\sigma_H]$. Then, failure by tooth breakage should be prevented by ensuring that bending stress σ_F less than or at most equal to the allowable bending stress $[\sigma_F]$.

For enclosed hard tooth surface gears, tooth breakage is a common failure phenomenon. Surface durability is equally important as gear teeth may experience numerous cycles of stresses. In order to provide adequate performance for both strengths, the gear teeth should be first designed according to $\sigma_F \leq [\sigma_F]$ and then checked by $\sigma_H \leq [\sigma_H]$.

For open gearings, wear happens before tooth breakage. The design approach is therefore similar to bending strength analysis, except 70–80% of the allowable bending stress is selected to consider wear effects.

8.6.6   Design Procedure and Guidelines

Gear design, like other design procedures, involves a series of approximations and iterations. Besides, the standard values for modules and pressure angle, plus limitation on tooth number, impose additional constraints. The following subsections provide the design procedure and guidelines to design a safe and long-lasting gear drive. Design cases are presented in Section 8.6.7.

8.6.6.1   Select Gear Type, Materials, Accuracy Grades, Heat Treatments and Manufacturing Methods

The selection of gear type, materials, accuracy grade and manufacturing process for a particular design scenario depends largely on many factors, including the gear layout, the transmitted power, the rotational speed, noise limitation and cost constraints. High alloy steel with high surface hardness usually results in a compact design but at a high cost. Also, higher quality and tighter tolerance gears with ground or shaved teeth improve load sharing, reduce dynamic loads and consequently lower stresses and improve surface durability, but also cost is high [3]. Therefore, compromise is indispensable while making design decisions. Moreover, because of its size, the possibility of undercut and more frequent contact, a pinion usually uses better and more expensive materials than gears.

8.6.6.2   Initial Selection of Design Variables

The number of teeth of a pinion, z_1

Generally speaking, gears with more teeth tend to run more smoothly and quietly than gears with fewer teeth. However, the number of pinion teeth should be as small as possible to keep a transmission system compact. But the possibility of interference or undercut is great with fewer teeth. For enclosed gearings, the number of pinion teeth is initially selected as $z_1 = 20$–40, while for open gearings, it is selected as $z_1 = 17$–20. The number of teeth of the meshing gear can be obtained by $z_2 = u z_1$.

Helix angle, β

A helix angle is specified for each helical gear design. Although a large helix angle benefits smooth and quiet operation, it may induce a large thrust load. A typical range of a helix angle is from 8° to 20°.

Face width factor, φ_d

Increasing face width reduces both bending stress and contact stresses and improves surface durability. However, the face width is normally not greater than the pitch diameter of pinion, as a wide face width increases possibility of misalignment and uneven load distribution along the gear teeth. The face width of pinion and gear can be obtained from

$$\begin{cases} b = \varphi_d d_1 \\ b_2 = b \\ b_1 = b + (5 \sim 10)mm \end{cases} \tag{8.87}$$

The selection of face width factor φ_d relates closely to how a gear is supported by bearings. For symmetric, asymmetric and cantilever support, face width factor φ_d is selected from the ranges of 0.9–1.4, 0.7–1.15 and 0.4–0.6, respectively [5]. Small values are for the design where both gears have hardness greater than 350HBW, while large values for at least one gear has hardness less than 350HBW.

8.6.6.3 Design by Gear Strength

(1) Design by gear teeth surface strength

After the initial values of design variables have been selected, the design process can proceed by applying loads and speeds imposed by design specifications to the specific design until both contact fatigue strength and bending fatigue strength are satisfied.

The diameter of pinion that is directly related to the gear size is determined to satisfy the requirement of surface contact strength against pitting. A larger pitch diameter usually leads to reduced contact stresses and improved surface durability but increased pitch line velocity, while a small pitch diameter may result in too few teeth to avoid interference. The smallest acceptable pinion diameter can be used as the first choice to keep the design compact.

(2) Design by gear teeth bending strength

The module that directly relates to tooth size is determined by satisfying bending strength requirement. A large module results in larger teeth and generates lower bending stresses. A standard acceptable value of module should be used to facilitate manufacturing and reduce cost.

(3) Design by both gear teeth surface strength and gear teeth bending strength

More often, gears are designed by both tooth surface fatigue strength $\sigma_H \le [\sigma_H]$ and bending fatigue strength $\sigma_F \le [\sigma_F]$ regardless of tooth surface hardness. A trial load factor K_t (usually $K_t = 1.2$–1.4) is used to calculate trial pitch diameter d_{1t} (or trial module m_{nt}). Then use d_{1t} to calculate pitch line velocity, determine dynamic factor K_v, transverse load factor K_a and face load factor K_β to obtain load factor K. If there is difference between the trial load factor K_t and K, the trial values of pitch diameter d_{1t} (or module m_{nt}) can be modified by

$$d_1 = d_{1t}\sqrt[3]{K/K_t} \tag{8.88}$$

or

$$m_n = m_{nt}\sqrt[3]{K/K_t} \tag{8.89}$$

The results from both strength analyses are compared and evaluated before the final design decision can be made.

8.6.6.4 Geometrical Calculation

Because of the requirement of standardization and manufacturing process, variables in gear transmission must be adjusted after the initial design. Basic rules are listed in Table 8.7.

Variables in gear transmission are constrained by each other. On the one hand, they have to meet the requirements listed in Table 8.7; on the other hand, they have to satisfy geometrical constraints for proper meshing. A reasonable compromise between relevant variables must be reached during design process.

For example, for a soft-tooth-surface helical gear in an enclosed gearing with a speed ratio $i = 3.243$, assume the module is obtained by

$$m_n \geq \sqrt[3]{\frac{2KT_1\cos^2\beta}{\varphi_d z_1^2} \cdot \frac{Y_{Fa}Y_{Sa}Y_\varepsilon Y_\beta}{[\sigma_F]}} = 2.31$$

A standard module is then selected as $m_n = 2.5\,\mathrm{mm}$. Select $z_1 = 24$, then $z_2 = iz_1 = 3.243 \times 24 = 77.832$, choose $z_2 = 78$. Assuming $\beta = 15°$, the centre distance is

$$a = \frac{m_n(z_1 + z_2)}{2\cos\beta} = 131.99$$

Round off the centre distance to 130 mm for the convenience of assembly and the helix angle is consequently changed to

$$\beta = \arccos\frac{m_n(z_1 + z_2)}{2a} = \arccos\frac{2.5 \times (24 + 78)}{2 \times 130} = 11.275°$$

If β is not within the range from $8°$ to $20°$, select z_1 and z_2 again and repeat this calculation.

Table 8.7 Basic rules for design variable selection.

Variables	Requirements	Variables	Requirements
Module m_n	Standard value	Centre distance a	0 or 5 at digit
Number of teeth z_1, z_2	Integer	Face width b	Integer
Pitch diameter d	Calculated value	Helix angle β	Calculated value

8.6.7 Design Cases

Example Problem 8.1

Design a pair of spur gears for a reducer to drive a conveyor, as shown in Figure E8.1. The electric motor rotates at 960 rpm in one direction. The gear reducer with a teeth ratio of $u = 4.0$ operates steadily, transmitting a power of $P = 10$ kW. The reducer is expected to work 16 hours daily, 300 days a year for 15 years.

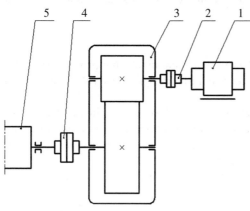

Figure E8.1 Illustration for Example Problem 8.1.

1. Motor   2. Coupling   3. Reducer

4. Coupling   5. Roller of conveyor

Solution

Steps	Computation	Results	Units
1. Select gear type, gear accuracy grades, materials and heat treatments	(1) Select spur gears according to the layout and requirement of the reducer. (2) Since conveyor is ordinary machinery whose speed is not very high, select gear accuracy grade 7 in ISO/GB standard by Table 8.5. (3) Material selection Select alloy steel 40Cr (AISI 5140) with hardness 280 HV for the pinion, heat treated by quenching and tempering and medium carbon steel 45 (AISI 1045) with hardness 210 HV for the gear, heat treated by quenching and tempering.	Spur gears Gear accuracy grade 7 40Cr for pinion 45 carbon steel for gear	
2. Select initial variables	(1) Select the number of teeth of the pinion z_1 Select $z_1 = 19$, the number of gear teeth is $z_2 = uz_1 = 4.0 \times 19 = 76$. So select $z_2 = 76$.		
3. Design by the tooth surface fatigue strength of the pinion	Using design formula Eq. (8.26) $$d_1 \geq \sqrt[3]{\frac{2KT_1}{\varphi_d} \cdot \frac{u\pm 1}{u} \cdot \left(\frac{Z_H Z_E}{[\sigma_H]}\right)^2}$$		

(continued)

Steps	Computation	Results	Units
(1) Decide the allowable contact stresses	1) Select the endurance limit of the pinion and gear from Table 8.6 Contact endurance limit of the pinion $\sigma_{Hlim1} = 740$ MPa Contact endurance limit of the gear $\sigma_{Hlim2} = 550$		
	2) The number of load cycles is calculated by Eq. (8.84) as $N_1 = 60n_1jL_h = 60 \times 960 \times 1 \times (16 \times 300 \times 15) = 4.15 \times 10^9$ $N_2 = 4.15 \times 10^9/4.0 = 1.04 \times 10^9$	$N_1 = 4.15 \times 10^9$ $N_2 = 1.04 \times 10^9$	
	3) Since the number of load cycles of the pinion and gear is greater than 5×10^7, select the life factor for contact stress slightly less than 1.0, as $K_{HN1} = 0.90$; $K_{HN2} = 0.95$. More precise data can be found in design handbooks or standards.		
	5) Calculate allowable contact stresses Select reliability of 99% and safety factor $S_H = 1.0$, the allowable contact stress is calculated by Eq. (8.85), as $[\sigma_{H1}] = \dfrac{K_{HN1}\sigma_{H\lim 1}}{S_H} = 0.90 \times 740 = 666 MPa$ $[\sigma_{H2}] = \dfrac{K_{HN2}\sigma_{H\lim 2}}{S_H} = 0.95 \times 550 = 522.5 MPa$	$[\sigma_{H1}] = 666$ $[\sigma_{H2}] = 522.5$	MPa MPa
(2) Decide values to be used in the calculation	1) Select a trial load factor $K_t = 1.3$.		
	2) The torque transmitted by the pinion is $T_1 = 9.55 \times 10^6 P_1/n_1 = 9.55 \times 10^6 \times 10/960 = 9.948 \times 10^4 N \cdot mm$	$T_1 = 9.948 \times 10^4$	N mm
	3) For symmetric supporting, select the face width factor φ_d as 1.0, as recommended in Section 8.6.6.	$\varphi_d = 1.0$	
	4) For the match of steel pinion and steel gear, the elastic coefficient is $Z_E = 189.8$ MPa$^{1/2}$. For a standard spur gear, the pressure angle is 20°, zone factor is $Z_H = 2.5$.	$Z_E = 189.8$ $Z_H = 2.5$	MPa$^{1/2}$
(3) Design calculation	1) Calculate trial pitch diameter of the pinion using the smaller allowable stress $d_{1t} \geq \sqrt[3]{\dfrac{2KT_1}{\varphi_d} \cdot \dfrac{u+1}{u} \cdot \left(\dfrac{Z_E Z_H}{[\sigma_H]}\right)^2}$ $= \sqrt[3]{\dfrac{2 \times 1.3 \times 9.948 \times 10^4}{1} \cdot \dfrac{4+1}{4} \cdot \left(\dfrac{189.8 \times 2.5}{522.5}\right)^2} mm$ $= 64.363 mm$		
	2) Calculate pitch line velocity, v $v = \dfrac{\pi d_{1t} n_1}{60 \times 1000} = \dfrac{\pi \times 64.363 \times 960}{60 \times 1000} = 3.23 m/s$		
	3) Calculate face width $b = \varphi_d \cdot d_{1t} = 1 \times 64.363 mm = 64.363 mm$		

(continued)

Steps	Computation	Results	Units
	4) Calculate the ratio of face width to tooth height b/h Module $m_{1t} = d_{1t}/z_1 = 64.363/19$ mm $= 3.388$ mm Tooth height $h = 2.25\, m_{1t} = 2.25 \times 3.388$ mm $= 7.621$ mm The ratio of $b/h = 64.363/7.621 = 8.445$		
	5) Calculate the load factor According to $v = 3.23\ \mathrm{m\,s^{-1}}$, gear accuracy grade 7, from Eqs. (8.13)–(8.16), the dynamic factor is found as $K_v = 1.10$. From Table 8.2, select transverse load factors $K_{Ha} = K_{Fa} = 1.0$. Select the application factor from Table 2.1 as $K_A = 1.0$. Determine the face load factor K_β from Table 8.3. For accuracy grade 7 and a symmetrically supported pinion, substitute the relevant value $K_{H\beta} = a_1 + a_2(1 + a_3\varphi_d^2)\varphi_d^2 + a_4 b$ $= 1.12 + 0.18(1 + 0.0 \times 1^2) \times 1^2 + 2.3$ $\times 10^{-4} \times 64.363 = 1.314$ From Eq. (8.18) $$N_0 = \frac{(b/h)^2}{1 + (b/h) + (b/h)^2}$$ $$= \frac{8.445^2}{1 + 8.445 + 8.445^2} = 0.883$$ From Eq. (8.17) $K_{F\beta} = (K_{H\beta})^{N_0} = 1.314^{0.883} = 1.27$ The load factor is then $K = K_A K_v K_{Ha} K_{H\beta} = 1.0 \times 1.10 \times 1.0$ $\times 1.314 = 1.445$	$K = 1.443$	
	6) The modified pitch diameter by the actual load factor K can be obtained from Eq. (8.88) as $d_1 = d_{1t}\sqrt[3]{K/K_t} = 64.363 \times \sqrt[3]{1.445/1.3}$ $= 66.671 mm$	$d_1 = 66.671$	mm
	7) Compute module m $m = \dfrac{d_1}{z_1} = \dfrac{66.671}{19} = 3.509 mm$		
4. Design by gear teeth bending strength	Using the design formula Eq. (8.29) $m \geq \sqrt[3]{\dfrac{2KT_1}{\varphi_d z_1^2} \cdot \dfrac{Y_{Fa} Y_{Sa}}{[\sigma_F]}}$		
(1) Decide the allowable bending stresses	1) Select the endurance limit of the pinion and gear from Table 8.6. Bending endurance limit of the pinion $\sigma_{Flim1} = 306$ MPa Bending endurance limit of the gear $\sigma_{Flim2} = 213$ MPa		

(*continued*)

Steps	Computation	Results	Units
	2) Since the numbers of load cycles of the pinion and gear are greater than 3×10^6 cycles, select the life factor for bending stress as $K_{FN1} = 0.85$, $K_{FN2} = 0.87$.		
	3) Calculate allowable bending stress Select safety factor $S_F = 1.4$, from Eq. (8.86), allowable bending stress is $[\sigma_{F_1}] = \dfrac{K_{FN1}\sigma_{F \lim 1}}{S_F} Y_{ST} = \dfrac{0.85 \times 306}{1.4} \times 2.0$ $= 371.6$ MPa $[\sigma_{F_2}] = \dfrac{K_{FN2}\sigma_{F \lim 2}}{S_F} Y_{ST} = \dfrac{0.87 \times 213}{1.4} \times 2.0$ $= 264.7$ MPa	$[\sigma_{F1}] = 371.6$ $[\sigma_{F2}] = 264.7$	MPa MPa
(2) Decide values to be used in the calculation	1) Compute the load factor K $K = K_A K_v K_{Fa} K_{F\beta} = 1 \times 1.10 \times 1.0 \times 1.27$ $= 1.397$	$K = 1.397$	
	2) Specify tooth form factor Y_{Fa} From Table 8.4, by interpolating, we have $Y_{Fa1} = 2.85$; $Y_{Fa2} = 2.232$		
	3) Specify stress correction factor Y_{sa} From Table 8.4, by interpolation, we have $Y_{sa1} = 1.54$; $Y_{sa2} = 1.762$		
	4) Compute and compare the values of $Y_{Fa} Y_{Sa}/[\sigma_F]$ for the pinion and gear $\dfrac{Y_{Fa1} Y_{Sa1}}{[\sigma_{F1}]} = \dfrac{2.85 \times 1.54}{371.6} = 0.0118$ $\dfrac{Y_{Fa2} Y_{Sa2}}{[\sigma_{F2}]} = \dfrac{2.232 \times 1.762}{264.7} = 0.0148$		
(3) Design calculation	$m \geq \sqrt[3]{\dfrac{2KT_1}{\varphi_d z_1^2} \cdot \dfrac{Y_{Fa} Y_{Sa}}{[\sigma_F]}} = \sqrt[3]{\dfrac{2 \times 1.397 \times 9.948 \times 10^4}{1 \times 19^2} \times 0.0148} = 2.25mm$ The module obtained by contact strength analysis is greater than that obtained by bending strength analysis. Since module relates closely to the tooth bending strength, while the pitch diameter of a gear is relevant to contact strength. Therefore, the module obtained from bending strength analysis is selected and round down to a standard value of 2.5 mm. The pitch diameter obtained from contact strength analysis $d_1 = 66.671$ mm is selected to calculated the number of teeth of the pinion, $z_1 = \dfrac{d_1}{m} = \dfrac{66.671}{2.5} = 26.67$ Select $z_1 = 26$. The number of teeth of the gear $z_2 = u z_1 = 4.0 \times 26 = 104$, select $z_2 = 104$. Thus, the designed gear can meet both contact and bending strength requirements.	$m = 2.5$ $z_1 = 26$ $z_2 = 104$	mm

(continued)

Steps	Computation	Results	Units
5. Geometrical calculation	(1) Pitch diameter $$d_1 = mz_1 = 26 \times 2.5 \text{ mm} = 65 \text{ mm}$$ $$d_2 = mz_2 = 104 \times 2.5 \text{ mm} = 260 \text{ mm}$$	$d_1 = 65$ $d_2 = 260$	mm
	(2) Centre distance $a = (d_1 + d_2)/2 = (65 + 260)/2 \text{ mm} = 162.5 \text{ mm}$	$a = 162.5$	mm
	(3) Face width of the gear $b = \varphi_d d_1 = 1 \times 65 mm = 65 mm$ Select $b_2 = 65$ mm, $b_1 = 70$ mm.	$b_1 = 70$ $b_2 = 65$	mm mm
6. Gear drawings	(omitted)		

Example Problem 8.2

Design a pair of soft tooth surface helical gears for the reducer in Example Problem 8.1, using the same design data.

Solution

Steps	Computation	Results	Units
1. Selection of gear type, gear accuracy grades, materials and heat treatments	1) Select helical gears according to the layout and requirement of the reducer. 2) Since conveyor is ordinary machinery whose speed is not very high, select gear accuracy grade 7 from Table 8.5. 3) Material selection Select alloy steel 40Cr (AISI 5140) with hardness 280 HV for the pinion, heat treated by quenching and tempering and medium carbon steel 45 (AISI 1045) with hardness 210 HV for the gear, heat treated by quenching and tempering.	Helical gears Gear accuracy grade 7 40Cr for pinion 45 carbon steel for gear	
2. Select initial variables	1) Select the number of teeth of the pinion z_1 Select $z_1 = 19$, the number of gear teeth is $z_2 = uz_1 = 4.0 \times 19 = 76$. 2) Select helical angle $\beta = 14°$.	$z_1 = 19$ $z_2 = 76$	
3. Design by the tooth surface fatigue strength of the pinion	From the design formula Eq. (8.56) $$d_1 \geq \sqrt[3]{\frac{2KT_1}{\varphi_d} \cdot \frac{u \pm 1}{u} \left(\frac{Z_H Z_E Z_\epsilon Z_\beta}{[\sigma_H]} \right)^2}$$		
(1) Decide the allowable contact stresses	1) Select the endurance limit of the pinion and gear from Table 8.6 Contact endurance limit of the pinion $\sigma_{Hlim1} = 740$ MPa Contact endurance limit of the gear $\sigma_{Hlim2} = 550$ MPa		

(*continued*)

Steps	Computation	Results	Units
	2) The number of load cycles is calculated by Eq. (8.84) as $N_1 = 60n_1jL_h = 60 \times 960 \times 1 \times (16 \times 300 \times 15) = 4.15 \times 10^9$ $N_2 = 4.15 \times 10^9/4.0 = 1.04 \times 10^9$	$N_1 = 4.15 \times 10^9$ $N_2 = 1.04 \times 10^9$	
	3) Since the number of load cycles of the pinion and gear are greater than 5×10^7, select life factor for contact stress slightly less than 1.0, as $K_{HN1} = 0.90$; $K_{HN2} = 0.95$. More precise data can be found in design handbooks or standards.		
	4) Calculate allowable contact stresses Select reliability of 99% and safety factor $S_H = 1.0$, the allowable contact stress is calculated by Eq. (8.85), as $[\sigma_{H1}] = \dfrac{K_{HN1}\sigma_{H\lim 1}}{S_H} = 0.90 \times 740 = 666 MPa$ $[\sigma_{H2}] = \dfrac{K_{HN2}\sigma_{H\lim 2}}{S^H} = 0.95 \times 550 = 522.5 MPa$	$[\sigma_{H1}] = 666$ $[\sigma_{H2}] = 522.5$	MPa MPa
(2) Decide values to be used in the calculation	1) Select a trial load factor $K_t = 1.3$		
	2) The torque transmitted by the pinion is $T_1 = 9.55 \times 10^6 P_1/n_1 = 9.55 \times 10^6 \times 10/960 = 9.948 \times 10^4 N \cdot mm$	$T_1 = 9.948 \times 10^4$	N·mm
	3) For symmetric supporting, select face width factor φ_d as 1.0.	$\varphi_d = 1.0$	
	4) For the match of steel pinion and steel gear, the elastic coefficient is $Z_E = 189.8 \, MPa^{1/2}$.	$Z_E = 189.8$	$MPa^{1/2}$
	5) For a standard helical gear, the pressure angle is $20°$, zone factor is calculated by $Z_H = \sqrt{\dfrac{2\cos\beta_b}{\sin\alpha_t\cos\alpha_t}}$ From Eq. (8.36) $\alpha_t = \arctan\left(\dfrac{\tan\alpha_n}{\cos\beta}\right) = \arctan\left(\dfrac{\tan 20°}{\cos 14°}\right)$ $= 20.56°$ and from Eq. (8.37) $\beta_b = \arctan(\tan\beta\cos\alpha_t) = \arctan(\tan 14° \cos 20.56°) = 13.13°$ Therefore, the zone factor is $Z_H = \sqrt{\dfrac{2\cos\beta_b}{\sin\alpha_t\cos\alpha_t}} = \sqrt{\dfrac{2\cos 13.13°}{\sin 20.56° \cos 20.56°}} = 2.433$	$Z_H = 2.433$	
	6) From Eq. (8.39), the transverse contact ratio is estimated as $\varepsilon_\alpha = \left[1.88 - 3.2\left(\dfrac{1}{z_1} + \dfrac{1}{z_2}\right)\right]\cos\beta$ $= \left[1.88 - 3.2\left(\dfrac{1}{19} + \dfrac{1}{76}\right)\right]\cos 14° = 1.62$ The face contact ratio is calculated by Eq. (8.40) as $\varepsilon_\beta = 0.318\varphi_d z_1 \tan\beta = 0.318 \times 1.0 \times 19 \tan 14° = 1.509$	$\varepsilon_\alpha = 1.62$ $\varepsilon_\beta = 1.509$	

(continued)

Steps	Computation	Results	Units
	7) Since $\varepsilon_\beta > 1$, substitute ε_β by $\varepsilon_\beta = 1$ in Eq. (8.52), the contact ratio factor is $$Z_\varepsilon = \sqrt{\frac{4\varepsilon_\alpha}{3}(1-\varepsilon_\beta)+\frac{\varepsilon_\beta}{\varepsilon_\alpha}}$$ $$= \sqrt{\frac{1}{1.62}} = 0.786$$	$Z_\varepsilon = 0.786$	
	1) From Eq. (8.53), the helix angle factor is $$Z_\beta = \sqrt{\cos\beta} = \sqrt{\cos 14°} = 0.985$$	$Z_\beta = 0.985$	
(3) Design calculation	2) Calculate pitch diameter of the pinion d_1 using the smaller allowable stress $$d_{1t} \geq \sqrt[3]{\frac{2KT_1}{\varphi_d} \cdot \frac{u\pm1}{u}\left(\frac{Z_H Z_E Z_\varepsilon Z_\beta}{[\sigma_H]}\right)^2}$$ $$= \sqrt[3]{\frac{2\times1.3\times9.948\times10^4}{1} \cdot \frac{4+1}{4} \cdot \left(\frac{2.433\times189.8\times0.786\times0.985}{522.5}\right)^2}$$ $$= 53.29 \text{ mm}$$		
	3) Calculate pitch line velocity, v $$v = \frac{\pi d_{1t} n_1}{60\times1000} = \frac{\pi\times53.29\times960}{60\times1000} = 2.67 m/s$$		
	4) Calculate face width $$b = \varphi_d d_{1t} = 1 \times 53.29 mm = 53.29 mm$$		
	5) Calculate the ratio of face width to tooth height b/h $$\text{Module } m_{nt} = \frac{d_{1t}\cos\beta}{z_1} = \frac{53.29\times\cos 14°}{19} = 2.72 mm$$ Tooth height $h = 2.25\, m_{nt} = 2.25 \times 2.72$ mm $= 6.12$ mm The ratio of $b/h = 53.29/6.12 = 8.71$		
	6) Calculate the load factor. From Eq. (8.12), the load factor is $$K = K_A K_v K_\alpha K_\beta$$ Determine the application factor from Table 2.1 as $K_A = 1.0$. According to $v = 2.67$ m s^{-1}, gear accuracy grade 7, from Eqs. (8.13)–(8.16), the dynamic factor is found as $K_v = 1.10$. From Table 8.2, select transverse load factor $K_{Ha} = K_{Fa} = 1.1$. Determine the face load factor K_β from Table 8.3. For accuracy grade 7 and a symmetrically supported pinion, $$K_{H\beta} = a_1 + a_2[1 + a_3\varphi_d^2]\varphi_d^2 + a_4 b$$ $$= 1.12 + 0.18 \times [1 + 0 \times 1^2] \times 1^2 + 2.3 \times 10^{-4} \times 53.29 = 1.312$$ From Eq. (8.18) $$N_0 = \frac{(b/h)^2}{1+(b/h)+(b/h)^2} = \frac{8.71^2}{1+8.71+8.71^2} = 0.887$$ From Eq. (8.17) $$K_{F\beta} = (K_{H\beta})^{N_0} = 1.312^{0.887} = 1.27$$ The load factor is then $$K = K_A K_v K_{Ha} K_{H\beta} = 1.0 \times 1.10 \times 1.1 \times 1.312$$ $$= 1.588$$	$K = 1.588$	

(continued)

Steps	Computation	Results	Units
	6) The modified pitch diameter from actual load factor K can be obtained from Eq. (8.88) as $d_1 = d_{1t}\sqrt[3]{K/K_t} = 53.29 \times \sqrt[3]{1.588/1.3} = 56.966$ mm		
	8) Compute module m $m_n = \dfrac{d_1 \cos\beta}{z_1} = \dfrac{56.966 \times \cos 14^\circ}{19} = 2.91mm$ Select $m_n = 3.0$ mm.	$m_n = 3.0$	mm
4. Geometrical calculation	1) Centre distance $a = \dfrac{(z_1 + z_2)m_n}{2\cos\beta} = \dfrac{(19 + 76) \times 3}{2 \times \cos 14^\circ} = 146.86mm$ Select $a = 145$ mm.	$a = 145$	mm
	2) Modified helix angle $\beta = \arccos\dfrac{(z_1 + z_2)m_n}{2a} =$ $\arccos\dfrac{(19 + 76) \times 3}{2 \times 145} = 10.65^\circ$ Since calculated β closes to the initially selected helix angle 14°, ε_α, Z_β and Z_H do not need to be modified.	$\beta = 10.65$	°
	3) Calculate the pitch diameters $d_1 = \dfrac{m_n z_1}{\cos\beta} = \dfrac{3 \times 19}{\cos 10.65^\circ} = 57.999mm$ $d_2 = \dfrac{m_n z_2}{\cos\beta} = \dfrac{3 \times 76}{\cos 10.65^\circ} = 231.996mm$	$d_1 = 57.999$ $d_2 = 231.996$	mm
	4) Face width of the gear $b = \varphi_d d_1 = 1 \times 57.999mm = 57.999mm$ Select $b_2 = 58$ mm, $b_1 = 63$ mm.	$b_1 = 63$ $b_2 = 58$	mm
5. Check gear teeth bending strength	From the design formula Eq. (8.62) $\sigma_F = \dfrac{KF_t}{bm_n}Y_{Fa}Y_{Sa}Y_\varepsilon Y_\beta \le [\sigma_F]$		
(1) Decide the allowable bending stresses	1) Select the endurance limit of the pinion and gear from Table 8.6 Bending endurance limit of the pinion $\sigma_{Flim1} = 306$ MPa Bending endurance limit of the gear $\sigma_{Flim2} = 213$ MPa		
	2) Since the numbers of load cycles of the pinion and gear are greater than 3×10^6 cycles, select the life factor for bending stress as $K_{FN1} = 0.85$, $K_{FN2} = 0.87$.		
	3) Calculate allowable bending stress Select safety factor $S_F = 1.4$, from Eq. (8.86), allowable bending stress is $[\sigma_{F1}] = \dfrac{K_{FN1}\sigma_{Flim1}}{S_F}Y_{ST} = \dfrac{0.85 \times 306}{1.4} \times 2.0$ $= 371.6$ MPa $[\sigma_{F2}] = \dfrac{K_{FN2}\sigma_{Flim2}}{S_F}Y_{ST} = \dfrac{0.87 \times 213}{1.4} \times 2.0$ $= 264.7MPa$	$[\sigma_{F1}] = 371.6$ $[\sigma_{F2}] = 264.7$	MPa MPa

(continued)

Steps	Computation	Results	Units
(2) Decide values to be used	1) Compute the load factor K $K = K_A K_v K_{F\alpha} K_{F\beta} = 1 \times 1.10 \times 1.1 \times 1.27$ $= 1.537$	$K = 1.537$	
	2) Virtual number of teeth $z_{v1} = \dfrac{z_1}{\cos^3 \beta} = \dfrac{19}{\cos^3 10.65°} = 20.01$ $z_{v2} = \dfrac{z_2}{\cos^3 \beta} = \dfrac{76}{\cos^3 10.65°} = 80.07$	$z_{v1} = 20.01$ $z_{v2} = 80.07$	
	3) Specify tooth form factor Y_{Fa} From Table 8.4, we have $Y_{Fa1} = 2.80$; $Y_{Fa2} = 2.22$	$Y_{Fa1} = 2.80$ $Y_{Fa2} = 2.22$	
	4) Specify stress correction factor Y_{sa} From Table 8.4, we have $Y_{sa1} = 1.55$; $Y_{sa2} = 1.77$	$Y_{sa1} = 1.55$ $Y_{sa2} = 1.77$	
	5) Specify contact ratio factor Y_ε From Eq. (8.60), $\cos \beta_b = \sqrt{1 - (\sin \beta \cos \alpha_n)^2} =$ $\sqrt{1-(\sin 10.65° \cos 20°)^2} = 0.9848$ From Eq. (8.59) $\varepsilon_{av} = \dfrac{\varepsilon_\alpha}{\cos^2 \beta_b} = \dfrac{1.62}{0.9848^2} = 1.6704$ From Eq. (8.58) $Y_\varepsilon = 0.25 + \dfrac{0.75}{\varepsilon_{av}} = 0.25 + \dfrac{0.75}{1.6704} = 0.699$	$Y_\varepsilon = 0.699$	
	6) Specify helix angle factor Y_β The face contact ratio is calculated from Eq. (8.40) $\varepsilon_\beta = 0.318 \varphi_d z_1 \tan \beta = 0.318 \times 1.0$ $\times 19 \tan 10.65° = 1.138$ Since $\varepsilon_\beta > 1$, substitute ε_β for $\varepsilon_\beta = 1.0$ in Eq. (8.61), the helix angle factor is $Y_\beta = 1 - \varepsilon_\beta \dfrac{\beta}{120°} = 1 - 1.0 \times \dfrac{10.65°}{120°} = 0.911$	$Y_\beta = 0.911$	
(3) Check bending strength	Since $F_t = \dfrac{2T_1}{d_1} = \dfrac{2 \times 9.948 \times 10^4}{57.999} = 3430.4N$ From Eq. (8.57) $\sigma_F = \dfrac{KF_t}{bm_n} Y_{Fa} Y_{Sa} Y_\varepsilon Y_\beta$ Therefore. $\sigma_{F1} = \dfrac{1.537 \times 3430.4}{58 \times 3.0} \times 2.8 \times 1.55 \times 0.699 \times 0.911$ $= 83.74 < [\sigma_{F1}]$ $\sigma_{F2} = \dfrac{1.537 \times 3430.4}{58 \times 3.0} \times 2.22 \times 1.77 \times 0.699 \times$ $0.911 = 75.82 < [\sigma_{F2}]$	The bending strength is enough.	
6. Gear drawings	(omitted)		

Example Problem 8.3

Design a pair of hard tooth surface helical gears for the reducer in Example Problem 8.1, using the same design data.

Solution

Steps	Computation	Results	Units
1. Selection of gear type, gear accuracy grades, materials and heat treatments	(1) Select helical gears according to the layout and requirement of the reducer. (2) Since the conveyor is ordinary machinery whose speed is not very high, select gear accuracy grade 7 from Table 8.5. (3) Material selection Select alloy steel 40Cr (AISI 5140) with hardness 550HV for both the pinion and gear, heat treated by flame hardening.	Helical gears Gear accuracy grade 7 40Cr for pinion and gear	
2. Select initial variables	(1) Select the number of teeth of the pinion z_1 Select $z_1 = 19$, the number of gear teeth is $z_2 = uz_1 = 4.0 \times 19 = 76$. So select $z_2 = 76$. (2) Select helical angle $\beta = 14°$.	$z_1 = 19$ $z_2 = 76$	
3. Design by gear teeth bending strength	Design formula by Eq. (8.63) $$m_n \geq \sqrt[3]{\frac{2KT_1\cos^2\beta}{\varphi_d z_1^2} \cdot \frac{Y_{Fa}Y_{Sa}Y_\varepsilon Y_\beta}{[\sigma_F]}}$$		
(1) Decide the allowable bending stresses	1) Select the endurance limit of the pinion and gear from Table 8.6 Bending endurance limit of the pinion and gear $\sigma_{Flim1} = \sigma_{Flim2} = 366$ MPa		
	2) The number of load cycles is calculated by Eq. (8.84) as $N_1 = 60n_1 jL_h = 60 \times 960 \times 1 \times (16 \times 300 \times 15) = 4.15 \times 10^9$ $N_2 = 4.15 \times 10^9/4.0 = 1.04 \times 10^9$	$N_1 = 4.15 \times 10^9$ $N_2 = 1.04 \times 10^9$	
	3) Since the numbers of load cycles of the pinion and gear are greater than 3×10^6 cycles, select the life factor for bending stress as $K_{FN1} = 0.85$, $K_{FN2} = 0.87$.		
	4) Calculate allowable bending stress Select safety factor $S_F = 1.4$, from Eq. (8.86), allowable bending stress is $[\sigma_{F1}] = \dfrac{K_{FN1}\sigma_{Flim1}}{S_F}Y_{ST} = \dfrac{0.85 \times 366}{1.4} \times 2.0$ $= 444.4$ MPa $[\sigma_{F2}] = \dfrac{K_{FN2}\sigma_{Flim2}}{S_F}Y_{ST} = \dfrac{0.87 \times 366}{1.4} \times 2.0$ $= 454.9$MPa	$[\sigma_{F1}] = 444.4$ $[\sigma_{F2}] = 454.9$	MPa MPa

(*continued*)

Steps	Computation	Results	Units
(2) Decide values to be used	1) Select a trial load factor $K_t = 1.3$.		
	2) The torque transmitted by the pinion is $T_1 = 9.55 \times 10^6 P_1/n_1 = 9.55 \times 10^6 \times 10/960$ $= 9.948 \times 10^4 N \cdot mm$	$T_1 = 9.948 \times 10^4$	N·mm
	3) For symmetric supporting, select the face width factor φ_d as 1.0.	$\varphi_d = 1.0$	
	4) Virtual number of teeth $z_{v1} = \dfrac{z_1}{\cos^3\beta} = \dfrac{19}{\cos^3 14°} = 20.799$ $z_{v2} = \dfrac{z_2}{\cos^3\beta} = \dfrac{76}{\cos^3 14°} = 83.196$	$z_{v1} = 20.799$ $z_{v2} = 83.196$	
	5) Specify tooth form factor Y_{Fa} From Table 8.4, by interpolating, we have $Y_{Fa1} = 2.771; Y_{Fa2} = 2.214$	$Y_{Fa1} = 2.771$ $Y_{Fa2} = 2.214$	
	6) 6) Specify stress correction factor Y_{sa} From Table 8.4, by interpolating, we have $Y_{Sa1} = 1.556; Y_{Sa2} = 1.773$	$Y_{Sa1} = 1.556$ $Y_{Sa2} = 1.773$	
	7) From Eq. (8.39), the transverse contact ratio is $\varepsilon_\alpha = \left[1.88 - 3.2\left(\dfrac{1}{z_1} + \dfrac{1}{z_2}\right)\right]\cos\beta$ $= \left[1.88 - 3.2\left(\dfrac{1}{19} + \dfrac{1}{76}\right)\right]\cos 14° = 1.62$ The face contact ratio is calculated from Eq. (8.40). $\varepsilon_\beta = 0.318\varphi_d z_1 \tan\beta = 0.318 \times 1.0 \times 19$ $\times \tan 14° = 1.509$	$\varepsilon_\alpha = 1.62$ $\varepsilon_\beta = 1.509$	
	8) Specify contact ratio factor Y_ε From Eq. (8.60), $\cos\beta_b = \sqrt{1 - (\sin\beta\cos\alpha_n)^2}$ $= \sqrt{1-(\sin 14° \cos 20°)^2} = 0.9738$ From Eq. (8.59) $\varepsilon_{av} = \dfrac{\varepsilon_\alpha}{\cos^2\beta_b} = \dfrac{1.62}{0.9738^2} = 1.7083$ From Eq. (8.58) $Y_\varepsilon = 0.25 + \dfrac{0.75}{\varepsilon_{av}} = 0.25 + \dfrac{0.75}{1.7083} = 0.689$	$Y_\varepsilon = 0.689$	
	9) Specify helix angle factor Y_β. Since $\varepsilon_\beta > 1$, substitute ε_β with $\varepsilon_\beta = 1.0$ in Eq. (8.61), the helix angle factor is $Y_\beta = 1 - \varepsilon_\beta \dfrac{\beta}{120°} = 1 - 1.0 \times \dfrac{14°}{120°} = 0.883$	$Y_\beta = 0.883$	

(continued)

Steps	Computation	Results	Units
	10) Compute and compare the values of $Y_{Fa}Y_{Sa}Y_{\varepsilon}Y_{\beta}/[\sigma_F]$ for the pinion and gear. $$\frac{Y_{Fa1}Y_{Sa1}Y_{\varepsilon}Y_{\beta}}{[\sigma_{F1}]} = \frac{2.771 \times 1.556 \times 0.689 \times 0.883}{444.4}$$ $= 0.00590$ $$\frac{Y_{Fa2}Y_{Sa2}Y_{\varepsilon}Y_{\beta}}{[\sigma_{F2}]} = \frac{2.214 \times 1.773 \times 0.689 \times 0.883}{454.9}$$ $= 0.00525$		
(3) Design calculation	1) Calculate trial modules $$m_n \geq \sqrt[3]{\frac{2KT_1\cos^2\beta}{\varphi_d z_1^2} \cdot \frac{Y_{Fa}Y_{Sa}Y_{\varepsilon}Y_{\beta}}{[\sigma_F]}}$$ $$= \sqrt[3]{\frac{2\times1.3\times9.948\times10^4\cos^2 14^\circ}{1.0\times19^2} \cdot \frac{2.771\times1.556\times0.689\times0.883}{444.4}}$$ $= 1.58mm$ Select $m_n = 2.0$.	$m_n = 2.0$	mm
4. Geometrical calculation	(1) Centre distance $$a = \frac{(z_1 + z_2)m_n}{2\cos\beta} = \frac{(19 + 76)\times 2}{2\times\cos 14^\circ} = 97.91mm$$ Select $a = 100$ mm. (2) Modified helix angle $$\beta = \arccos\frac{(z_1 + z_2)m_n}{2a} = \arccos\frac{(19 + 76)\times 2}{2\times 100}$$ $= 18.19^\circ$ Select $\beta = 18.19^\circ$. Since calculated β is close to the initially selected helix angle 14°, Y_{Fa}, Y_{Sa}, Y_{ε} and Y_{β} do not need to be modified. $$d_1 = \frac{m_n z_1}{\cos\beta} = \frac{2\times 19}{\cos 18.19^\circ} = 39.999mm$$ $$d_2 = \frac{m_n z_2}{\cos\beta} = \frac{2\times 76}{\cos 18.19^\circ} = 159.996mm$$ (3) Face width of the gear $b = \varphi_d d_1 = 1\times 39.999mm = 39.999mm$ Select $b_2 = 40$ mm, $b_1 = 45$ mm.	$a = 100$ $\beta = 18.19^\circ$ $d_1 = 39.999$ $d_2 = 159.996$ $b_1 = 45$ $b_2 = 40$	mm mm mm mm mm
5. Modify the load factor	1) Calculate pitch line velocity v $$v = \frac{\pi d_1 n_1}{60\times 1000} = \frac{\pi \times 39.999\times 960}{60\times 1000} = 2.01m/s$$ 2) Calculate the ratio of face width and tooth height b/h. Tooth height $h = 2.25m_n = 2.25\times 2 = 4.5$ mm The ratio of $b/h = 40/4.5 = 8.88$		

(continued)

Steps	Computation	Results	Units
	3) Calculate the load factor. From Eq. (8.12), the load factor is $K = K_A K_v K_\alpha K_\beta$ Determine the application factor from Table 2.1 as $K_A = 1.0$. According to $v = 2.01 \text{ m s}^{-1}$, gear accuracy grade 7, from Eqs. (8.13)–(8.16), the dynamic factor is found as $K_v = 1.10$. From Table 8.2, select transverse load factor $K_{Ha} = K_{Fa} = 1.2$. Determine the face load factor K_β from Table 8.3. Initially, we can assume $K_{H\beta} \le 1.34$ to start calculation. Due to limited data, we use data for gear accuracy grade 6 and a symmetrically supported pinion, $K_{H\beta} = a_1 + a_2(1 + a_3\varphi_d^2)\varphi_d^2 + a_4 b$ $= 1.05 + 0.26 \times (1 + 0 \times 1^2) \times 1^2 + 1.6 \times 10^{-4} \times 40 = 1.316$ From Eq. (8.18) $N_0 = \dfrac{(b/h)^2}{1 + (b/h) + (b/h)^2} = \dfrac{8.88^2}{1 + 8.88 + 8.88^2}$ $= 0.889$ From Eq. (8.17) $K_{F\beta} = (K_{H\beta})^{N_0} = 1.316^{0.889} = 1.27$ The load factor is then $K = K_A K_v K_{Fa} K_{F\beta} = 1.0 \times 1.10 \times 1.2 \times 1.27 = 1.676$ 4) The module is modified as $m_n = m_{nt}\sqrt[3]{K/K_t} = 1.58 \times \sqrt[3]{\dfrac{1.676}{1.3}} = 1.72$ We still select $m_n = 2.0 \text{ mm}$.	$K = 1.676$	
6. Check tooth surface fatigue strength	The tooth surface fatigue strength is calculated by Eq. (8.55) $\sigma_H = Z_H Z_E Z_\epsilon Z_\beta \sqrt{\dfrac{KF_t}{bd_1} \cdot \dfrac{u \pm 1}{u}} \le [\sigma_H]$		
(1) Decide the allowable contact stresses	1) Select the endurance limit of the pinion and gear from Table 8.6. Contact endurance limit of the pinion and gear $\sigma_{Hlim1} = \sigma_{Hlim2} = 1180 \text{ MPa}$ 2) Since the numbers of load cycles of the pinion and gear are greater than 5×10^7, select the life factor for contact stress slightly less than 1.0, as $K_{HN1} = 0.90$; $K_{HN2} = 0.95$. 3) Calculate allowable contact stresses. Select reliability of 99% and safety factor $S_H = 1$, the allowable contact stress is calculated by Eq. (8.85), as $[\sigma_{H1}] = \dfrac{K_{HN1}\sigma_{H\lim 1}}{S_H} = 0.9 \times 1180 MPa$ $= 1062 MPa$ $[\sigma_{H2}] = \dfrac{K_{HN2}\sigma_{H\lim 2}}{S^H} = 0.95 \times 1180 MPa$ $= 1120 MPa$	$[\sigma_{H1}] = 1062$ $[\sigma_{H2}] = 1120$	MPa MPa

(continued)

Steps	Computation	Results	Units
(2) Decide values to be used in the calculation	1) Calculate the load factor $K = K_A K_v K_{H\alpha} K_{H\beta} = 1.0 \times 1.10 \times 1.2 \times 1.316$ $= 1.737$	$K = 1.737$	
	2) For the match of steel pinion and steel gear, the elastic coefficient is $Z_E = 189.8\,\text{MPa}^{1/2}$.	$Z_E = 189.8$	$\text{MPa}^{1/2}$
	3) For a standard helical gear, the pressure angle is $20°$, zone factor is calculated by $Z_H = \sqrt{\dfrac{2\cos\beta_b}{\sin\alpha_t \cos\alpha_t}}$ From Eq. (8.36) $\alpha_t = \arctan\left(\dfrac{\tan\alpha_n}{\cos\beta}\right) = \arctan\left(\dfrac{\tan 20°}{\cos 18.19°}\right)$ $= 20.96°$ And from Eq. (8.37) $\beta_b = \arctan(\tan\beta\cos\alpha_t) = \arctan(\tan 18.19°$ $\cos 20.96°) = 17.06°$ Therefore, the zone factor is $Z_H = \sqrt{\dfrac{2\cos\beta_b}{\sin\alpha_t \cos\alpha_t}} = \sqrt{\dfrac{2\cos 17.06°}{\sin 20.96° \cos 20.96°}} = 2.392$	$Z_H = 2.392$	
	4) From Eq. (8.39), the transverse contact ratio is $\varepsilon_\alpha = \left[1.88 - 3.2\left(\dfrac{1}{z_1} + \dfrac{1}{z_2}\right)\right]\cos\beta$ $= \left[1.88 - 3.2\left(\dfrac{1}{19} + \dfrac{1}{76}\right)\right]\cos 18.19° = 1.586$ The face contact ratio is calculated from Eq. (8.40) $\varepsilon_\beta = 0.318\varphi_d z_1 \tan\beta = 0.318 \times 1.0$ $\times 19\tan 18.19° = 1.988$	$\varepsilon_\alpha = 1.586$ $\varepsilon_\beta = 1.988$	
	5) Since $\varepsilon_\beta > 1$, substitute $\varepsilon_\beta = 1$ in Eq. (8.52), the contact ratio factor is $Z_\varepsilon = \sqrt{\dfrac{4\varepsilon_\alpha}{3}(1 - \varepsilon_\beta) + \dfrac{\varepsilon_\beta}{\varepsilon_\alpha}}$ $Z_\varepsilon = \sqrt{\dfrac{1}{1.586}} = 0.794$	$Z_\varepsilon = 0.794$	
	6) From Eq. (8.53), the helix angle factor is $Z_\beta = \sqrt{\cos\beta} = \sqrt{\cos 18.19°} = 0.975$	$Z_\beta = 0.975$	
	7) Compute tangential force F_t $F_t = \dfrac{2T_1}{d_1} = \dfrac{2 \times 9.948 \times 10^4}{39.999} = 4974.12N$		
	8) Check the tooth surface fatigue strength $\sigma_H = Z_H Z_E Z_\varepsilon Z_\beta \sqrt{\dfrac{KF_t}{bd_1} \cdot \dfrac{u\pm1}{u}}$ $= 2.392 \times 189.8 \times 0.794 \times 0.975 \times$ $\sqrt{\dfrac{1.737\times 4974.12}{40\times 39.999} \cdot \dfrac{4\pm1}{4}} = 913.1 \leq [\sigma_H]$	Contact strength is enough	
7. Gear drawings	(omitted)		

Table 8.8 Comparison of design results.

Variables	Spur gears Example Problem 6.1	Helical gears (soft tooth surface) Example Problem 6.2	Helical gears (hard tooth surface) Example Problem 6.3
z_2/z_1	104/26	76/19	76/19
m (m_n)	2.5	3	2
d_1	65	57.99	39.999
d_2	260	231.99	159.996
a	162.5	145	100
b_2/b_1	65/70	58/63	40/45
β	0	10.65°	18.19°

The design results of three cases are summarized in Table 8.8 for comparison. For the same design task using helical gears, especially hard tooth surface helical gears, one can obtain a more compact design than using spur gears or soft tooth surface helical gears.

8.7 Structural Design of Gears

The selection of gear structure is relevant to many factors, such as geometrical dimensions, materials, manufacturing methods and costs. The most important factor is diameters. For simplicity and low cost, small size pinions are often forged integrally with a shaft to form a gear shaft, as shown in Figure 8.12a. When the diameter of addendum circle is less than 160 mm, gears may be made with a solid hub from a forged blank, see Figure 8.12b.

Large gears with addendum diameter d_a within the range of 200–500 mm, or with d_a three times the shaft diameter, often have a forged or machined web with holes as indicated in Figure 8.12c. Even larger gears with d_a within the range of 400–1000 mm can use a cast spoked gear, see Figure 8.12d. Excess material is removed from gears in Figure 8.12c,d to save material and also to minimize the total gear inertia. The detailed dimension of gear structure is determined by manufacturing process and empirical formulas and can be referred to in design handbooks [5, 9].

Gears are usually mounted to a shaft by either a key, double keys or a spline, depending on the transmitted power, rotational speed and the strength of keys. Because of stress concentration in keyways, gear blanks must accommodate a radial distance of at least four modules from the top of keyseat to the outside of the gear.

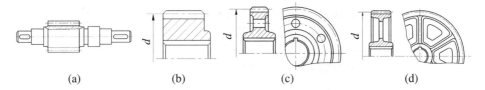

| (a) | (b) | (c) | (d) |

Figure 8.12 Gear structures.

8.8 Lubrication and Efficiency

When a pair of gear teeth mesh with each other, both rolling and sliding occur between contact surface of gear teeth, except at the pitch point where pure rolling occurs [6]. The sliding may cause friction, wear, significant heat and power losses.

To reduce friction and wear, a continuous film of lubricant should be maintained between the mating tooth surfaces to prevent direct metal-to-metal contact of meshing teeth. To keep operating temperature at an acceptable level, a sufficient flow rate and adequate quantity of oil should be provided to the meshing surface to remove frictional heat. Adequate lubrication and cooling capacity is critical for smooth operation, power transmission and gear life.

Gears operate under a diversity of conditions and lubrication methods vary accordingly. For light loads, low speeds, low power transmission, intermittent operation and open gearings, oil can, drip oiler or periodically supplied grease may be applied to the mating tooth surfaces. When gears operate in an enclosed housing at pitch line velocity less than 12 m s^{-1}, large gears may dip into an oil supply sump at the bottom of gear case and carry the oil to the mesh. For even high speeds and high capacity gearing systems, positive oil circulation systems are often required, using a pump to draw oil from the sump and deliver it at a controlled rate to the meshing teeth.

Power losses due to friction in spur gears typically range between 0.5–1.0% of the transmitted power, and 1.0–2.0% in helical and bevel gears. The power transmission efficiency depends on materials, surface characteristic, pitch line velocity and lubrication. In general, gear drives have high power transmission efficiency with long service life and high reliability.

References

1 Budynas, R.G. and Nisbett, J.K. (2011). *Shigley's Mechanical Engineering Design*, 9e. New York, USA: McGraw-Hill.
2 Juvinall, R.C. and Marshek, K.M. (2011). *Fundamentals of Machine Component Design*, 5e. New York: Wiley.
3 Mott, R.L. (2003). *Machine Elements in Mechanical Design*, 4e. Prentice Hall.
4 Hindhede, U., Zimmerman, J.R., Hopkins, R.B. et al. (1983). *Machine Design Fundamentals: A Practical Approach*. New York: Wiley.
5 Pu, L.G. and Ji, M.G. (2006). *Mechanical Design*, 8e. Beijing: Higher Education Press.
6 Collins, J.A. (2002). *Mechanical Design of Machine Elements and Machines: A Failure Prevention Perspective*, 1e. New York: Wiley.
7 American Gear Manufacturers Association. Standard ANSI/AGMA 2001-D04. Fundamental Rating Factors and Calculation Methods for Involute Spur and Helical Gear Teeth. (Metric Edition). American Gear Manufacturers Association, Alexandria, VA, 2004.
8 ISO 6336–1:2006 Calculation of load capacity of spur and helical gears. Part 1: Basic principles, introduction and general influence factors. Switzerland: International Organization for Standards, 2006
9 Wen, B.C. (2015). *Machine Design Handbook*, 5e, vol. 2. Beijing: China Machine Press.

10 Jelaska, D. (2012). *Gears and Gear Drives*, 1e. New York: Wiley.
11 Gere, J.M. and Timoshenko, S.P. (1996). *Mechanics of Materials*, 4e. CL Engineering.
12 Qiu, X.H. (1997). *Mechanical Design*, 4e. Beijing: Higher Education Press.
13 Huang, X.K. and Zheng, W.W. *Theory of Machines and Mechanisms*. Beijing: People's Education Press.
14 ISO 6336–2:2006 Calculation of load capacity of spur and helical gears. Part 2: Calculation of surface durability (pittings). Switzerland: International Organization for Standards, 2006.
15 ISO 6336–3:2006 Calculation of load capacity of spur and helical gears. Part 3: Calculation of tooth bending strength. Switzerland: International Organization for Standards, 2006.
16 ISO 6336–5:2003 Calculation of load capacity of spur and helical gears. Part 5: Strength and quality of materials. International Organization for Standards, 2003.

Problems

Review Questions

1 Discuss the main failure modes of gear drives, the failing mechanism, location and prevention measures.

2 Which factors affect the face load factor K_β? How could one improve uneven load distribution along the face width?

3 What are the main variables to be decided in gear design and how should one select them?

4 For a pair of gears, how could one decide which gear is prone to fail due to pitting and which gear is prone to fail because of tooth breakage?

5 How could one increase the contact strength and bending strength of a cylindrical gear?

Objective Questions

1 In a gear drive, both the pinion and gear are made of medium carbon steel. The pinion is heat treated by quenching and tempering and the gear by normalizing. The contact stress on the tooth surface of the pinion and gear is____.
   (a) $\sigma_{H1} > \sigma_{H2}$
   (b) $\sigma_{H1} = \sigma_{H2}$
   (c) $\sigma_{H1} < \sigma_{H2}$
   (d) hard to decide

2 Assuming the pitch diameter is constant in a spur gear drive, the larger the _____, the higher the bending strength of the gear.

(a) modulus
(b) torque transmitted
(c) number of the teeth of a gear
(d) load factor

3 In order to improve the contact fatigue strength of tooth surface, we can _____.
(a) reduce the pressure angle α
(b) increase the centre distance a
(c) reduce the face width factor φ_d
(d) increase the gear modulus m

4 For a pair of gears with hardness less than 350HBW, we usually _____.
(a) select the hardness of the pinion tooth surfaces < the hardness of the gear tooth surfaces
(b) select the hardness of the pinion tooth surfaces = the hardness of the gear tooth surfaces
(c) select the hardness of the pinion tooth surfaces > the hardness of the gear tooth surfaces
(d) have no restriction on the hardness of the pinion and gear tooth surfaces

5 In a cylindrical gear transmission, if the pitch diameters of the pinion and gear are kept constant and the module is reduced, which of the following transmission qualities will be improved? _____
(a) The gear tooth bending strength
(b) The gear tooth contact strength
(c) The gear transmission stability
(d) None of them

Calculation Questions

1 A pair of enclosed spur gears have $z_1 = 20$, $z_2 = 80$, $m = 3$ mm and face width factor of $\varphi_d = 1.0$. The pinion rotates at $n_1 = 970$ rpm. The allowable contact stress of the pinion and gear are $[\sigma_{H1}] = 700$ MPa, $[\sigma_{H2}] = 500$ MPa, respectively. The load factor is $K = 1.5$, zone factor is $Z_H = 2.5$, elastic coefficient is $Z_E = 189.8$ MPa$^{1/2}$. Determine the power that can be transmitted by tooth surface contact strength analysis.

2 The variables of a pair of standard spur gears are listed in the table as

Gear	m	z	b	Y_{Fa}	Y_{Sa}	$[\sigma_F]$	$[\sigma_H]$
Units	mm		mm			MPa	MPa
1	4	19	65	2.85	1.54	410	650
2	4	50	60	2.34	1.70	380	570

Answer the following questions:

(a) Which gear is more likely to fail because of pitting? And which gear is more likely to fail because of tooth breakage?

(b) Decide the value of face width factor φ_d.

(c) If the load factor is $K = 1.3$, calculate the maximum torque to be transmitted by tooth bending strength analysis.

(d) If the load factor is $K = 1.3$, elastic coefficient $Z_E = 189.8\,\text{MPa}^{1/2}$ and zone factor $Z_H = 2.5$, calculate the maximum torque to be transmitted by tooth contact strength analysis.

3 The variables of a pair of gears are listed here:

$z_1 = 20$, $z_2 = 40$, $m = 2\,\text{mm}$, $b = 40\,\text{mm}$, $Y_{Fa1} = 2.97$, $Y_{Fa2} = 2.35$, $Y_{Sa1} = 1.52$, $Y_{Sa2} = 1.68$, $Z_H = 2.5$, $Z_E = 189.8(\text{MPa})^{1/2}$, $K_1 = K_2 = 1.3$, $[\sigma_{H1}] = 500\,\text{MPa}$, $[\sigma_{H2}] = 470\,\text{MPa}$, $[\sigma_{F1}] = 390\,\text{MPa}$, $[\sigma_{F2}] = 370\,\text{MPa}$. Please do the following:

a) Compare the contact strength and bending strength of two gears and predict the failure modes of each gear.

b) Calculate the maximum torque that can be transmitted by this pair of gears.

4 In the gear box in Figure P8.1, the variables for the first pair of spur gears are: $z_1 = 20$, $z_2 = 100$, $m_1 = 2.5\,\text{mm}$ and $b_2 = 45\,\text{mm}$; The variables for the second pair of spur gears are: $z_3 = 23$, $z_4 = 77$, $m_3 = 3\,\text{mm}$ and $b_4 = 60\,\text{mm}$. For the two pairs of gears, the allowable stresses of the pinions are $[\sigma_{H1}] = [\sigma_{H3}] = 500\,\text{MPa}$ and the allowable stresses of the gears are $[\sigma_{H2}] = [\sigma_{H4}] = 420\,\text{MPa}$. Select load factor $K = 1.3$ and elastic coefficient $Z_E = 189.8\,\text{MPa}^{1/2}$. Determine:

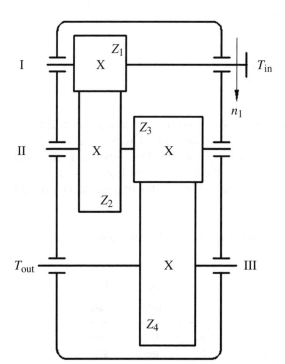

Figure P8.1 Illustration for Calculation Question 4.

a) the allowable output torque T_{output} by contact strength calculation;

b) if after heat treatment, the allowable stresses of the pinions increase to $[\sigma_{H1}]' = [\sigma_{H3}]' = 650\,\text{MPa}$ and the allowable stresses of the gears increase to $[\sigma_{H2}]' = [\sigma_{H4}]' = 600\,\text{MPa}$, by what percentage will the allowable output torque T_{output} calculated by contact strength analysis be increased by?

Design Problems

1   The input power for a pair of helical gears is $P_1 = 5.5\,\text{kW}$. The pinion and gear rotate at $n_1 = 480\,\text{rpm}$ and $n_2 = 150\,\text{rpm}$, respectively. The initially selected variables are: the number of teeth of the pinion $z_1 = 28$, helix angle $\beta = 12°$, face width ratio $\varphi_d = 1.1$. The calculated pitch diameter by tooth surface fatigue strength analysis is $d_1 = 70.13\,\text{mm}$. Decide the normal module m_n, centre distance a, helix angle β and face width of the pinion b_1 and gear b_2.

2   A reducer composes of a pair of bevel gears and helical gears, as shown in Figure P8.2. The rotational speed of the small bevel gear is $n_1 = 960\,\text{rpm}$ and the transmitted power is $P_1 = 7.5\,\text{kW}$. The number of teeth of bevel gears are $z_1 = 25$, $z_2 = 79$ and the large end module is $m = 3\,\text{mm}$. The number of teeth of helical gears are $z_3 = 21$, $z_4 = 80$. Normal module is $m_n = 4\,\text{mm}$ and normal pressure angle is $\alpha_n = 20°$. Centre distance is $a = 210\,\text{mm}$. The face width of the bevel gear and helical gear are 35 mm and 45 mm, respectively. Please decide on the following:

Figure P8.2 Illustration for Design Problem 2.

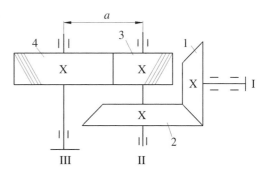

a) The cone distance of the bevel gear R, pitch cone angle δ_1 and δ_2, mean diameter of small bevel gear d_{m1};

b) The helix angle β and pitch diameter of the pinion and gear, d_3, d_4.

c) The rotational direction of the small bevel gear, so that the thrust forces acting on the intermediate shaft are of the opposite direction.

d) The forces acting on the gear 2 and gear 3 and indicate the force directions.

3   Design a pair of spur gears to drive a reciprocating compressor, which works 8 h daily, 300 days each year for 10 years. The pair of spur gears is to transmit a power of 10 kW while the pinion rotates at 960 rpm. The gear ratio is $u = 4.0$. Consider using both soft tooth surface and case-hardened carburized surface gears. Compare the design results.

4   Design a pair of helical gears using data in Design Problem 3. Consider using both soft tooth surface and case-hardened carburized surface gears.

Structure Design Problems

1   While designing a pair of spur gears, the face width of pinion is usually wider than that of gear by 5–10 mm. The reason for this is _____.
    (a) to ensure the strength of pinion is greater than that of gear
    (b) to ensure the pinion and the gear have a similar strength
    (c) to facilitate assembly and to ensure contact length
    (d) to ensure stable transmission and to increase efficiency

2   The helix angle direction and rotational direction of gear 1 in a gear reducer are shown in Figure P8.3. Gear 2 has a normal module of $m_{n2} = 2$ mm, 61 teeth and a helix angle of $\beta_2 = 12.5°$; Gear 3 has a normal module of $m_{n3} = 4$ mm and 19 teeth. Do the following:

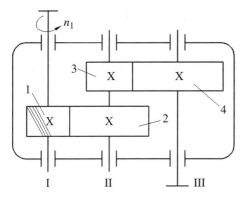

Figure P8.3 Illustration for Structure Design Problem 2.

(a) To reduce the thrust force on shaft II, determine the helix directions of gears 2 and 3;
(b) Indicate the forces acting on gear 2 and 3;
(c) To ensure shaft II carries no thrust load, decide on the helix angle of gear 3.

3   Design two layouts of gears in a reducer. In the reducer, the input shaft and output shaft are concentric. The rotational speed of the input and output shafts are of the same and opposite directions, respectively.

4   A pair of spur gears and a pair of helical gears are to be used in the gear reducer shown in Figure P8.4. Should the pair of helical gears be arranged in the first stage or in the second stage of power transmission?

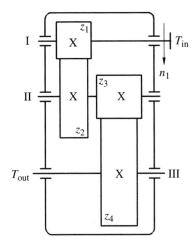

Figure P8.4 Illustration for Structure Design Problem 4.

5   Power is to be transmitted from an electric motor to a conveyor through a belt drive and a two-stage gear reducer. A pair of spur gears and a pair of helical gears are to be used in the gear reducer. Propose at least two designs and compare them.

CAD Problems

1   Write a flow chart for the gear drive design process.

2   Develop a program to implement a user interface similar to Figure P8.5 and complete Example Problem 8.1.

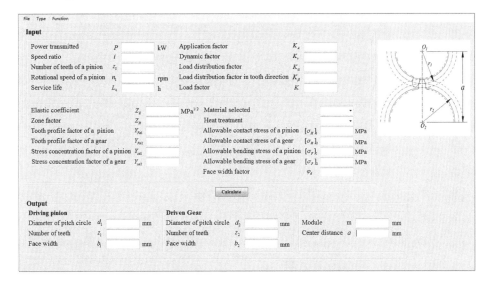

Figure P8.5 Illustration for CAD Problem 2.

9

Wormgear Drives

Nomenclature

a	centre distance, mm
c	clearance, mm
d	reference diameter, mm
d_a	diameter of addendum circle, mm
d_f	diameter of dedendum circle, mm
E	elastic modulus, MPa
F_a	axial force, N
F_n	normal force, N
F_r	radial force, or separating force, N
F_t	tangential force, or transmitted force, N
f_v	equivalent coefficient of friction
I	moment of inertia, mm^4
i	speed ratio, velocity ratio
j	number of gear teeth meshing per revolution
K	load factor
K_A	application factor

K_v	dynamic factor
K_β	face load factor
K_{FN}	bending life factor
K_{HN}	contact life factor
L	total contact length, mm
L_h	design life in hours, h
l	span between two bearings, mm
m	module, mm
m_a	axial module, mm
m_n	normal module, mm
m_t	transverse module, mm
N	number of cycles
n	rotational speed, rpm
P	transmitted power, kW
p_a	axial pitch, mm
p_t	transverse circular pitch, mm
p_z	lead, mm
q	worm diameter factor, worm quotient
S	housing external surface area, m^2

Analysis and Design of Machine Elements, First Edition. Wei Jiang.
© 2019 John Wiley & Sons Singapore Pte. Ltd. Published 2019 by John Wiley & Sons Singapore Pte. Ltd.
Companion website: www.wiley.com/go/Jiang/analysis_of_machine_elements

T	transmitted torque, N·mm	β	helix angle, °
t_a	ambient air temperature, °C	γ	lead angle, °
t_o	oil operating temperature, °C	η	efficiency
u	teeth ratio, gear ratio	η_1	meshing efficiency
v	pitch line velocity, m s^{-1}	η_2	bearing efficiency
v_s	sliding velocity, m s^{-1}	η_3	efficiency due to lubricating oil
x	profile shift coefficient		churning
Y_{Fa}	tooth form factor	ρ	radius of curvature, mm
Y_{Sa}	stress correction factor	σ_F	bending stress, MPa
Y_β	helix angle factor	$[\sigma_F]$	allowable bending stress, MPa
Y_ε	contact ratio factor	$[\sigma_F]'$	basic allowable bending stress,
$[y]$	allowable deflection, mm		MPa
z_1	number of thread of a worm	σ_H	contact stress, MPa
z_2	number of wormgear teeth	$[\sigma_H]$	allowable contact stress, MPa
z_{v2}	virtual number teeth of a	$[\sigma_H]'$	basic allowable contact stress, MPa
	wormgear	φ_v	equivalent friction angle, °
Z_E	elastic coefficient, MPa$^{1/2}$,	ω	angular velocity, rad s^{-1}
	(N mm^{-2})$^{1/2}$		
α	pressure angle, °	*Subscripts*	
α_a	axial pressure angle, °	1	worm
α_n	normal pressure angle, °	2	wormgear
α_s	heat transfer coefficient,	a	axial plane
	W/(m^2 °C)	t	transverse plane
α_t	transverse pressure angle, °		

9.1 Introduction

9.1.1 Applications, Characteristics and Structures

Wormgear drives, or worm gearings, are used to transmit motion and power between crossed shafts, usually at a right angle [1]. The drive consists of a worm and a wormgear. The worm resembles a power screw thread on a high-speed shaft, driving the wormgear whose appearance is similar to that of a helical gear. Wormgear drives are widely used in machine tools and in automotive and many other machines.

Wormgear drives can achieve a high-speed ratio ($i = 5$–80) in a compact design. They provide smooth and quite operation. However, the power transmission efficiency is usually less than 90%, which is far lower than other types of gear drives. This is due to the frictional loss caused by sliding between meshing teeth surfaces. Sliding also makes it necessary to use expensive antifriction materials for wormgears. Besides, self-locking will occur if a lead angle is less than the equivalent friction angle.

9.1.2 Types of Wormgear Drives

According to the worm profile, there are three basic types of wormgear drives, that is, cylindrical, toroidal and spiroid worm (or cone worm) gear drives, as shown in Figure 9.1.

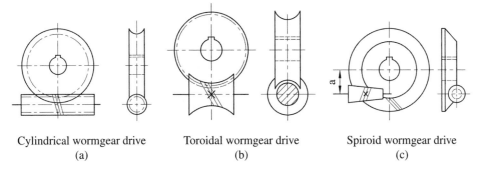

Cylindrical wormgear drive (a)   Toroidal wormgear drive (b)   Spiroid wormgear drive (c)

Figure 9.1 Types of wormgear drives.

9.1.2.1 Cylindrical Wormgear Drives

A cylindrical worm usually mates with a toroidal wormgear having teeth that are throated, wrapping partially around the worm, as shown in Figure 9.1a. It is a single enveloping worm gearing. The contact between worm thread and wormgear teeth is along a line and the power transmission capacity is quite good [2]. According to the tooth profile, which relates closely to the manufacturing methods, cylindrical worms can be classified as Archimedean worms, straight-sided normal worms and involute helicoid worms. Archimedean worms (ZA worms) are the most common type. They have threads that are straight-sided in the axial section. In the transverse section, the thread forms an Archimedean spiral. Therefore, in the central plane that goes through the worm axis and perpendicular to the wormgear axis, the Archimedean worm gearing resembles a helical gear and rack drive. Straight-sided normal worms (ZN worm) have threads that are straight-sided in a normal section. In the transverse section the thread forms a prolate involute. Involute helicoid worms (ZI worm) have an involute form in the transverse section [3]. This book will focus on discussing Archimedean worms.

9.1.2.2 Toroidal Wormgear Drives

When a worm profile is also throated to envelop a wormgear, it is a toroidal wormgear drive, or a double enveloping wormgear drive, as shown in Figure 9.1b. In a toroidal wormgear drive, 3–11 wormgear teeth are typically in contact with the toroidal worm, depending upon the ratio. The increased number of wormgear teeth that are in contact with worm significantly increases load carrying capacity. However, toroidal worms are difficult to manufacture, and the precision alignment of worm and wormgear is critical.

9.1.2.3 Spiroid Wormgear Drives

Compared with the hypoid gear drive mentioned in Section 8.1.2, which has a relatively small shaft offset, a spiroid wormgear drive has larger offsets and the spiroid worm resembles a tapered worm, as shown in Figure 9.1c.

9.1.3 Geometry and Terminology

The central plane is defined as the plane going through the axis of the worm and perpendicular to the axis of the wormgear. In the central plane, the meshing of an Archimedean worm and a wormgear is similar to the meshing of a rack and a helical gear, as shown in

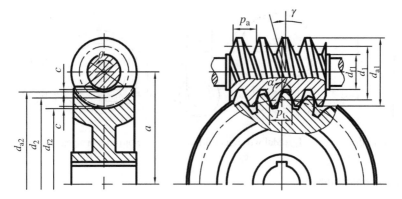

Figure 9.2 Geometry and terminology of a cylindrical. Adapted from Wen 2015.

Figure 9.2. Thus, standard module and pressure angle are defined in this plane. Parameters in the central plane are calculated similar to those of gear drives.

9.1.3.1 Module *m* and Pressure Angle *a*

For the proper meshing of a wormgear drive, the axial module of a worm should be equal to the transverse module of the mating wormgear, and both are the same as a standard module, that is,

$$m_{a1} = m_{t2} = m \tag{9.1}$$

Similarly, the axial pressure angle of worm, the transverse pressure angle of mating wormgear, and standard pressure angle are identical, that is,

$$\alpha_{a1} = \alpha_{t2} = \alpha \tag{9.2}$$

For Archimedean worms, the axial pressure angle is the standard value, that is $\alpha_a = 20°$, while for straight-sided normal worms and involute helicoid worms, the normal pressure angle is the standard value, that is $\alpha_n = 20°$. The relationship between axial pressure angle α_a and normal pressure angle α_n is [4]

$$\tan \alpha_a = \frac{\tan \alpha_n}{\cos \gamma} \tag{9.3}$$

9.1.3.2 The Worm Diameter d_1 and Worm Diameter Factor *q*

Because a wormgear is cut by a hob acting as a worm, the reference diameter of the hob should be identical to the reference diameter of the worm, so that the fabricated wormgear can mesh properly with the worm. To limit the number of hobs for the same module, worm diameter factor *q* is specified for the convenience of manufacturing management. The relationship between standard module *m*, worm reference diameter d_1 and worm diameter factor *q* is expressed as

$$d_1 = mq \tag{9.4}$$

9.1.3.3 The Number of Threads of Worm z_1 and the Number of Wormgear Teeth z_2

Worms have a single thread, as in a typical screw, or multiple threads. The number of threads in a worm is frequently referred as the number of starts. In practice, commonly used numbers of starts are 1, 2, 4 or 6, which is usually selected according to the speed ratio and efficiency. Similar to a spur or helical gear, the number of wormgear teeth z_2 is related to the speed ratio as $z_2 = iz_1$.

9.1.3.4 Worm Lead Angle γ and Wormgear Helix Angle β

Axial pitch p_a is the axial distance from a point on one thread to the corresponding point on the adjacent thread. For proper meshing, the axial pitch of worm p_a must be equal the transverse circular pitch of wormgear p_t in the central plane. The lead of a worm p_z is the axial distance that a point on a worm would move as the worm is rotated one revolution. For a multi-threaded worm, the worm lead is the product of the number of thread and axial pitch, that is, $p_z = z_1 p_a$, similar to a thread shown in Figure 3.2b.

Lead angle γ is the angle between the tangent to the worm thread at reference diameter and the plane normal to the worm axis. The worm lead angle γ must equal the wormgear helix angle β with the same hand for a 90° shaft angle for proper meshing [5]. The lead angle can be obtained from a simple triangle that would be formed if the thread of worm is unwrapped from the reference cylinder, which is similar to the calculation of thread lead angle illustrated in Figure 3.2b. The relationship between lead angle γ, lead p_z and worm reference diameter d_1 are expressed as

$$\tan \gamma = \frac{p_z}{\pi d_1} = \frac{z_1 p_a}{\pi d_1} = \frac{z_1 \pi m}{\pi m q} = \frac{z_1}{q} \tag{9.5}$$

9.1.3.5 Profile Shift Coefficient x

Wormgears are cut by a straight hob with a radial feeding movement between the hob and wormgear blank. To adapt to a predetermined centre distance, the hob can move radially from its normal position towards or away from the wormgear, resulting the reduction or enlargement of wormgear tooth dimensions [3]. The profile shift coefficient is calculated by [4]

$$x_2 = \frac{a}{m} - \frac{d_1 + d_2}{2m} \tag{9.6}$$

9.1.3.6 Centre Distance a

As with a spur or helical gear, the reference diameter of a wormgear is related to its module and number of teeth. Therefore, the centre distance is calculated by

$$a = \frac{1}{2}(d_1 + d_2 + 2x_2 m) = \frac{1}{2}m(q + z_2 + 2x_2) \tag{9.7}$$

Table 9.1 shows typical matching parameters common to wormgears and cylindrical worms. More data can be found in references [6, 7].

Table 9.1 Matching parameters of worms and wormgears [6, 7].

Centre distance a, (mm)	Module m, (mm)	Reference diameter d_1, (mm)	$m^2 d_1$, (mm³)	Number of thread z_1	Number of teeth of a wormgear z_2	Profile shift coefficient x_2
50	1	18	18	1	82	0.000
	1.25	22.4	35	1	62	0.040
	1.6	20	51.2	1	51	−0.500
				2		
				4		
	2	22.4	89.6	1	39	−0.100
				2		
				4		
	2.5	28	175	1	29	−0.100
				2		
				4		
				6		
100	2	35.5	142	1	82	0.125
	2.5	28	175		70	−0.600
		45	281.25		62	0.000
	3.15	35.5	352.25	1	53	−0.3889
				2		
				4		
	4	40	640	1	41	−0.500
				2		
				4		
	5	50	1250	1	31	−0.500
				2		
				4		
				6		
160	3.15	56	555.6	1	83	0.4048
	4	40	640		70	0.000
		71	1136		62	0.125
	5	50	1250	1	53	0.500
				2		
				4		
	6.3	63	2500.47	1	41	−0.1032
				2		
				4		
	8	80	5120	1	31	−0.500
				2		
				4		
				6		

(Continued)

Table 9.1 (Continued)

Centre distance a, (mm)	Module m, (mm)	Reference diameter d_1, (mm)	$m^2 d_1$, (mm³)	Number of thread z_1	Number of teeth of a wormgear z_2	Profile shift coefficient x_2
200	4	71	1136	1	82	0.125
	5	50	1250		70	0.000
		90	2250		62	
	6.3	63	2500.47	1	53	0.246
				2		
				4		
	8	80	5120	1	41	−0.500
				2		
				4		
	10	90	9000	1	31	0.000
				2		
				4		
				6		

9.2 Working Condition Analysis

9.2.1 Kinematic Analysis

9.2.1.1 Speed Ratio i and Gear Ratio u

Like gear drives, the speed ratio of a worm gearing is defined as the ratio of the rotational speed of driving worm and wormgear, and the gear ratio is the ratio of wormgear teeth to the number of worm threads. The relationship between speed ratio i and gear ratio u is defined as

$$i = \frac{n_1}{n_2} = \frac{z_2}{z_1} = u \tag{9.8}$$

Obviously, single-thread worms have a high-speed ratio. However, they are comparatively inefficient in power transmission. Single-thread worms are preferably used for fine tuning, or for self-locking to hold a mechanism at a preset position.

9.2.1.2 Sliding Velocity Analysis

Because of the kinematics of contact between meshing gear teeth as they pass through contact zone, a component of sliding motion exists in worm gear drives. Figure 9.3 shows the relationship between worm pitch line velocity v_1, wormgear pitch line velocity v_2 and sliding velocity v_s. Vectorially [8],

$$\overrightarrow{v}_1 = \overrightarrow{v}_2 + \overrightarrow{v}_s$$

Consequently, the sliding velocity v_s is calculated by

$$v_s = \sqrt{v_1^2 + v_2^2} = \frac{v_1}{\cos \gamma} \tag{9.9}$$

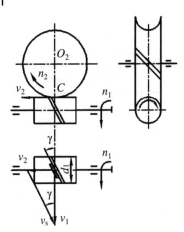

Figure 9.3 Sliding velocity analysis.

The high-sliding velocity between the worm thread and wormgear teeth will cause friction, wear and heat leading to the failure of teeth surface.

9.2.2 Forces on Worm and Wormgear Teeth

Forces acting on a worm and wormgear are usually considered at a pitch point and can be conveniently resolved into three mutually perpendicular components. They are tangential force F_t, radial force F_r and axial force F_a, as illustrated in Figure 9.4. The force directions on a worm and wormgear are determined similarly to that of helix gears, considering the direction of both worm thread and worm rational speed. The paired forces are in opposite directions according to the action and reaction principle.

The tangential force acting on a worm is computed by torque, power and rotational speed, with the direction opposite to the worm pitch line velocity at pitch point.

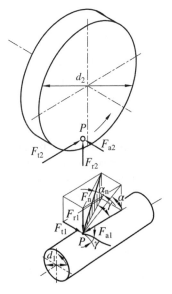

Figure 9.4 Force analysis of a wormgear drive.

The tangential force acting on the worm F_{t1} equals to the axial force on the wormgear F_{a2} for the usual 90° shaft angle, that is,

$$F_{t1} = \frac{2T_1}{d_1} = F_{a2} \tag{9.10}$$

The radial force on the worm and wormgear are equal, directing to the centre of worm and wormgear, respectively, separating the mating worm and wormgear.

$$F_{r1} = F_{r2} = F_{t2} \tan \alpha \tag{9.11}$$

The axial force acting on the worm F_{a1} equals to the tangential force on the wormgear F_{t2} for the usual 90° shaft angle. The direction of axial force follows the Right- or Left-Hand Rule. The magnitude is calculated by

$$F_{a1} = F_{t2} = \frac{2T_2}{d_2} \tag{9.12}$$

Ignoring friction, the normal force can be regarded as acting on the pitch point in the normal plane, expressed as

$$F_n = \frac{F_{a1}}{\cos \gamma \cos \alpha_n} = \frac{F_{t2}}{\cos \gamma \cos \alpha_n} = \frac{2T_2}{d_2 \cos \gamma \cos \alpha_n} \tag{9.13}$$

Because the output power P_2 is less than the input power P_1 due to power losses caused by friction, the relation between the output power P_2 and input power P_1 can be expressed as $P_2 = \eta P_1$, where η is transmission efficiency. This relation can be rewritten as $T_2 \omega_2 = \eta T_1 \omega_1$. Thus, we have the relations between the input torque T_1 and output torque T_2 as

$$T_2 = T_1 i \eta \tag{9.14}$$

9.2.3 Potential Failure Modes

Similar to previously discussed gearings, the principal failure modes in an enclosed wormgear drive are fatigue pitting and tooth breakage. Both are observed mainly in wormgears made of bronzes after extended service.

Different from previously discussed gearings where the motion of one tooth relative to the mating tooth is primarily rolling, there is an inherently sliding motion between worm threads and wormgear teeth, as illustrated in Figure 9.3. Therefore, friction and wear play an important role in the performance of wormgear drives.

An enclosed worm gearing operating at a high speed or under a heavy load with inadequate lubrication often shows a failure mode of scuffing. The high sliding velocity between mating surfaces of worm and wormgear, combined with high pressure, will generate high temperature and consequently reduce the viscosity of lubricant. The thinned oil film thickness due to reduced lubricant viscosity will result in the direct contact of worm and wormgear teeth and eventually cause scuffing.

The principal failure modes for open wormgear drives are tooth wear and breakage. Wear varies in a wide range depending on assembly, operation and lubrication. Inaccurate assembly, frequent starting and stopping, insufficient lubrication, accelerate wear process. Breakage can be observed mainly after severe wear. As a common scenario, only the teeth of wormgear are broken.

9.3 Load Carrying Capacities

The determination of load carrying capacity for wormgear drives is more complicated than for other types of gear drives due to complex profile geometry. Besides, wormgear drive capacity is often limited not only by fatigue strength but also by thermal capacity.

However, since an Archimedean wormgear drive resembles the meshing of helical gear and rack in the central plane, it is reasonable to predict contact and bending fatigue strength similar to those for helical gears, with factors introduced to account for their differences.

9.3.1 Tooth Surface Fatigue Strength Analysis

Usually, hardened steel worms operate with soft bronze wormgears. As worm threads are inherently strong and robust, the surface fatigue strength analysis is only performed on wormgear teeth.

Similar to other gearings, contact stress calculation on wormgears is carried out on the basis that the stress at pitch point does not vary greatly from the maximum stress at other points of engagement, but is more convenient to be determined. The contact stress calculation starts with the Hertz formula,

$$\sigma_{H\max} = \sqrt{\frac{KF_n}{L} \cdot \frac{\left(\frac{1}{\rho_1} \pm \frac{1}{\rho_2}\right)}{\pi \left(\frac{1-\mu_1^2}{E_1} + \frac{1-\mu_2^2}{E_2}\right)}}$$

where

K – load factor, $K = K_A K_\beta K_v$;

K_A – application factor. Since wormgear drives operate smoothly, low to moderate values within the recommended range from Table 2.1 can be selected for worm gearing design;

K_β – face load factor. For stable load, select $K_\beta = 1.0$; For shock and vibration load, select $K_\beta = 1.1$–1.3 [6];

K_v – dynamic factor. For pitch line velocity of a wormgear less than 3 m s^{-1}, select $K_v = 1.0$–1.1; otherwise, select $K_v = 1.1$–1.2;

Z_E – elastic coefficient. For hardened steel worms mating with cast tin bronze, cast aluminium bronze, cast irons and nodular cast iron, the elastic coefficients are selected as 155, 156, 162 and 181.4 $(\text{N mm}^{-2})^{1/2}$, respectively [6].

F_n – normal force, calculated by Eq. (9.13);

ρ_1, ρ_2 – radius of curvature at pitch point, mm; Since a wormgear drive resembles a helical gear and rack at the central plane, $\rho_1 = \infty$, $\rho_2 \approx d_2 \sin\alpha / 2\cos\gamma$.

L – total contact length, mm.

After substituting these variables into Hertz formula, the contact stress at the mating surface of the wormgear is estimated as [6]

$$\sigma_H = Z_E \sqrt{\frac{9.4KT_2}{d_1 d_2^2}} \leq [\sigma_H] \tag{9.15}$$

Based on the required capability of wormgears to operate without significant damage from pitting, it gives a design formula from Eq. (9.15) as

$$m^2 d_1 \geq 9.4 K T_2 \left(\frac{Z_E}{z_2 [\sigma_H]} \right)^2 \tag{9.16}$$

where $[\sigma_H]$ is allowable contact stress of a wormgear, which will be introduced in Section 9.4.3.1.

9.3.2 Tooth Bending Strength Analysis

According to the Lewis formula, which is applied to a helical gear in Eq. (8.57), the bending stress at the root fillet of wormgear teeth can be similarly expressed as

$$\sigma_F = \frac{K F_{t2}}{\widehat{b_2} m_n} Y_{Fa2} Y_{Sa2} Y_\epsilon Y_\beta = \frac{2 K T_2}{\widehat{b_2} d_2 m_n} Y_{Fa2} Y_{Sa2} Y_\epsilon Y_\beta \tag{9.17}$$

where
$\widehat{b_2}$ – curved length of wormgear tooth.
m_n – normal module, $m_n = m \cos \gamma$, mm;
Y_ϵ – contact ratio factor, select $Y_\epsilon = 0.667$;
Y_β – helix angle factor, calculated by $Y_\beta = 1 - \dfrac{\gamma}{120°}$;

Incorporating these into Eq. (9.17), the following equation can be obtained and used for rough estimates of bending strength as [6]

$$\sigma_F = \frac{0.666 K T_2}{d_1 d_2 m} Y_{Fa2} Y_{Sa2} Y_\beta \leq [\sigma_F] \tag{9.18}$$

Where wormgear tooth form factor Y_{Fa} and stress correction factor Y_{Sa} can be found from Table 8.4 according to the virtual number of teeth $z_{v2} = z_2 / \cos^3 \gamma$ and profile shift coefficient x_2 (limit to $x_2 = \pm 0.5$ in the table). More data can be found in design handbooks [6] for wormgears with other profile shift coefficients. $[\sigma_F]$ is the allowable bending stress of wormgear, which will be introduced in Section 9.4.3.2.

The computed values of tooth bending stresses from the left side of Eq. (9.18) are compared with the fatigue strengths of wormgear material for strength evaluation. The design formula can be derived from Eq. (9.18) as

$$m^2 d_1 \geq \frac{0.666 K T_2}{z_2 [\sigma_F]} Y_{Fa2} Y_{Sa2} Y_\beta \tag{9.19}$$

Combined with the value estimated by Eq. (9.16), the design variables for the worm and wormgear can be decided from Table 9.1.

9.3.3 Rigidity Analysis

The shaft carrying a worm must be sufficiently rigid to limit deflection at pitch point. Otherwise, the worm and wormgear could not mesh properly. Since a worm usually integrates with a shaft, the dedendum diameter of worm d_{f1} is used as shaft diameter for rigidity analysis. Therefore, the deflection and rigidity of worm can be calculated by [10]

$$y = \frac{\sqrt{F_{t1}^2 + F_{r1}^2}}{48 E I} l^3 \leq [y] \tag{9.20}$$

where

I – moment of inertia,$I = \frac{\pi d_{f1}^4}{64}$;

l – span between two bearings, usually $l = 0.9d_2$, where $d_2 = mz_2$;

$[y]$ – allowable deflection, usually select $[y] = (0.001-0.0025)\, d_1$, where d_1 is reference diameter of the worm.

9.3.4 Efficiency and Thermal Capacity

9.3.4.1 Efficiency of Wormgear Drives

Efficiency is defined as the ratio of output power to input power. Compared with other gearing systems, wormgear drive has a much lower mechanical efficiency, mainly due to friction loss.

The efficiency of enclosed wormgear drive η is determined by meshing efficiency η_1, bearing efficiency η_2 and efficiency η_3 due to churning of lubricating oil, which gives

$$\eta = \eta_1 \eta_2 \eta_3 \tag{9.21}$$

The meshing efficiency η_1 is calculated by [11]

$$\eta_1 = \frac{\tan \gamma}{\tan(\gamma + \varphi_v)} \tag{9.22}$$

where

φ_v – equivalent friction angle, which can be obtained from

$$\varphi_v = \arctan f_v \tag{9.23}$$

where equivalent coefficient of friction f_v can be estimated by sliding velocity from Figure 9.5. When the lead angle is equal to or less than the equivalent friction angle, that is, $\gamma \leq \varphi_v$, self-locking will occur [1]. Self-locking can be utilized when it is desired to hold a mechanism at a preset position.

The product of bearing efficiency η_2 and efficiency η_3 is usually within the range of 0.95–0.96. Thus, the total efficiency of an enclosed wormgear drive η is

$$\eta = (0.95 \sim 0.96)\frac{\tan \gamma}{\tan(\gamma + \varphi_v)} \tag{9.24}$$

Both lead angles and equivalent friction angles, or equivalent coefficients of friction, affect meshing efficiency η_1. Figure 9.6 illustrates the relationship graphically. Obviously, the larger the lead angle, the higher the efficiency. From the definition of lead angle in Eq. (9.5), it is clear that the number of threads has major effect on lead angle. To obtain a high efficiency, a multiple-threaded worm should be used. However, the efficiency increases with lead angle only until approximately 45°, beyond that, the increase of lead angle will have negative effect on efficiency and may also cause difficulties in worm fabrication. Therefore, commonly used numbers of thread are limited to 1, 2, 4 and 6 only, and the efficiency for the initial design can be correspondingly selected as 0.7, 0.8, 0.9 and 0.95.

9.3.4.2 Thermal Analysis

In a worm gearing, heat is generated by teeth meshing, bearings rolling and oil churning. The heat must be dissipated to the ambient atmosphere at such a rate that oil operating temperature is kept below 80°C. Otherwise, viscosity decreases severely and lubricant deteriorates rapidly, which will greatly reduce the lubricity of oil.

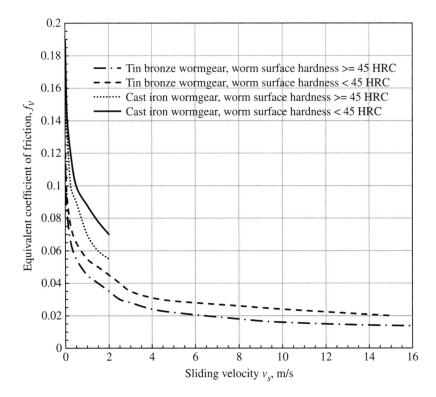

Figure 9.5 Equivalent coefficient of friction f_v versus sliding velocity v_s in wormgear drives [6].

Figure 9.6 Meshing efficiency η_1 as a function of lead angle and coefficient of friction.

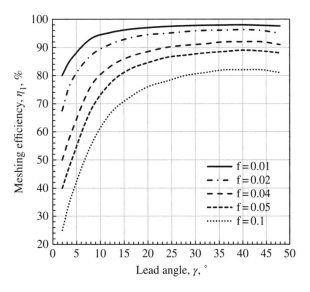

The rate of heat generated by friction during worm operation is

$$H_1 = 1000P(1 - \eta)$$

The rate of heat removed by dissipation from the free surface of housing is

$$H_2 = \alpha_s S(t_o - t_a)$$

To keep thermal balance, the amount of heat developed by the continuous operation of wormgear drive should be equal to the heat removed by dissipation from the free surface of housing within the same time, that is, $H_1 = H_2$, we have

$$t_o = t_a + \frac{1000P(1 - \eta)}{\alpha_s S} \leq 80^\circ C \tag{9.25}$$

The required area for conventional housing design may be roughly estimated from

$$S = \frac{1000P(1 - \eta)}{\alpha_s(t_o - t_a)} \text{ m}^2 \tag{9.26}$$

where

α_s – heat transfer coefficient of housing surface, $\alpha_s = (8.15\text{–}17.45) \text{ W m}^{-2} \, ^\circ\text{C}$ [4]. A larger value is for good ventilation;

t_o – oil operating temperature, $t_o \leq 80^\circ\text{C}$, normally $t_o = 60\text{–}70^\circ\text{C}$;

t_a – ambient air temperature, $t_a = 20^\circ\text{C}$;

$t_o - t_a$ – temperature rise.

If the oil operating temperature exceeds 80°C, or the dissipated surface area is insufficient, various means of cooling the housing or the lubricant oil may be used. For example, cooling fins can be incorporated to increase housing surface area; or fans and water coils can be used together to achieve cooling by increasing heat transfer coefficients.

9.4 Design of Wormgear Drives

9.4.1 Introduction

The task and objective of wormgear drive design are similar to those of gear drives. The provided design specifications include transmitted power, the rotational speed of worm and wormgear or gear ratio and operating conditions.

The design of a wormgear drive needs to determine worm types, accuracy grade levels and the arrangement of worm and wormgear. It also involves the selection of tentative material candidates, heat treatments, as well as initial variables like the number of worm threads. These initially selected design variables must satisfy both strength and thermal criteria.

Like the design of other machine elements, iteration in wormgear drive design is also unavoidable as information provided is scarce for solving unknowns directly. Besides, the values of some variables must be assumed and a trial solution be performed.

9.4.2 Materials and Heat Treatments

In a wormgear drive, sliding is far more extensive than in spur gears, and wormgears are more vulnerable to sliding friction. Materials, therefore, are required to have low coefficients of friction, sufficient strength and enough durability. A hardened ground steel worm and a nonferrous wormgear are a good combination to reduce friction and wear.

As a general rule of thumb, worms are usually made of medium carbon steel or alloy steel. For high speeds and heavy loads, alloy steels carburized or quenched to a hardness of 56–62 HRC are used. For low speeds and light loads, medium carbon steels that are heat treated to a hardness of 220–300 HBW are preferred.

Bronzes are the most common nonferrous metals used for wormgears. Bronze wormgears are able to wear-in and thus increase contact area [5]. The low elasticity modulus of bronzes provides greater tooth deflection and improves load sharing between teeth. Bronze gears also have good corrosion and wear resistance and low friction coefficients.

Typically used bronzes for wormgears are phosphor or tin bronze, lead bronze and aluminium iron bronze [1, 2, 6]. Most bronzes are cast, but wrought forms are also available. The choice of bronzes needs to take into account sliding velocity. If the sliding velocity is high, cast tin bronze, lead bronze, is recommended. If the sliding velocity is medium ($\leq 6 \, \mathrm{m \, s^{-1}}$) and load is heavy, cast aluminium–iron bronze is better. For low speeds and light loads, grey cast irons could be selected.

9.4.3 Allowable Stresses

9.4.3.1 Allowable Contact Stresses

If wormgears are made of bronze with strength limits less than 300 MPa, the main failure mode will be contact fatigue. The allowable contact stress is decided by [6]

$$[\sigma_H] = K_{HN}[\sigma_H]' \tag{9.27}$$

The basic allowable contact stresses $[\sigma_H]'$ at $N = 10^7$ is obtained from Table 9.2. At other numbers of cycles, contact life factor K_{HN} is introduced by [4]

$$K_{HN} = \sqrt[8]{\frac{10^7}{N}} \tag{9.28}$$

When $N > 25 \times 10^7$, use $N = 25 \times 10^7$ and when $N < 2.6 \times 10^5$, use $N = 2.6 \times 10^5$. The number of cycles is calculated by

$$N = 60 \, jn_2 L_h \tag{9.29}$$

where n_2 is the rotational speed of the wormgear, L_h is design life in hours and j is the number of meshings of a wormgear tooth per revolution.

If wormgears are made of grey cast iron or high strength bronze, the load carrying capacity of wormgear drive is decided by surface strength against scuffing. Due to limited data, a tentative contact calculation is used instead. The allowable contact stresses are selected from Figure 9.7, considering both mating worm hardness and sliding velocities.

9.4.3.2 Allowable Bending Stresses

Similar to the allowable contact stress, the allowable bending stress is obtained from

$$[\sigma_F] = K_{FN}[\sigma_F]' \tag{9.30}$$

where bending life factor is [4]

$$K_{FN} = \sqrt[9]{\frac{10^6}{N}} \tag{9.31}$$

Table 9.2 Basic allowable stresses of wormgears [4, 6].

Wormgear materials	Materials designation					Casting methods Sliding velocity limits v_s (m s^{-1})	Allowable contact stresses at $N = 10^7$ [σ_H]', MPa Worm surface hardness		Allowable bending stresses at $N = 10^6$ [σ_F]', MPa	
	ASTM/ AISI No.	BS No.	DIN No.	GB No.	ISO No.		≤45 HRC	>45 HRC	One side meshing	Both sides meshing
Cast tin bronze	C90700	PB4		ZCuSn 10P1	CuSn10P	Sand mould ($v_s \leq 12$)	160	180	45	30
						Metal mould ($v_s \leq 25$)	220	240	63	40
Cast lead bronze	C83600	LG2	G-CuSn5 ZnPb	ZCuSn5 Pb5Zn5	CuPb5 Sn5Zn5	Sand mould ($v_s \leq 10$)	110	130	30	22
						Metal mould ($v_s \leq 12$)	130	145	32	26
Cast aluminium iron bronze	C95200	AB1	G-CuAl 10Fe	ZCuAl 10Fe3	CuAl 10Fe3	Sand mould ($v_s \leq 10$)	See Figure 9.7		80	60
						Metal mould ($v_s \leq 10$)			90	70
Grey cast iron	No. 25	EN-GJL- 150	GG15	HT150	150	Sand mould ($v_s \leq 2$)			40	28
	No. 35 No. 40	EN-GJL- 250	GG25	HT250	250	Sand mould ($v_s \leq 2$–5)			48	34

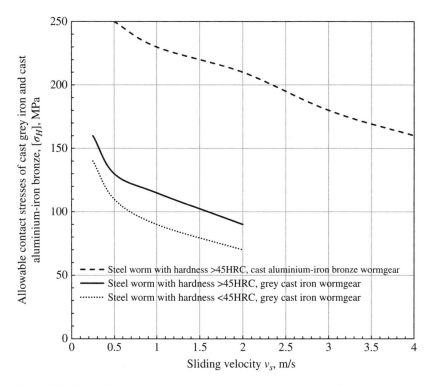

Figure 9.7 Allowable contact stresses of grey cast iron and cast aluminium–iron bronze [6].

When $N > 25 \times 10^7$, use $N = 25 \times 10^7$ and when $N < 10^5$, use $N = 10^5$. The basic allowable bending stresses $[\sigma_F]'$ at $N = 10^6$ can be obtained from Table 9.2.

9.4.4 Design Criteria

In general, the load carrying capacities of most wormgear drives are limited by pitting, wear and thermal capacity. Since the strength of worm teeth is stronger than that of a wormgear, failures often happen on wormgear surfaces. Thus, the calculation of working capacity applies only to wormgear teeth. The computed value of both contact stress and tooth bending stress are compared with fatigue strength of wormgear materials to evaluate strength. Furthermore, the continuous rated capacity of wormgear drive is often limited by temperature rise and oil temperature must not exceed 80°C for satisfactory operation.

Therefore, enclosed wormgear drives are designed by surface contact strength, that is, $\sigma_H \leq [\sigma_H]$, checked by bending strength, that is, $\sigma_F \leq [\sigma_F]$, and oil temperature is limited to $t_o \leq 80°C$ by thermal capacity analysis. For open wormgear drives, wear caused by significant sliding velocity is the primary failure mode. Open wormgear drives are thus designed by bending strength $\sigma_F \leq [\sigma_F]$, with an assumed number of wormgear teeth greater than 80.

9.4.5 Design Procedure and Guidelines

The design procedure for wormgear drives is similar to that of gear drives. Here, we list guidelines particular to wormgear drive design. A design case is given next to illustrate the design procedure.

9.4.5.1 The Arrangement of Wormgear Drive and Selection of Accuracy Grade Levels

When a wormgear drive is lubricated by bath lubrication, the worm is to be placed over the wormgear if the pitch line velocity of worm is greater than $5\,\mathrm{m\,s^{-1}}$ and below or by the side of wormgear if pitch line velocity less than $10\,\mathrm{m\,s^{-1}}$ [3]. There is no specific arrangement requirement for spay lubrication. For general application, worms are usually selected at 6–8 accuracy grade levels and one grade lower for wormgears.

9.4.5.2 The Selection of the Number of Worm Threads z_1 and the Number of Wormgear Teeth z_2

Considering transmission efficiency and fabrication capability, the commonly used numbers of threads are limited to 1, 2, 4 and 6. The number of teeth of meshing gear are obtained by $z_2 = uz_1$. To avoid undercut, as well to improve transmission stability, the number of wormgear teeth should be greater than 28. Under constant reference diameter of the wormgear, increasing the number of wormgear teeth consequently reduces the module, which eventually reduces the bending strength of the wormgear teeth. On the other hand, if the module keeps constant, the increase in the number of wormgear teeth will produce a large reference diameter, increasing the span between two bearings supporting the worm and ultimately reducing worm rigidity. In summary, the number of wormgear teeth should be within the range of 28–80.

9.4.5.3 Design by Wormgear Strength

After initial design variables have been selected, a wormgear drive can be designed in either order to satisfy both contact fatigue strength and bending fatigue strength. The final design variables are selected from Table 9.1 by the calculated m^2d according to Eq. (9.16) or Eq. (9.19).

9.4.6 Design Cases

Example Problem 9.1

Design a wormgear drive to transmit a power of 9 kW at a worm speed of 1460 rpm. The desired speed ratio is $i = 20$. The wormgear drive works steadily with a life of $L_h = 10\,000$ hours.

Steps	Computation	Results	Units
1. Selection of worm type and materials	(1) Select Archimedean worms (ZA).	ZA worm	
	(2) Select quenched medium carbon steel with hardness of 45–55 HRC for the worm. Select cast tin bronze ZCuSn10Pl for the wormgear.		

Steps	Computation	Results	Units
2. Design by surface contact strength	From Eq. (9.16), the design formula by contact strength is $$m^2 d_1 \geq 9.4 KT_2 \left(\frac{Z_E}{z_2 [\sigma_H]} \right)^2$$ 1) Determine the torque acting on the wormgear T_2 Select $z_1 = 2$, efficiency $\eta = 0.8$, then $$T_2 = 9.55 \times 10^6 \frac{P_2}{n_2} = 9.55 \times 10^6 \frac{P_1 \eta}{n_1/i} =$$ $$9.55 \times 10^6 \times \frac{9 \times 0.8}{1460/20} = 0.94$$ $$\times 10^6 N \cdot mm$$ 2) Specify load factor K For a stable work load, select application factor as $K_A = 1.15$ from Table 2.1. Select the face load factor as $K_\beta = 1.0$; dynamic factor is $K_v = 1.05$ for low speed and minor impact. The load factor is then $$K = K_A K_\beta K_v = 1.15 \times 1.0 \times 1.05 = 1.21$$ 3) Determine elastic factor Z_E Since hardened steel worms mating with a cast tin bronze wormgear, select $Z_E = 155 \text{ MPa}^{1/2}$. 4) Specify allowable contact stress $[\sigma_H]$ Since the wormgear material is cast tin bronze ZCuSnl0P1, metal mould casting and the hardness of worm surface is greater than 45 HRC, from Table 9.2, select the basic allowable contact stress as $[\sigma_H]' = 268 \text{ MPa}$. The number of cycles $$N = 60 j n_2 L_h = 60 \times 1 \times \frac{1460}{20} \times 10000 = 4.38 \times 10^7$$ Contact life factor $$K_{HN} = \sqrt[8]{\frac{10^7}{N}} = \sqrt[8]{\frac{10^7}{4.38 \times 10^7}} = 0.8314$$ Then the allowable contact stress is $$[\sigma_H] = K_{HN}[\sigma_H]' = 0.8314 \times 268 = 222.8 \text{ MPa}$$ 5) Calculate the value of $m^2 d_1$ $$m^2 d_1 \geq 9.4 KT_2 \left(\frac{Z_E}{z_2 [\sigma_H]} \right)^2 = 9.4 \times 1.21 \times 0.94 \times$$ $$10^6 \left(\frac{155}{2 \times 20 \times 222.8} \right)^2 = 3234$$ From Table 9.1, select $m = 8$ mm, $d_1 = 80$ mm, $a = 200$ mm.	$a = 200$ $m = 8$ $d_1 = 80$	mm mm mm
3. Parameters and dimensions of the worm and wormgear	1) Parameters and dimensions of the worm From Eq. (9.4), the worm diameter factor is $$q = \frac{d_1}{m} = 10$$ From Eq. (9.5), lead angle is $$\gamma = \arctan \left(\frac{z_1}{q} \right) = \arctan \frac{2}{10} = 11°18'36''$$ $$= 11.31°$$	$q = 10$ $\gamma = 11.31$	°

Steps	Computation	Results	Units
	2) Parameters and dimensions of the wormgear The number of wormgear teeth $z_2 = 41$; Profile shift coefficient $x_2 = -0.5$; Check the speed ratio $$i = \frac{z_2}{z_1} = \frac{41}{2} = 20.5$$ The error of speed ratio $$\frac{20.5 - 20}{20} = 2.5\%$$ It is acceptable. Reference diameter of the wormgear $d_2 = mz_2 = 8 \times 41 = 328\,\text{mm}$	$z_1 = 2$ $z_2 = 41$ $x_2 = -0.5$ $d_2 = 328$	mm
4. Check the bending strength of wormgear teeth	Check the bending strength of the wormgear teeth by Eq. (9.18) $$\sigma_F = \frac{0.666KT_2}{d_1 d_2 m} Y_{Fa2} Y_{Sa2} Y_\beta \le [\sigma_F]$$ Virtual number of teeth $$z_{v2} = \frac{z_2}{\cos^3 \gamma} = \frac{41}{\cos^3 11.31^\circ} = 43.48$$ From Table 8.4, by interpolating, we have wormgear tooth form factor $Y_{Fa2} = 2.87$, and stress correction factor $Y_{Sa2} = 1.47$. Helix angle factor $$Y_\beta = 1 - \frac{\gamma}{120^\circ} = 1 - \frac{11.31^\circ}{120^\circ} = 0.906$$ Specify basic allowable bending stress from Table 9.2 for ZCuSnl0P1 wormgear as $[\sigma_F]' = 56\,\text{MPa}$. Bending life factor $$K_{FN} = \sqrt[9]{\frac{10^6}{N}} = \sqrt[9]{\frac{10^6}{4.38 \times 10^7}} = 0.657$$ Allowable bending stress is $[\sigma_F] = K_{FN}[\sigma_F]' = 56 \times 0.657 = 36.8\,\text{MPa}$ The bending stress is $$\sigma_F = \frac{0.666KT_2}{d_1 d_2 m} Y_{Fa2} Y_{Sa2} Y_\beta$$ $$= \frac{0.666 \times 1.21 \times 0.94 \times 10^6}{80 \times 328 \times 8}$$ $\times 2.87 \times 1.47 \times 0.906 = 13.79\ \text{MPa}$ $\sigma_F < [\sigma_F]$ Therefore, bending strength is satisfactory.	$Y_{Fa2} = 2.87$ $Y_{Sa2} = 1.47$ $Y_\beta = 0.906$	
5. Thermal analysis	Omitted		
6. Structural design	Omitted		

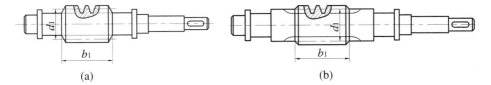

Figure 9.8 Worm structure.

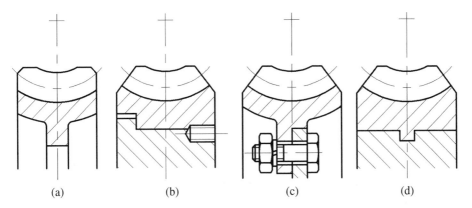

Figure 9.9 Structure of wormgears.

9.5 Structural Design of Wormgear Drives

Worms are usually integrated with a shaft, fabricated either by turning or milling, as shown in Figure 9.8 parts a and b, respectively, followed by case hardening and grinding or polishing.

The structure of a wormgear is largely dependent on the size. When the diameter is less than 100 mm, a bronze wormgear can be casted as a whole solid, as shown in Figure 9.9a. Greater diameter wormgears are often composed of a toothed rim made from expensive bronze mounted on the centre or hub of less expensive materials. For relatively small sizes, the bronze toothed rim is press-fitted on a steel or cast iron central part with interference, assisted by 6–12 pin connections, as shown in Figure 9.9b. For greater structures or easily worn wormgears, the bronze rim has a flange secured by a number of bolts to the hub of wormgear, as shown in Figure 9.9c. In the case of batch manufacturing, a bimetallic design, as shown in Figure 9.9d, is used.

9.6 Lubrication of Wormgear Drives

Because of high sliding velocities and associated frictional heat, lubrication is extremely important for wormgear drives. Poor lubrication results in low efficiency, excessive wear and even scuffing.

The selection of lubricant viscosity and lubrication methods for wormgear drives are similar to those for gear drives. They are selected by sliding velocity and load character-istics. For enclosed wormgear drives, high viscosity lubricant and oil bath lubrication are usually chosen for heavy load applications with sliding velocities less than $5 \, \text{m s}^{-1}$.

As sliding velocities increase, low viscosity lubricant and oil spray with pressure are preferred. A higher viscosity lubricant or grease can be used for open wormgear drives.

References

1 Budynas, R.G. and Nisbett, J.K. (2011). *Shigley's Mechanical Engineering Design*, 9e. New York, USA: McGraw-Hill.
2 Mott, R.L. (2003). *Machine Elements in Mechanical Design*, 4e. Prentice Hall.
3 Jelaska, D. (2012). *Gears and Gear Drives*, 1e. New York, NY: Wiley.
4 Pu, L.G. and Ji, M.G. (2006). *Mechanical Design*, 8e. Beijing: Higher Education Press.
5 Juvinall, R.C. and Marshek, K.M. (2011). *Fundamentals of Machine Component Design*, 5e. New York, NY: Wiley.
6 Wen, B.C. (2015). *Machine Design Handbook*, 5e, vol. 1&2. Beijing: China Machine Press.
7 Zhang, X., Wang, Z., Ji, D. et al. GB10085-88(1990). *Basic Parameters of Cylindrical Worm Gears*. Beijing: Standard Press of China.
8 Hibbeler, R.C. (2006). *Engineering Mechanics-Dynamics*, 11e. Prentice Hall.
9 Hindhede, U., Zimmerman, J.R., Hopkins, R.B. et al. (1983). *Machine Design Fundamentals: A Practical Approach*. New York, NY: Wiley.
10 Gere, J.M. and Timoshenko, S.P. (1996). *Mechanics of Materials*, 4e. CL Engineering.
11 Huang, X.K. and Zheng, W.W. (1981). *Theory of Machines and Mechanisms*. Beijing: People's Education Press.

Problems

Review Questions

1 For power transmission, why should the number of teeth of a wormgear be within the range of 28–80?

2 Under what condition can a worm be placed on the top of or below a wormgear?

3 What will be the effect of sliding velocities on the worm transmission under poor lubrication or under sufficient lubrication?

4 To improve heat dissipation capability, should the fan in a wormgear drive be mounted on the worm shaft or on the wormgear shaft?

Objective Questions

1 The most effective method of improving the efficiency of a wormgear drive is____.
   (a) to increase the modulus m
   (b) to increase the number of thread of a worm
   (c) to increase the worm diameter factor q
   (d) to reduce the worm diameter factor q

2 The worm reference diameter d_1 could not be calculated by _____ (γ is lead angle)
- (a) $d_1 = mq$
- (b) $d_1 = 2a - d_2$
- (c) $d_1 = \frac{z_1 m}{\tan \gamma}$
- (d) $d_1 = mz_1$

3 In a wormgear drive, not only is the module standardized, but also the reference diameter of the worm d_1. The purpose of such regulation is _____.
- (a) to facilitate assembly
- (b) to reduce the number of tools required to fabricate worms
- (c) to reduce the number of tools required to fabricate wormgears and to facilitate the standardization of tools
- (d) to improve the precision of manufacturing

4 The purpose of heat balance calculation for a worm gearing is to control temperature in order to prevent___
- (a) the deterioration of lubricants and gluing of the wormgear surface
- (b) the decrease of worm mechanical properties
- (c) the decrease in transmission efficiency
- (d) the annealing of the wormgear

Calculation Questions

1 A worm gearing set is shown in Figure P9.1. The driving worm rotates anticlockwise. Indicate the direction of radial forces, tangential forces and axial forces that each worm and wormgear is subjected to. Show the rotational directions of each shaft. Determine the hand helix of the wormgears and draw them on the sketch.

Figure P9.1 Illustration for Calculation Question 1.

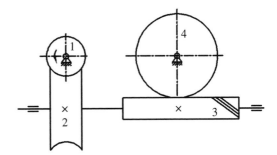

2 A power transmission composes of a pair of bevel gears, a worm and a wormgear, as shown in Figure P9.2. The rotational direction of the output bevel gear ω_4 is also shown in the figure.
- (a) To counteract the forces on the intermediate shaft, determine the helix direction of the wormgear, and the rotational direction of the worm.
- (b) Indicate tangential and axial forces acting on each element.

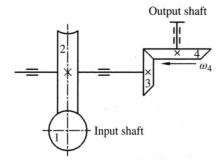

Output shift   **Figure P9.2** Illustration for Calculation Question 2.

3   A wormgear with 41 teeth and module of $m = 5$ mm mates with a double-threaded worm with a diameter of $d_1 = 50$ mm. Assume the coefficient of friction is $f = 0.1$. Determine (1) speed ratio, (2) diameter of wormgear, (3) centre distance, (4) lead angle of the worm and whether the wormset is self-locking and (5) efficiency.

4   The input power for a wormgear reducer is $P = 3$ kW with a total efficiency of $\eta = 0.8$. The heat dissipation area is 1 m^2 and heat transfer coefficient is $\alpha_s = 15 \text{ W m}^{-2}\ °\text{C}$. Assume the ambient temperature is 20°C. It is required that the oil temperature does not exceed 80°C at heat balance. Please check whether the lubricant oil sump temperature meets the requirements.

5   A 10 kW, a 970 rpm electric motor is to drive a lift through a right-handed wormgear reducer, as shown in Figure P9.3. The reducer has a double-threaded worm and a 60-tooth wormgear, with a module of $m = 8$ mm. The worm has an axial pitch of 25.12 mm and a reference diameter of 64 mm.

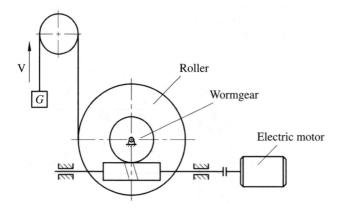

Figure P9.3 Illustration for Calculation Question 5.

Assume the friction coefficient between the worm and wormgear is 0.1. Calculate or determine the following:
(a) When the lift is going up, indicate the rotational direction of the electric motor;
(b) Indicate the direction of forces acting on the worm;

(c) Determine the lead angle of the worm and whether the worm set is self-locking;

(d) Calculate the meshing efficiency;

(e) Calculate the forces acting on the wormgear;

(f) Calculate sliding velocity between worm and wormgear;

(g) Determine the power delivered to the lift.

Design Problems

1   A double-threaded steel worm rotates at 1460 rpm, meshing with a 53-tooth wormgear transmitting 20 kW to the output shaft. The wormgear drive works 8 h a day steadily for 10 years. Design the wormgear drive.

2   A double-threaded steel worm with a module of $m = 6.3$ mm rotates at 1460 rpm, meshing with a 53-tooth wormgear. The worm is hardened steel and the gear is sand-cast bronze. The wormgear set works steadily 8 h a day for 10 years. Decide the maximum power to be transmitted by the wormgear drive.

3   A wormgear drive is utilized to reduce the speed of a 1460 rpm motor driving the worm down to an output wormgear shaft speed of approximately 100 rpm and provides 7.5 kW power to the driven machine. The application factor is 1.25. Design a wormgear drive and specify the nominal required power rating of the driving motor.

4   Three designs, that is, using a single-thread worm, a double-thread worm and a four-thread worm, are proposed for a wormgear drive to produce a speed ratio of 20 when the wormgear rotates at 90 rpm. Worms in all the three designs have a module of 5 mm, and a worm reference diameter of 50 mm. The worms are hardened steel and the wormgears are chilled bronze. Compare these designs and compute the rated output torque, satisfying both bending strength and pitting resistance requirements.

Structure Design Problems

1   Find the errors in Figure P9.4 and correct them.

2   Find the errors in Figure P9.5 and correct them.

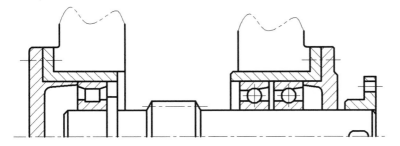

Figure P9.4 Illustration for Structure Design Problem 1.

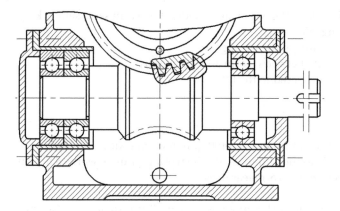

Figure P9.5 Illustration for Structure Design Problem 2.

CAD Problems

1 Write a flow chart for the wormgear drive design process.

2 Develop a program to implement a user interface similar to Figure P9.6 and complete Example Problem 1.

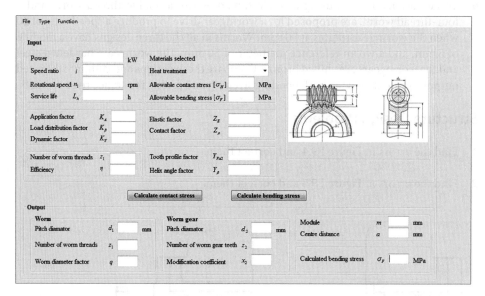

Figure P9.6 Illustration for CAD Problem 2.

10

Shafts

Nomenclature

d	shaft diameter, mm
d_v	equivalent diameter of a stepped shaft, mm
E	elastic modulus, MPa
F	force, N
G	shear modulus of elasticity, MPa
g	acceleration of gravity, mm/s^2
I	moment of inertia, mm^4
J	polar moment of inertia, mm^4
K	cantilever length, mm
k	shaft stiffness or shaft spring rate, N/m
L	calculated length of a shaft, mm

l	span, mm
M	resultant bending moment, N mm
M_{ca}	equivalent bending moment, N mm
M_H	horizontal bending moment, N mm
M_V	vertical bending moment, N mm
m	mass, kg
m_n	normal module, mm
n	rotational speed, rpm
n_c	critical speed, rpm
P	transmitted power, kW
S	safety factor
S_{ca}	calculated safety factor
S_σ	safety factor in tension

Analysis and Design of Machine Elements, First Edition. Wei Jiang.
© 2019 John Wiley & Sons Singapore Pte. Ltd. Published 2019 by John Wiley & Sons Singapore Pte. Ltd.
Companion website: www.wiley.com/go/Jiang/analysis_of_machine_elements

$[S]$	allowable safety factor	$[\sigma_{-1}]$	allowable bending stress, MPa
T	torque, N mm	τ	shear stress, MPa
W	section modulus, mm^3	τ_{-1}	endurance limit in shear at $r = -1$,
W_T	polar section modulus, mm^3		MPa
w_i	weight of ith mass, N	τ_s	yield strength in shear, MPa
y	deflection, mm	τ_T	torsional shear stress, MPa
$[y]$	allowable deflection, mm	$[\tau]$	allowable shear stress, MPa
z	number of segments in a stepped shaft	φ	unit length angular deflection, $°$ m^{-1}
α	correction coefficient	$[\varphi]$	allowable unit length angular
θ	slope, rad		deflection, $°$ m^{-1}
$[\theta]$	allowable slope, rad	ω_n	fundamental natural frequency,
σ	normal stress, MPa		rad s^{-1}
σ_{-1}	endurance limit at stress ratio of -1, MPa	*Subscripts*	
		i	values of ith segment in a stepped
σ_a	stress amplitude, MPa		shaft
σ_b	ultimate tensile strength, MPa	*max*	maximum value
σ_{ca}	design stress, MPa	S	static state
σ_m	mean stress, MPa	σ	normal stress
σ_s	yield strength in tension, MPa	τ	shear stress

10.1 Introduction

10.1.1 Applications, Characteristics and Structures

Shafts are important elements in virtually all types of machinery. They are supported by bearings, rotate and transmit power and thereby torque. A typical application of shafts is input, output or intermediate shafts supporting gears in gear reducers.

Unlike standard elements, shafts are designed for specific applications individually, considering the mounted elements, operating loads and service conditions. Power transmission elements, such as gears, belt pulleys, chain sprockets, couplings and so on are attached to the shafts by keys, splines, pins and other devices. When shafts rotate with power transmission elements, power and rotational motion are transmitted from one element to other rotating parts of shaft system [1–3].

10.1.2 Types of Shafts

A shaft is a long cylindrical element loaded torsionally, transversely and/or axially as a machine operates. Consequently, shafts can be classified as transmission shafts, axles and spindles according to the loads they carry [4]. Transmission shafts carry both bending and torsional moments, and are usually stepped, as shown in Figure 10.1a. They support gears, pulleys and so on and often transmit power. Axles carry bending moments only. They are used to support rotating wheels, pulleys and the like. An axle can turn with a wheel as a unit, like a railway car axle, as illustrated in Figure 10.1b, or is nonrotating,

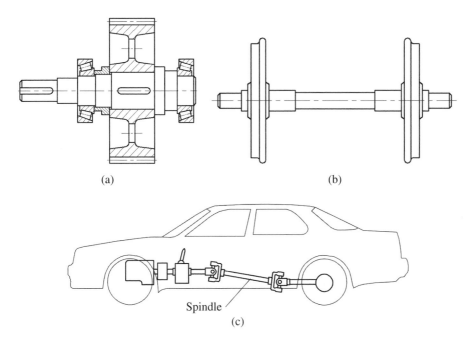

(a)   (b)

Spindle

(c)

Figure 10.1 Types of shafts classified by loading.

like the axle of the front wheel of a bicycle. Spindles take torsional moments only, like the one in an automobile, as shown in Figure 10.1c.

According to the shapes of axis, shafts include straight shafts, with either different diameters or constant diameters and crankshafts. Stepped shafts have different diameters and are usually used as transmission shafts, while constant diameter shafts work as axles or spindles. Crankshafts are used to convert reciprocating motion into rotary motion or vice versa. Most shafts are solid and hollow shafts are used to save on weight. Also, flexible shafts are used to transmit power along a curved path when space is limited or the axes of the power source and driven machine are not aligned with each other, as shown in Figure 10.2.

Figure 10.2 Flexible shaft.

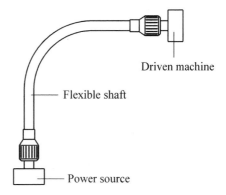

Driven machine

Flexible shaft

Power source

10.2 Working Condition Analysis

10.2.1 Force Analysis

The loads on a shaft arise from its mounted elements. To simplify force analysis, the distributed loads on mounted elements, such as gears, belt pulleys, chain sprockets and bearings, are treated as concentrated forces acting at the midpoint of element width, as the width of the element is small compared to the total length of the shaft. The weight of the shaft and the attached elements are also neglected, as they are small compared to the applied loads.

In a typical transmission shaft shown in Figure 10.3a, power transmission elements, like gears, apply tangential, radial and axial forces F_t, F_r, F_a, on the shaft. The forces are generally not all in the same plane and bending moment diagrams are in mutually perpendicular planes, as shown in Figure 10.3b,c. Their vector sum produces the resultant bending moment in Figure 10.3d. Since the shaft rotates, the phase angle of moments is not important [2].

While transmitting power, a shaft is inherently subjected to a torsional moment or torque. The torque developed from one power transmission element must balance the torque from other elements, usually through a portion of a shaft, as illustrated by the torque distribution diagram in Figure 10.3e. Torque keeps a constant value at stable operation.

Apparently, shafts are subjected to the combination of axial, transverse shear, bending and torsional loads. These loads may be static or fluctuating during operation, depending on the specific application. After visualizing the forces, torques and bending moments in the shaft, the critical sections, as well as stresses and strength at these critical locations, can be determined.

10.2.2 Stress Analysis

When a rotating shaft transmits a constant unidirectional torque, the produced torsional shear stress is greatest on the shaft outer surface. These torsional shear stresses are usually steady, but may sometimes fluctuate, depending on applications.

Transverse loads from power transmission elements generate bending moments, result in completely reversed cyclic bending stresses. The bending stress is greatest on the outer surfaces. Transverse loads may also result in transverse shear stresses, which are normally small and can be neglected in stress analysis [5].

In addition, axial loads, such as those generate from helical gears, may produce either normal tensile or compressive stresses on the shaft. However, they are usually negligibly constant small stresses, and can be neglected when bending stresses are present in a shaft [2].

These stresses may exist simultaneously in a shaft. Therefore, the maximum shear stress theory or maximum distortion energy theory [5] are used to appraise stresses for the shaft analysis and design.

10.2.3 Deflection and Rigidity

A shaft deforms due to the loads applied by the elements it carries. Excessive bending or torsional deflections may cause misalignments in gear meshes, leading to an uneven load

(a)

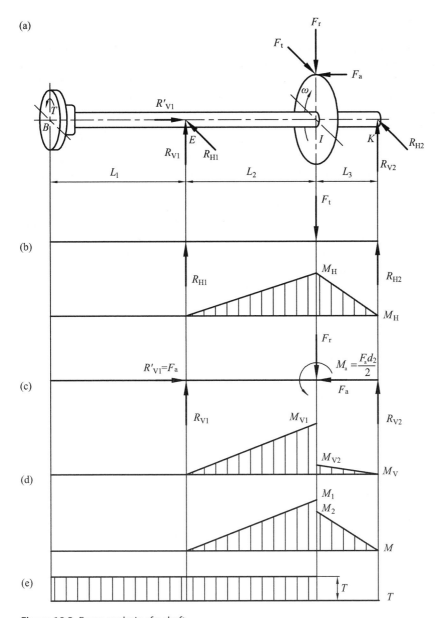

Figure 10.3 Force analysis of a shaft.

distribution along tooth width, which may result in malfunction, interference, excessive wear, vibration, noise or heat generation [6]. Furthermore, the greater the deflection, either bending or torsional, the lower the corresponding critical speed. Therefore, the rigidity or stiffness of shafts, both in bending and torsion, must be analysed. The deflection must be kept within specified limits to ensure satisfactory operation of mounted elements.

10.2.4 Rotating Shaft Dynamics

Shafts are nearly always part of a dynamic system. When a shaft operates at a speed close to its natural frequency, vibration amplitudes may suddenly increase and may destroy the system. Shafts may vibrate in transverse, torsional and axial directions. Axial vibration rarely happens during shaft operation and will not be discussed here.

Due to imperfection in manufacturing and assembly, the mass centre of a shaft rotating system seldom coincides exactly with the centre of rotation. Consequently, as the shaft rotates, eccentricity causes a centrifugal force deflection [2]. The greater the eccentricity, the larger the centrifugal force deflection and the severer the transverse vibration. When the shaft transverse frequency is close to or identical to the resonance of frequency, transverse resonance happens.

Torsional vibration may be excited when fluctuating torques are encountered due to periodic changes of transmitted power. If the frequency of torsional input and torsional natural frequencies of the shaft coincide, torsional resonance may damage the shaft.

Therefore, the operating speeds of shaft should avoid close to the critical speeds to prevent violent vibration or resonance, preferably below the lowest critical speed by a factor of 2–3 [6]. For a shaft with an operating speed higher than the lowest critical speed, it is important to ensure a quick through resonance during the startup and shut down cycle.

10.2.5 Potential Failure Modes

Considering the rotation of shaft, the bending stress induced by transverse loads from power transmission elements is completely reversed stress. The transverse shear stress produced by transverse loads may be completely reversed stress. The torsional shear stress generated by torques may be steady or fluctuating, depending on applications. The axial stress arises from helical gears or preloaded bearings, which are usually steady but can fluctuate sometimes. Obviously, fatigue is an important potential failure mode for power transmission shafts.

Furthermore, excessive bending or torsional deflections may lead misalignment in gear meshes and bearings, hamper gear performance and cause undesirable noise [3]. Finally, in high speed mechanical systems, shafts operate close to critical speeds may excite intolerable vibration. The increased vibration amplitudes may destroy the shaft system. Therefore, resonance of vibration is also one of the potential failure modes.

10.3 Load Carrying Capacities

10.3.1 Strength Analysis

As discussed before, stresses on a rotating shaft may involve torsional shear stress, transverse shear stress, bending stress or axial stress components, any or all of which may be fluctuating stresses. In general, shaft design must base on multiaxial states of stress produced by fluctuating loads [6]. However, in practice, strength analysis is mostly based on simple loading conditions. If a shaft is or is mainly subjected to a stable torque, that is, a spindle, a torsional strength calculation is performed. If a shaft is subjected to steady bending moments only, that is, an axle, a bending strength calculation is used. If a shaft is subjected to both bending moments and torques, that is, a transmission shaft, the combination of bending and torsion approach is used and fatigue strength analysis is performed when necessary. Static strength is calculated at peak loads to avoid excessive plastic deformation if a shaft is subjected to overload impact. The structure of the shaft needs to be modified if the strength does not meet requirements. After strength analysis, either the minimum shaft diameter to successfully support the loads or the safety factor for a specific design can be determined.

10.3.1.1 Torsional Strength Analysis

A spindle carries an operating torque only. The torsional shear stress at the critical section should not exceed the allowable shear stress, predicted by

$$\tau_T = \frac{T}{W_T} = \frac{9.55 \times 10^6 \frac{P}{n}}{0.2d^3} \leq [\tau] \tag{10.1}$$

where $W_T = \pi d^3/16$ is the polar section modulus.

For a transmission shaft carrying both bending moments and torques, we can assume the shaft is subjected to torsion only by reducing the allowable shear stress to consider the bending effect. The following equation obtained from Eq. (10.1) can then be used to estimate minimum diameter d to start the design iteration as

$$d \geq \sqrt[3]{\frac{9.55 \times 10^6}{0.2[\tau]}} \sqrt[3]{\frac{P}{n}} \tag{10.2}$$

The allowable shear stresses $[\tau]$ for low to medium carbon steels, stainless steels and alloy steels are within the range of 20–40, 15–25 and 40–52 MPa [7], respectively. When a shaft is subjected to torsion only, select a large value of $[\tau]$, while if a shaft is subjected to both bending and torsion, select a small value of $[\tau]$.

If the shaft has a keyseat or two keyseats, the resultant estimated diameter should be increased by approximately 3–5% or 7–10%, respectively [7]. Then round down the diameter to the next available size compatible with the mating elements.

10.3.1.2 Combination of Torsional and Bending Strength Analysis

The strength analysis for a transmission shaft subjected to combined torsional and bending loads follows the procedures next:

(1) Determine the forces exerted on the shaft by gears, belt sheaves or chain sprockets and resolve them in the horizontal and vertical planes, see Figure 10.3a–c. Determine the supporting reaction point on the shaft. The location of effective reaction point depends on the type of bearings and their layout, which is usually at the mid-width of radial bearings and should be found out in bearing catalogues for angular contact bearings and special bearings if the span is short. Solve for the reactions on all supporting bearings in each plane.
(2) Draw bending moment diagrams in the horizontal (Figure 10.3b) and vertical (Figure 10.3c) planes, and combine orthogonal moments as vectors to obtain resultant moments along the shaft by $M = \sqrt{M_H^2 + M_V^2}$, as shown in Figure 10.3d.

(3) Determine the transmitted torque and develop a torque diagram, usually between the midpoint of torque transmission elements, see Figure 10.3e;

(4) After the bending moment and torque have been decided, the maximum shear stress theory can be used to calculate design stress by

$$\sigma_{ca} = \sqrt{\sigma^2 + 4\tau^2}$$

Since the bending stress in the shaft is completely reversed stress as shafts rotate while torsional shear stress is usually not, a correction coefficient α is introduced to consider the difference in stress ratios between bending and torsional shear stresses. The design stress is modified as

$$\sigma_{ca} = \sqrt{\sigma^2 + 4(\alpha\tau)^2} \tag{10.3}$$

If torsional stress is a static stress, select $\alpha = 0.3$; if torsional stress is repeated stress or unknown, select $\alpha = 0.6$; if torsional stress is also a completely reversed stress, select $\alpha = 1.0$. For circular shafts, the bending stress is [5]

$$\sigma = \frac{M}{W} \tag{10.4}$$

and torsional shear stress is [5]

$$\tau = \frac{T}{W_T} = \frac{T}{2W} \tag{10.5}$$

where W and W_T are section modulus and polar section modulus, respectively.

Substitute Eqs. (10.4) and (10.5) to Eq. (10.3), the design stress for rotating, solid circular shafts at the critical cross section is

$$\sigma_{ca} = \frac{M_{ca}}{W} = \frac{\sqrt{M^2 + (aT)^2}}{0.1d^3} \le [\sigma_{-1}] \tag{10.6}$$

where M_{ca} is the equivalent bending moment, and allowable bending stress $[\sigma_{-1}]$ can be found in Table 10.1.

Since both loads and cross sections vary along the shaft, it is necessary to calculate and evaluate stresses at several critical sections. Possible critical locations of maximum stress may be decided by a combined effect of diameter variation and high torque and bending moments.

When a shaft is subjected to bending only, that is, an axle, no torque is applied. Substitute $T = 0$ in Eq. (10.6) so the stress can then be calculated.

10.3.1.3 Fatigue Strength Analysis

To evaluate fatigue strength of a shaft, stress variation, diameter changes, stress concentrations and surface conditions of shaft must be taken into account. The fatigue strength is usually calculated at critical locations on the outer surface, where the stress magnitude is large and where stress concentrations exist [2]. Normally, several critical sections along a stepped shaft are selected, and the stresses at each critical point are calculated and compared with the allowable values to ensure safety. The fatigue strength analysis of combined stresses introduced in Eq. (2.50) is duplicated here as

$$S_{ca} = \frac{S_\sigma S_\tau}{\sqrt{S_\sigma^2 + S_\tau^2}} \ge [S] \tag{10.7}$$

Table 10.1 Shaft materials and their mechanical properties [1, 2, 4, 7, 8].

	Materials designation				Heat treatment	Hardness	Ultimate strengths σ_b	Yield strength σ_s	Endurance limit in bending σ_{-1}	Endurance limit in shear τ_{-1}	Allowable bending stress $[\sigma_{-1}]$	Applications
	ASTM/AISI No.	BS No.	GB No.	ISO No.		HBW	MPa					
Carbon steels	1020	S235JR	Q235A	Fe360A		120	440	240	180	105	130	Low loads, unimportant applications
	1045	060A42 C45E	45	C45E4	N QT	195 236	600 650	300 360	240 270	140 155	172 194	Most widely used
Alloy steels	5120	527A20	20Cr	20Cr4	C Q T	56~62 HRC	850 650 650	550 400 400	375 280 280	215 160 160	230 170 170	High strength and toughness applications
	5140	530A40 530M40	40Cr	41Cr4	QT	264	750	550	350	200	214	Heavy load applications
	3140	640M40	40CrNi		QT	285	1000	800	485	280	296	Important shaft
		905M39	38Cr MoAl	41CrAl Mo74	QT	229	990	850	495	285	237	High strength, high wear resistance
Stainless steels	410	410S21	12Cr13	X12Cr13	QT	202	600	420	275	155	168	Corrosion environment
	304	X5CrNi 18-10	06Cr19 Ni10	X5CrNi 18-10	A		568	276	228	132	152	Corrosion environment
	302	X10CrNi 18-8	12Cr18 Ni9	X10CrNi 18-8	Q	192	550	220	205	120	145	High and low temperature, corrosion environment
Nodular iron	80-60-03	600/3	QT600-3	600-3		233	600	420	215	185		Complex structures
	120-90-02	800/2	QT800-2	800-2		290	800	480	290	250		

Note: N = normalizing, T = tempering, C = Carbonizing, Q = Quenching, A = Annealing.

If a shaft is subjected to normal stress only, the safety factor is calculated by

$$S_\sigma = \frac{\sigma_{-1}}{K_\sigma \sigma_a + \psi_\sigma \sigma_m} \geq [S] \tag{10.8}$$

And if a shaft is subjected to shear stress only, the safety factor is

$$S_\tau = \frac{\tau_{-1}}{K_\tau \tau_a + \psi_\tau \tau_m} \geq [S] \tag{10.9}$$

The allowable safety factor [S] is selected between 1.3 and 2.5, depending on the properties of material and application. The variables in Eqs. (10.8) and (10.9) have been introduced in Chapter 2.

10.3.1.4 Static Strength Analysis

When a shaft is subjected to an overload impact, the load peak value is used to calculate static strength to avoid excessive plastic deformation by [7]

$$S_{Sca} = \frac{S_{S\sigma} S_{S\tau}}{\sqrt{S_{S\sigma}^2 + S_{S\tau}^2}} \geq [S_S] \tag{10.10}$$

where

S_{Sca} – calculated static safety factor.

$[S_S]$ – allowable static safety factor. The allowable static safety factor is selected between 1.2 and 2.2 corresponding to the variation of σ_s/σ_b from 0.45 to 0.9.

$S_{S\sigma}$ – static safety factor when a shaft is subjected to bending and axial stresses only.

$$S_{S\sigma} = \frac{\sigma_s}{\sigma_{max}} \tag{10.11}$$

$S_{S\tau}$ – static safety factor when a shaft is subjected to torsional stress only.

$$S_{S\tau} = \frac{\tau_s}{\tau_{max}} \tag{10.12}$$

where

σ_s, τ_s – yield strength in tension and shear, respectively, MPa, usually $\tau_s = (0.55\text{–}0.62)\sigma_s$;

σ_{max}, τ_{max} – maximum bending and shear stresses at critical section, MPa;

10.3.2 Rigidity Analysis

Deflection and rigidity analysis usually follows strength analysis after the geometry of the shaft has been determined. Shaft deflection, both linear and angular, should be checked at locations where mating elements like gears and bearings are mounted, especially for slender shafts.

10.3.2.1 Bending Deflections and Slopes

The calculations of deflections and slopes for a stepped shaft are more complicated since both moment and cross-sectional moment of inertia change along the shaft. A stepped shaft can be treated as a uniform diameter shaft with an equivalent diameter of d_v, and then use the formula in *Mechanics of Materials* [5] to calculate the deflection y and slope θ. The equivalent diameter of a stepped shaft can be obtained from [4]

$$d_v = \sqrt[4]{\frac{L}{\sum\limits_{i=1}^{z} \frac{l_i}{d_i^4}}} \tag{10.13}$$

where

l_i, d_i –length and diameter of the ith segment in a stepped shaft;

L –calculated length of stepped shaft. When a load is applied between bearings, $L = l$; for an end loaded cantilever beam, $L = l + K$, where l is the span and K is the cantilever length;

z –number of segments in the stepped shaft;

The calculation of slope and deflection are based on successive integration of differential equations for the elastic beam deflection curve expressed as [5, 6]

$$\frac{d^2y}{dx^2} = \frac{M}{EI} \tag{10.14}$$

$$\theta = \frac{dy}{dx} = \int \frac{M}{EI} dx \tag{10.15}$$

$$y = \frac{d\theta}{dx} = \int\int \frac{M}{EI} dxdx \tag{10.16}$$

The integration may be performed either analytically, graphically or numerically. For a shaft of given length and loading, the bending deflection is inversely proportional to the product EI, as indicated in Eq. (10.16). Therefore, the effective way to increase the rigidity of a shaft is to increase the diameter of shaft.

The bending rigidity criteria are then

$$y \leq [y] \tag{10.17}$$

$$\theta \leq [\theta] \tag{10.18}$$

The allowable misalignment of a shaft is determined by the requirements of mounted specific gears or bearings by checking gear or bearing catalogues. As a rough guideline, the allowable bending deflections for a transmission shaft is $(0.0003–0.0005)l$, where l is span between bearings. When a shaft supports gears, the allowable bending deflections is selected as $(0.01–0.03)m_n$, where m_n is the normal module of mounted gears. The allowable slopes should be checked at locations where gears and bearings are mounted. At the cross section where a gear is mounted, the allowable slopes should be within $0.001–0.002$ rad. The allowable slopes for the shaft where a deep groove ball bearing, a cylindrical roller bearing and a tapered roller bearing, is mounted should be less than 0.005, 0.0025 and 0.0016 rad, respectively [7].

10.3.2.2 Torsional Deflections

Torsional rigidity is less important unless in some special applications. If a shaft has uniform diameter over its whole length, the unit length angular deflection may be readily calculated from formula in *Mechanics of Materials* [5], repeated here as

$$\varphi = \frac{180}{\pi} \times \frac{T}{GJ} \times 10^3 = 5.73 \times 10^4 \frac{T}{GJ} \tag{10.19}$$

For a stepped shaft with individual cylinder length l_i and torque T_i, the unit length angular deflection can be estimated from [4]

$$\varphi = 5.73 \times 10^4 \frac{1}{LG} \sum_{i=1}^{z} \frac{T_i l_i}{J_i} \tag{10.20}$$

where J is polar moment of inertia of circular cross section of a shaft, $J = \pi d^4/32$, mm⁴;

L –shaft length subjected to torsion, mm;
T_i, l_i, J_i –torque, length and polar moment of inertia of the ith segment in a stepped shaft.

Similar, the torsional rigidity criterion is

$$\varphi \leq [\varphi] \tag{10.21}$$

where $[\varphi]$ is allowable unit length angular deflection that depends on the shaft application. For general transmission shafts, select $[\varphi] = 0.5\text{--}1°\ \text{m}^{-1}$; for precision shafts, select $[\varphi] = 0.25\text{--}0.5°\ \text{m}^{-1}$ [7].

10.3.3 Critical Speed Analysis

Resonance happens when a shaft reaches a critical speed. At the critical speed, the shaft is unstable, with deflections increasing enough to break the shaft. To avoid the resonance of vibrations, critical speeds need to be determined especially for high speed shafts.

A shaft may have the first, second, third … critical speeds. Critical speed is usually referred to as the lowest shaft speed that excites a resonant condition in the system [9]. For a single mass shaft with deflection y_0 shown in Figure 10.4, the fundamental natural frequency, ω_n, is estimated by a spring model, as

$$\omega_n = \sqrt{\frac{k}{m}} = \sqrt{\frac{g}{y_0}} \tag{10.22}$$

and the critical speed, n_c, is

$$n_c = \frac{30}{\pi} \sqrt{\frac{k}{m}} = \frac{30}{\pi} \sqrt{\frac{g}{y_0}} \tag{10.23}$$

where k is shaft stiffness or shaft spring rate. It equals weight divided by deflection $k = mg/y_0$.

For a multiple mass shaft with deflection y_i at each mass w_i shown in Figure 10.5, the critical speed is estimated by [10]

$$n_c = \frac{30}{\pi} \sqrt{\frac{g \sum_{i=1}^{z} w_i y_i}{\sum_{i=1}^{z} w_i y_i^2}} \tag{10.24}$$

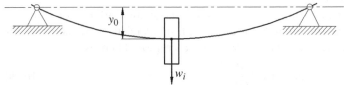

Figure 10.4 Single mass shaft system.

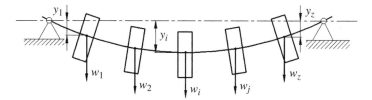

Figure 10.5 Multiple mass shaft system.

w_i – weight of ith mass, N;
y_i – deflection of ith mass centre from centreline of rotation, mm.

A stepped shaft can be first partitioned into segments and placing its weight at the segment centroid and then employing Eq. (10.24) to calculate the critical speed. Detailed derivation of these equations for critical speeds can be found in books on mechanical vibrations [3, 10].

10.4 Design of Shafts

10.4.1 Introduction

Shaft design aims to specify reasonable dimensions to ensure shafts satisfy operational requirements. The design process has much interdependence on the design of mounted elements, such as gears, bearings and so on. Therefore, shaft design must consider the initial analysis and design of these elements simultaneously. Similar to the design of other elements, shaft design involves the load carrying capacity analysis discussed previously and structural design.

Structural design is a flexible and complicated process and depends greatly on specific applications. It involves the specification of shaft geometries so that it is compatible with mounted elements. Many factors, including mounting, locating and manufacturing, need to be considered to ensure a secured location of each element and reliable power transmission. Although there is no absolute rule for shaft structural design, the following sections aim to provide general guidelines. Structural design is an important task in shaft design, as it affects the performance, costs and assembly of shafts.

10.4.2 Materials and Heat Treatments

Considering the potential failure modes introduced in Section 10.2.5, candidate materials for power transmission shafts should have good strength, especially fatigue strength, high stiffness and, in some applications, good wear and corrosion resistance. Other factors, such as cost, weight and machinability, also need to be considered.

Most power transmission shafts are made of low- or medium-carbon steels, either hot-rolled or cold-drawn. Low- or medium-carbon steels are commonly chosen as they are at reasonable price, less sensitive to stress concentration and can be easily heat treated. If higher strength is required, low alloy steels may be selected using appropriate heat treatment, such as quenching or tempering to achieve desired properties. Surface hardening, including carburizing, nitriding and case hardening, can be used to increase strength and wear resistance of shafts. For forged shafts, such as automotive crankshafts, high-strength nodular cast iron is frequently selected because of shock absorption properties and low cost. If shafts operate in a corrosive environment or at an elevated temperature, stainless steel or titanium alloy are required in spite of high costs or great difficulty with fabrication [6].

If strength is a critical issue in shaft design, a higher strength material like heat-treated alloy steel may be a proper selection; on the contrary, if deflection is of principle concern, the size of cross section rather than material strength should be increased to improve rigidity, as stiffness is represented by elastic modulus, which is essentially constant for all steels [2]. Table 10.1 lists commonly used shaft materials and their mechanical properties. Because of variations in composition, heat treatments or formation, data listed in the table are the averages for specimens with diameter less than 100 mm. Designers should consult suppliers or arrange tests for material properties for critical element design.

10.4.3 Design Criteria

As introduced before, the main failure modes in a shaft are fatigue fracture, excessive deflection and critical speed resonance. A good design indicates that a shaft should have enough strength to prevent fracture failure, sufficient rigidity to avoid excessive deflection and acceptable critical speed to prevent resonance.

Generally, most shafts require strength analysis by satisfying strength criteria selected from Eqs. (10.6)–(10.10), depending on stress states. Slender shafts need to check rigidity by satisfying rigidity criteria listed in Eqs. (10.17), (10.18) and (10.21), depending on mounted elements. High speed shafts require critical speed analysis by Eqs. (10.23) or (10.24) to ensure the fundamental natural frequency is significantly above the operating frequency to avoid violent vibration.

10.4.4 Design Procedure and Guidelines

In shaft design practice, the two tasks of shaft design, that is, load carrying capacity analysis and structural design, alternate with each other. A general design procedure is suggested next:

1. Select materials and heat treatments, determine mechanical properties, that is, the ultimate strength, yield strength, endurance limits and so on.
2. Determine the initial minimum diameter of the shaft
3. Design preliminary structure
   - Based on the functional requirement, specify the location of each element to be mounted on the shaft, including bearings;
   - Start from the minimum shaft diameter, determine the diameter and axial length of each segment of stepped shaft according to the width of the hub of mounted elements and the space between them. The diameters of each shaft segment should be integers. The size and tolerance of shaft diameters should be compatible with mounted elements. The overall length of shaft should be kept small to reduce bending moments and deflections;
   - Propose an initial structure of shaft, considering factors about manufacturing and assembly; Specify design details such as fillet radii, shoulder heights, keyseat dimensions and tolerances.

4. Perform force analysis, stress analysis, rigidity analysis and critical speed analysis if required.
5. Finalize structural design by modifying initial design if strength, rigidity and/or critical speed do not meet criteria; repeat previous steps if required.
6. Produce drawings.

10.4.5 Structural Design of Shafts

Structural design determines the shape and dimensions of a stepped shaft. Geometrical variations in a shaft, such as shoulders, grooves and keyseats, rely on the layout of mounted elements. An ideal shaft design should facilitate manufacturing and assembly, as well as location of mating elements. Sufficient strength and rigidity should be guaranteed in shaft design. The following issues greatly affect shaft structure and need to be addressed during the structural design process.

10.4.5.1 Measures to Increase Shaft Strength and Rigidity

The layout of shaft mating elements, for example, gears and bearings, must be specified in the early design stage to facilitate producing torque and bending moment diagrams. Appropriate layout of power transmitting elements could reduce the loads on a shaft and ultimately increase strength. As illustrated in Figure 10.6, the maximum torque on the shaft is $T_1 + T_2$ in Figure 10.6a, while T_1 in Figure 10.6b.

Shafts should be kept as short, stiff and light, as possible. Bearings are better placed on either side of power transmitting elements to provide stable supports and close to the power transmitting elements to minimize bending moments.

Stress concentration greatly affects fatigue strength of shaft. Therefore, geometrical discontinuities, such as shoulders, keyseats and retaining ring grooves, should be away from high stress regions. If this is not possible, use generous radii for fillets and relief grooves and good surface finishes. Local surface enhancement processes, such as shot peening, cold rolling or heat treatment, like carburizing, nitriding, case hardening and so on, are recommended for these regions.

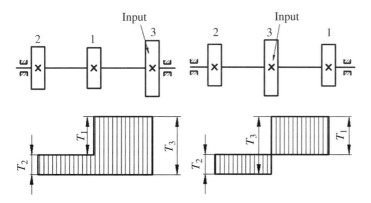

Figure 10.6 Layouts of power transmitting elements on shafts. (a) Poor and (b) good.

10.4.5.2 Locating and Fastening Elements on a Shaft

To prevent relative movement between elements and a shaft, each element must be located accurately on the shaft, both axially and circumferentially and securely held in position during operation to ensure reliable power transmission.

(1) Axially locating and fastening elements on a shaft

Elements must be held in position along shafts, especially for thrust loads producing elements, such as helical or bevel gears, or tapered roller bearings. Accurate axial positioning of gears or bearings requires a shoulder, spacer, locknut and lockwasher, retaining ring, collar and screw, tapered surface and so on.

A shoulder is a change in the shaft diameter against which to locate an element carrying large thrust loads. The use of shaft shoulders is a simple, convenient and reliable method for axially locating elements in a stepped shaft. Shoulders include locating shoulders and non-locating shoulders.

For a locating shoulder, the diameter of shaft shoulder must be sufficient to ensure solid seating, usually with a height of $h = (0.07–0.1)d$. Shoulder rings shown in Figure 10.7a have the same function as shoulders, usually with a width of $b \geq 1.4\ h$. When a locating shoulder supports a bearing, the radial dimension of mating shaft shoulders and housing shoulders must be of sufficient size so as to provide adequate surfaces to support bearings, as illustrated in Figure 10.7b. The desirable shoulder diameters for seating bearings on a shaft and a housing are specified in bearing catalogues or design handbooks [7]. Such specifications not only provide a secure surface against which to locate the bearing, but also offer enough space for pullers (see Figure 11.12) to access bearings for convenient disassembly.

The fillet radius at a shoulder needs to be sized to avoid interference with the chamfer of mating elements [1]. Normally, the fillet radius is the maximum permissible radius on the shaft, smaller than that of fillet or chamfer on the mating elements, that is, $r < R$ in Figure 10.7a, or $r < C$ in Figure 10.7c. A larger radius on the shaft would not permit mating elements to sit tightly against the shoulder. Stress concentration depends on the ratio of two shaft diameters and on the fillet radius produced. Of course, the larger the fillet radius, the smaller the stress concentration.

The height of non-locating shoulder is usually selected as $h = 1–2$ mm for the convenience of manufacturing and assembly.

A spacer is a simple and reliable axial locating element, suitable for positioning adjacent elements with a short interval. The length of wheel hub should be 2–3 mm longer than the mating shaft segment to guarantee a reliable axial location of power transmission elements, as shown in Figure 10.8a. Spacers are not suitable for high speed shafts due to clearance fit between spacer and shaft.

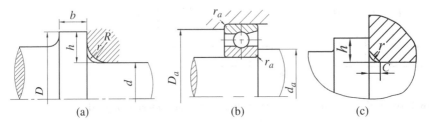

(a)   (b)   (c)

Figure 10.7 Dimensions of locating shoulders.

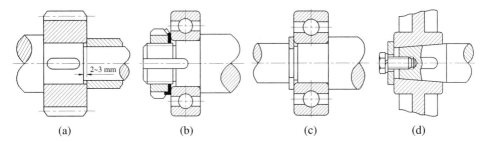

Figure 10.8 Axially locating and fastening elements on shafts.

Locknuts and lockwashers in Figure 10.8b are suitable for locating elements at the end of shaft. It can carry large thrust force, but the screw thread may reduce the strength of shaft. A retaining ring (or snap ring) is frequently used to hold an element in place axially on a shaft or in a housing bore. As shown in Figure 10.8c, a retaining ring is installed in the groove cut in the shaft after the bearing is retained in place. Since the groove can be regarded as two sharp-filleted shoulders positioned close together, stress concentration is fairly high [1]. Therefore, retaining rings are located where stresses are reasonably low.

A tapered surface between shaft and mounted element is often used on the overhanging end of shaft. A collar and screw are then used to lock the mounted element tightly to the shaft, as shown in Figure 10.8d. This approach is convenient for assembly and disassembly and can carry large thrust loads. When thrust loads are small, it may be feasible to rely on press fits to maintain an axial location [2].

(2) Circumferentially locating and fastening elements on a shaft

While transmitting torque and rotatory motion, it is necessary to prevent relative rotation between mounted elements and shafts. Keys, splines, setscrews, pins, interference fit and tapered fit, can be used to circumferentially locating and fastening elements on a shaft.

Power transmission elements like gears, pulleys, sprockets and so on are attached to a shaft by keys or splines to transmit medium to high torques, while setscrews (Figure 3.3e), interference fit (Figure 10.8c) and pins (Figure 4.7b) are used for light service. Both keys and splines can be made with a reasonably loose slip fit to permit the hub to slide axially along the shaft. They are the most effective and economical means for torque transmission. Setscrews are inexpensive, yet not suitable for applications where loosening would impose a safety hazard. Interference fit is commonly used in bearing assembly. However, it has a strict requirement for the manufacturing of the mating surface and it is inconvenient to assemble and disassemble elements.

10.4.5.3   Machinability and Assemblability of Shafts

The structure of shaft should facilitate manufacture, assembly, measurement and maintenance. The dimensions of fillet radius, chamfers, keyways, grooves and suchlike on different shaft segments with similar diameters has to be consistent. The keyways on different segments of a shaft should be on the same generatrix.

A grinding-relief groove should be provided for the grinded shaft segment to prevent the grinding operation from going all the way to the shoulder. Similarly, a relief groove should be designed for screw fabrication, as shown in Figure 10.9.

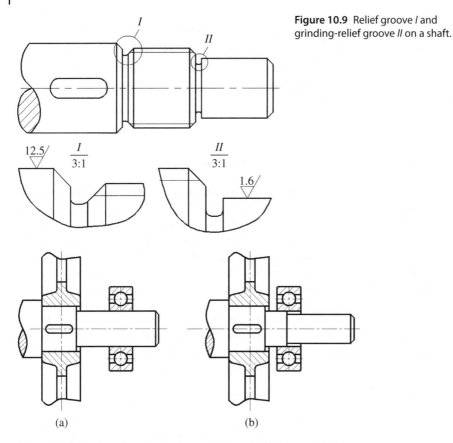

Figure 10.9 Relief groove *I* and grinding-relief groove *II* on a shaft.

Figure 10.10 Design of mating shaft segment length. (a) Poor and (b) better.

When elements are to be press-fit on a shaft, as the bearing shown in Figure 10.10, the length of mating shaft should be shortened and the diameter of the rest shaft segment be slightly reduced to allow the bearing to slide easily over the shaft up to its final position. Such a design reduces manufacturing and assembly cost as it requires a close tolerance only for a short distance.

Assembly sequence also affects shaft structure. During shaft design, several assembly possibilities are normally proposed, compared, ranked and selected. As an example, in Figure 10.11a, the assembly sequence on the left of shoulder ring is left rolling contact bearing 3, bearing cover 2 and half coupling 1. On the right, the assembly sequence is

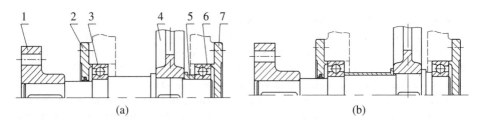

Figure 10.11 Shaft structure comparison.

gear 4, spacer 5, right rolling contact bearing 6 and bearing cover 7. The axial space between bearing cover 2 and half coupling 1 is to provide access space for dismounting bearing cover 2 to provide the bearing with lubricant. Figure 10.11b gives a similar design, yet it requires a longer spacer.

10.4.6 Design Cases

Example Problem 10.1

Design the output shaft III of a double-reduction, helical gear reducer in Figure E10.1. The power to be transmitted by the output shaft is $P_3 = 9.5\,\mathrm{kW}$. The output shaft rotates at a rotational speed of $n_3 = 120\,\mathrm{rpm}$. The helical gear on the output shaft has the number of teeth of $z = 108$ with modulus $m_n = 3\,\mathrm{mm}$, a helix angle of $\beta = 12.4°$ and face width of $B = 50\,\mathrm{mm}$. The safety factor is selected as 1.5. In Figure E10.1, the relevant dimension details are $A = 15\,\mathrm{mm}$, $B_1 = 40\,\mathrm{mm}$, $C = 20\,\mathrm{mm}$, $B_2 = 50\,\mathrm{mm}$ and $S = 8\,\mathrm{mm}$.

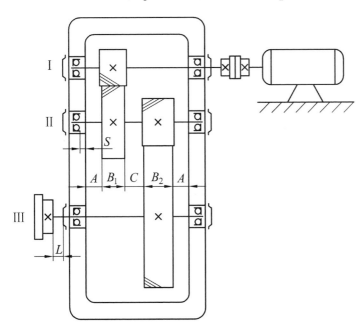

Figure E10.1 Illustration for Example Problem 10.1.

Solution

Steps	Computation	Results	Units
1. Select shaft material and heat treatment	Select medium-carbon steel 45 for the shaft, temper treated. From Table 10.1, the allowable stress is $[\sigma_{-1}] = 172\,\mathrm{MPa}$.	$[\sigma_{-1}] = 172$	MPa

(continued)

Steps	Computation	Results	Units
2. Determine minimum diameter of the shaft	According to Section 10.3.1.1, the allowable shear stress $[\tau]$ for medium-carbon steels is within the range of 20~40 MPa. Select a value of 35 MPa. From Eq. (10.2), the minimum acceptable diameter of the shaft is estimated as $$d \geq \sqrt[3]{\frac{9.55 \times 10^6 \frac{P}{n}}{0.2[\tau]}} = \sqrt[3]{\frac{9.55 \times 10^6}{0.2 \times 35} \times \frac{9.5}{120}} = 47.62\text{mm}$$		

3. Preliminary structural design

(1) Decide the layout of elements on the shaft	Two possible layouts of elements on the shaft are proposed, as shown in Figure 10.11. After comparison, select Figure 10.11a.		
(2) Shaft elements positioning	Axial positioning: The half coupling is positioned by the end cap on the left and the shoulder on the right. The left bearing seats against the shoulder and is located in place by the left bearing cover; while the right bearing by the spacer and the right bearing cover. The gear seats against the shoulder ring on the left and have the spacer to located it in place on the right. Circumferential positioning: Both the gear and half coupling are circumferentially located on the shaft by parallel keys.		
(3) Determine the diameter and length of each segment of the shaft	Shaft segment *I*: The initially estimated minimum diameter of shaft is the location where a coupling is to be installed. The length and diameter of this segment should be compatible with the selected coupling. An important parameter for selecting a coupling is torque. The torque transmitted by the shaft is $$T = 9.55 \times 10^6 \frac{P_3}{n_3} = 9.55 \times 10^6 \times \frac{9.5}{120} = 7.56 \times 10^5 \text{N} \cdot \text{mm}$$ Considering the variation of load during operation, select application factor $K_A = 1.3$ from Table 2.1, then the calculated torque T_{ca} is $$T_{ca} = K_A T = 1.3 \times 7.56 \times 10^5 = 9.83 \times 10^5 \text{N} \cdot \text{mm}$$ Select elastomeric pin coupling LX3 from GB/T 5014-2003 [11], whose nominal torque is 1 250 000 N mm. The aperture of the half coupling $d_I = 48$ mm, the total length of the half coupling is 112 mm and the mating length of hub bore is 84 mm. Therefore, select $d_I = 48$ mm, $L_I = 82$ mm.	$d_I = 48$ $L_I = 82$	mm mm

(*continued*)

Steps	Computation	Results	Units
	Shaft segment *II*:	$d_{II} = 58$	mm
	To meet the requirement of axial location of the half coupling, the right side of shaft segment *I* needs a shoulder with a height of approximate $h = (0.07{\sim}0.1)\,d$. Therefore, the diameter of segment *II* is $d_{II} = 58$ mm.	$L_{II} = 50$	mm
	The total width of the bearing cover is 20 mm by structural design. According to the requirement of assembly and disassembly of the bearing cover to facilitate grease addition for the left bearing, select $L_{II} = 50$ mm.		
	Shaft segment *III*:	$d_{III} = 65$	mm
	According to $d_{II} = 58$ mm, a pair of angular contact ball bearing 7313C is tentatively selected to carry both radial and axial forces, with dimensions of $d \times D \times T = 65$ mm $\times$ 140 mm $\times$ 33 mm [12]. Therefore, $d_{III} = d_{VII} = 65$ mm, $L_{III} = 33$ mm.	$L_{III} = 33$	mm
	Shaft segment *VI*:	$d_{VI} = 70$	mm
	Select the diameter of the shaft mating with the gear as $d_{VI} = 70$ mm. A spacer is used between the right side of gear and the right bearing. Since the width of the gear hub is $B_2 = 50$ mm, select $L_{VI} = 48$ mm.	$L_{VI} = 48$	mm
	Shaft segment *V*:	$d_V = 80$	mm
	A shoulder ring is used to locate the gear. The height of the shoulder ring should be $h > 0.07d$. Therefore, select $h = 5$ mm. The diameter of the shoulder ring is $d_V = 80$ mm. The width of the should ring should be $b \geq 1.4\,h$, select $L_V = 10$ mm.	$L_V = 10$	mm
	Shaft segment *IV*:	$d_{IV} = 77$	mm
	A shoulder is used for locating the left bearing axially. The height for the location shoulder for bearing 7313C is specified by bearing catalogues as 6 mm [12]. Therefore, $d_{IV} = 77$ mm.	$L_{IV} = 73$	mm
	Refer to Figure E10.1, $L_{IV} = S + A + B_1 + C - L_V = 8 + 15 + 40 + 20 - 10 = 73$mm		
	Shaft segment *VII*:	$d_{VII} = 65$	mm
	The right angular contact ball bearing 7313C is mounted on shaft segment *VII*. Refer to Figure , therefore, $L_{VII} = T + S + A + B_2 - L_{VI} = 33 + 8 + 15 + 50 - 48 = 58$mm	$L_{VII} = 58$	mm
	Thus, the diameters and lengths of each shaft segment are tentatively decided.		

(*continued*)

Steps	Computation	Results	Units
(4) Circumferentially locating elements on the shaft	According to d_{VI} and Table 4.1, the dimension of a key is selected as $b \times h \times l = 20 \times 12 \times 40$ mm. Meanwhile, to ensure better concentricity between the gear hub and shaft, the fit between the gear hub and the shaft is selected as H7/n6. Similarly, a parallel key used to connect the half coupling and the shaft is selected as $14 \times 9 \times 70$ mm, and the fit is H7/k6. A light interference fit is selected between the bearing and shafts. The tolerance of shaft diameter is selected as m6.	Key for gear $20 \times 12 \times 40$ H7/n6 Key for coupling $14 \times 9 \times 70$ H7/k6	
(5) Determine the shaft fillets and chamfers	Select the shaft chamfer as 2 mm. The fillet radii of shaft shoulders are shown in Figure E10.2. The initially designed shaft structure is shown in Figure E10.2.		
4. Force analysis			
(1) Determine the forces acting on the gear	Since the reference diameter of the low-speed stage gear is $$d_2 = \frac{m_n z}{\cos \beta} = \frac{3 \times 108}{\cos 12.4^\circ} = 331.739 \text{mm}$$ $$F_t = \frac{2T}{d_2} = \frac{2 \times 7.56 \times 10^5}{331.739} = 4558\text{N}$$ Therefore $$F_r = \frac{F_t \tan \alpha_n}{\cos \beta} = \frac{4558 \times \tan 20^\circ}{\cos 12.4^\circ} = 1699\text{N}$$ $$F_a = F_t \tan \beta = 4558 \times \tan 12.4^\circ = 1002\text{N}$$ The directions of tangential force F_t, radial force F_r and axial force F_a are shown in Figure E10.3.	$F_t = 4558$ $F_r = 1699$ $F_a = 1002$	N N N
(2) Forces on the shaft	Forces acting on the gear are assumed to act at the midpoint of the face width. Select the value for the effective reaction point as $a = 27.4$ mm from the bearing catalogues for bearing 7313C [12]. According to the previous structural design, decide the spans of $L_1 = 118.4$ mm, $L_2 = 113.6$ mm, $L_3 = 53.6$ mm. Resolve the forces in two perpendicular planes, usually horizontal plane and vertical plane. Perform free body diagram analysis for the reaction forces at all supporting bearings in each plane. Therefore, Reaction forces in horizontal plane are $R_{H1} = 1461$ N, $R_{H2} = 3097$ N Reaction forces in horizontal plane are $R_{V1} = 1539$ N, $R_{V2} = 160$ N	$R_{H1} = 1461$ $R_{H2} = 3097$ $R_{V1} = 1539$ $R_{V2} = 160$	N N N N

(continued)

Steps	Computation	Results	Units
(3) Bending moment diagram and torque diagram of the shaft	Produce bending moment diagrams to determine the distribution of bending moments in the shaft. Moment in horizontal plane $M_H = 165\,999\,\text{N mm}$ Moment in vertical plane $M_{V1} = 174\,830\,\text{N mm}, M_{V2} = 8576\,\text{N mm}$ Total moment M $M_1 = \sqrt{165999^2 + 174830^2} = 241083\,\text{N·mm}$ $M_2 = \sqrt{165999^2 + 8576^2} = 166220\,\text{N·mm}$ Torque T $T = 756\,000\,\text{N mm}$ The bending moment diagram and torque diagram of the shaft is shown in Figure 10.3.	$M_H = 165\,999$ $M_{V1} = 174\,830$ $M_{V2} = 8576$ $M_1 = 241\,083$ $M_2 = 166\,220$ $T = 756\,000$	N mm N mm N mm N mm N mm N mm

5. Strength analysis

Steps	Computation	Results	Units
(1) Check the shaft strength by the combination of torque and bending moment	Check the strength at dangerous cross section I that bears the greatest bending moment and torque. Since the shaft rotates in one direction, the torsional shear stress is regarded as a repeated stress, select $\alpha = 0.6$. Calculate the stress at the dangerous cross section according to the equivalent moment as $$\sigma_{ca} = \frac{\sqrt{M_1^2 + (\alpha T)^2}}{W} = \frac{\sqrt{241083^2 + (0.6 \times 756000)^2}}{0.1 \times 70^3}$$ $= 14.98\,\text{MPa}$ Since $\sigma_{ca} < [\sigma_{-1}]$, the shaft is safe.	$\sigma_{ca} < [\sigma_{-1}]$	
(2) Check fatigue strength of the shaft	Fatigue strength analysis is performed at locations where diameter changes, the torque and bending moment are large and stress concentration occurs. Therefore, we check fatigue strength at the right side of the cross section H. Bending moment M and bending stress are: $$M = M_1 \frac{113.6 - 25}{113.6} = 188028\,\text{MPa}$$ $$\sigma = \frac{M}{W} = \frac{188028}{0.1 \times 70^3} = 5.48\,\text{MPa}$$ Torque T and torsional shear stress are: $T = 756\,000\,\text{N mm}$ $$\tau = \frac{M}{W_T} = \frac{756000}{0.2 \times 70^3} = 11.0\,\text{MPa}$$ Axial force F and tensile stress are: $F_a = 1002\,\text{N}$ $$\sigma = \frac{F}{\frac{\pi}{4}d^2} = \frac{1002}{\frac{\pi}{4} \times 70^2} = 0.26\,\text{MPa}$$ From Table 10.1, we have $\sigma_b = 600\,\text{MPa}$, $\sigma_{-1} = 240\,\text{MPa}$, $\tau_{-1} = 140\,\text{MPa}$.	$S_{ca} > [S]$	

(continued)

Steps	Computation	Results	Units
	Select the mean stress influence factor as $\psi_\sigma = 0.1$ and $\psi_\tau = 0.05$ by referring to Chapter 2. The detailed calculation of combined influence factors can refer to Example Problem 2.2 as $K_\sigma = 2.95$, $K_\tau = 2.33$. Therefore, the safety factors on right side of cross section H are:		

$$S_\sigma = \frac{\sigma_{-1}}{K_\sigma \sigma_a + \psi_\sigma \sigma_m} = \frac{240}{2.95 \times 5.48 + 0.1 \times 0.26} = 14.82$$

$$S_\tau = \frac{\tau_{-1}}{K_\tau \tau_a + \psi_\tau \tau_m} = \frac{140}{2.33 \times \frac{11}{2} + 0.05 \times \frac{11}{2}} = 10.70$$

From Eq. (2.50)

$$S_{ca} = \frac{S_\sigma S_\tau}{\sqrt{S_\sigma^2 + S_\tau^2}} = \frac{14.82 \times 10.70}{\sqrt{14.82^2 + 10.70^2}} = 8.68 > 1.5$$

So the strength on right side of cross section H is enough.

| 6. Produce a working drawing of the shaft | Specify the final dimensions of the shaft considering design details such as tolerances, fillet radii, shoulder heights and keyseat. Produce a working drawing, as shown Figure E10.4. | | |

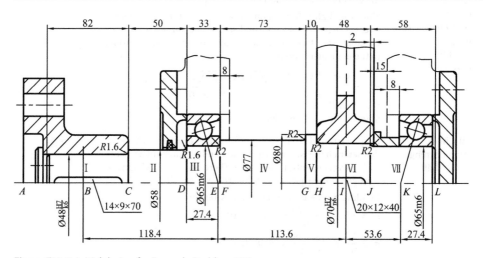

Figure E10.2 Initial design for Example Problem 10.1.

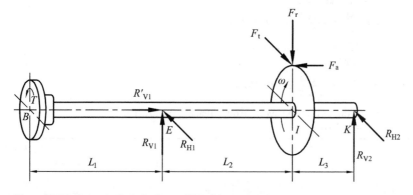

Figure E10.3 Force analysis for Example Problem 10.1.

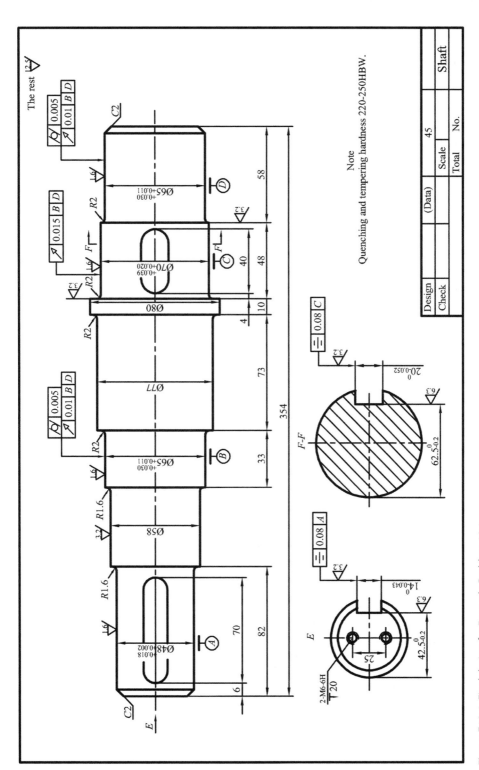

Figure E10.4 Final design for Example Problem 10.1.

References

1 Mott, R.L. (2003). *Machine Elements in Mechanical Design*, 4e. Prentice Hall.
2 Budynas, R.G. and Nisbett, J.K. (2011). *Shigley's Mechanical Engineering Design*, 9e. New York: McGraw-Hill.
3 Juvinall, R.C. and Marshek, K.M. (2011). *Fundamentals of Machine Component Design*, 5e. New York: Wiley.
4 Pu, L.G. and Ji, M.G. (2006). *Mechanical Design*, 8e. Beijing: Higher Education Press.
5 Gere, J.M. and Timoshenko, S.P. (1996). *Mechanics of materials*, 4e. CL Engineering.
6 Collins, J.A. (2002). *Mechanical Design of Machine Elements and Machines: A Failure Prevention Perspective*, 1e. New York: Wiley.
7 Wen, B.C. (2015). *Machine Design Handbook*, 5e, vol. 3. Beijing: China Machine Press.
8 Wen, B.C. (2015). *Machine Design Handbook*, 5e, vol. 1. Beijing: China Machine Press.
9 Hindhede, U., Zimmerman, J.R., Hopkins, R.B. et al. (1983). *Machine Design Fundamentals: A Practical Approach*. New York: Wiley.
10 Géradin, M. and Rixen, D.J. (2015). *Mechanical Vibrations: Theory and Application to Structural Dynamics*, 3e. New York: Wiley.
11 GB/T 5014-2003 (2003). *Pin Coupling Elastomer*. Beijing: Standardization Administration of the People's Republic of China.
12 Gong, Y.P., Tian, W.L., Zhang, W.H., and Huang, Q.B. (2008). *Project Design of Mechanical Design*. Beijing: Science Press.

Problems

Review Questions

1   According to the load a shaft carries, shafts include three types. What types are the front shaft, middle shaft and back shaft of a bicycle?

2   How should one decide on the input shaft and output shaft in a gear reducer?

3   A transmission shaft made of medium-carbon steel has an equivalent diameter of d_v. The shaft is supported by a pair of bearings with a spur gear mounted in the middle. The maximum deflection and twist angles are y and φ, respectively. Would the following measures reduce the deflection and twist angle? And which method is the most effective?
   (1) Increase the equivalent diameter to $2d_v$
   (2) Reduce 50% of the span
   (3) Use alloy steel

4   Why are most transmission shafts stepped? What should be considered while deciding the diameter and length of each segment of a shaft?

5 List the factors that affect the fatigue strength of a shaft. If a shaft does not have sufficient fatigue strength, what measures can be taken to satisfy the strength requirement?

Objective Questions

1 When a shaft carries combined bending and torsional stresses, a factor α is introduced in the stress calculation because _____.
   (a) a deviation exists between theoretical and experimental results
   (b) the torsional shear stress maybe not a completely reversed stress
   (c) the bending stress maybe not a completely reversed stress
   (d) the shaft has stress concentration

2 The purpose of increase a fillet radius in shoulder design is _____.
   (a) to ensure a reliable circumferential location
   (b) to ensure a reliable axial location
   (c) to facilitate manufacture
   (d) to reduce stress concentration and to improve fatigue strength of shaft

3 Which of the following could *not* effectively improve the rigidity of a shaft? _____
   (a) Use high-strength alloy steels.
   (b) Change the structure of shaft.
   (c) Change the diameter of shaft.
   (d) Change the bearing position of shaft.

4 In Figure P10.1, the width of the element installed on the shaft is B. The length of the mating shaft segment is L. In order to locate the element axially on the shaft securely, the relationship between B and L should be_____
   (a) $B < L$
   (b) $B = L$
   (c) $B > L$
   (d) there is no relationship between B and L

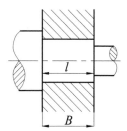

Figure P10.1 Illustration for Objective Question 4.

Calculation Questions

1   The total weight of a railway freight car and the goods it carries is $G = 800\,\text{kN}$. The carriage is supported by eight wheels installed on four axles and the load acting on each axle is shown in Figure P10.2. The distance between the loading point and rail centre line is $s = 220\,\text{mm}$. Select a load factor of $K_A = 1.2$. The axle is made from tempered medium-carbon steel, with an allowable stress of $[\sigma_{-1b}] = 75\,\text{MPa}$. Decide the diameter of the axle at wheel installation.

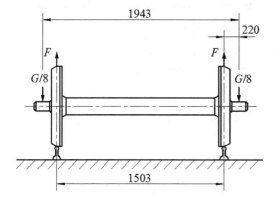

Figure P10.2 Illustration for Calculation Question 1.

2   A machined shaft shoulder is shown in Figure P10.3. The small diameter d is 60 mm, the large diameter D is 72 mm and the fillet is $r = 6\,\text{mm}$. The shoulder carries a bending moment of $103\,000\,\text{N}\,\text{mm}$ and a steady torsion moment of $741\,000\,\text{N}\,\text{mm}$. The heat-treated carbon steel shaft has an ultimate strength of $\sigma_b = 600\,\text{MPa}$, endurance limit in tension $\sigma_{-1} = 275\,\text{MPa}$ and in shear $\tau_{-1} = 155\,\text{MPa}$, respectively. Determine the fatigue safety factor of the shaft.

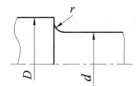

Figure P10.3 Illustration for Calculation Question 2.

3   The shaft in Figure P10.4 is driven by a driver B at a speed of $n = 300\,\text{r}\,\text{min}^{-1}$. The input power is $P_B = 10\,\text{kW}$. The output power from the driven gears are $P_A = 6\,\text{kW}$, $P_C = 4\,\text{kW}$, respectively. The allowable shear stress of the shaft is $[\tau] = 40\,\text{MPa}$ and the allowable unit length angular deflection is $[\varphi] = 1°\,\text{m}^{-1}$. The shear modulus of the shaft is $G = 8 \times 10^4\,\text{MPa}$. Decide the shaft diameter by strength and rigidity analysis, respectively.

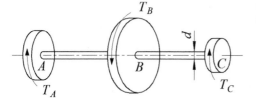

Figure P10.4 Illustration for Calculation Question 3.

4   A shaft with a diameter of 60 mm rotates at 1250 rpm. Three discs are mounted on the shaft, with weight of $w_1 = 10$ N, $w_2 = 25$ N and $w_3 = 7$ N, respectively, as shown in Figure P10.5. The deflection at the corresponding mounting locations are $y_1 = 0.305$ mm, $y_2 = 0.457$ mm and $y_3 = 0.178$ mm, respectively. Estimate the fundamental critical speed and comment on its acceptability.

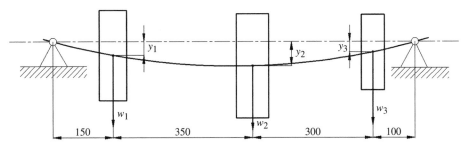

Figure P10.5 Illustration for Calculation Question 4.

Design Problems

1   Design the output shaft in the gear reducer in Figure P10.6. The rotational speed of the shaft is 80 rpm. The power transmitted by the shaft is $P = 3.15$ kW. The parameters of the gear are: $m_n = 3$ mm, helix angle $\beta = 12°$, the number of teeth $z = 94$ and the face width $b = 72$ mm.

Figure P10.6 Illustration for Design Problem 1.

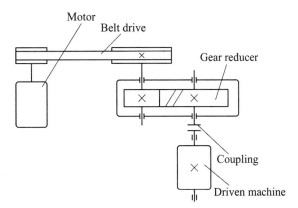

2   A spur pinion and a helical gear is mounted on a shaft carried by two bearings in Figure P10.7. The radial loads on both gears are in the same plane and are 3000 N for the spur pinion and 1000 N for the helical gear. A thrust load of 500 N on the helical gear is carried by the right bearing. The shaft rotates at a speed of 1000 rpm. Design the shaft. Draw a sketch to scale of the shaft showing all proposed dimensions.

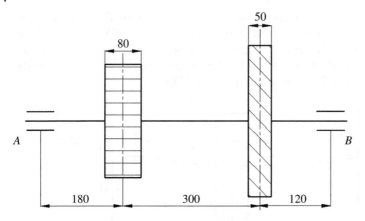

Figure P10.7 Illustration for Design Problem 2.

3   In a double-reduction gear reducer in Figure P10.8, the input shaft is driven by a motor through a belt drive, transmitting a power of 25 kW at a speed of 750 rpm. Gears in the reducer have parameters as $z_1 = 21$, $z_2 = 77$, $m_{n1} = 3$ mm, $\beta_1 = 11.478°$ and face width $b_2 = 80$ mm; $z_3 = 23$, $z_4 = 74$, $m_{n3} = 4$ mm, $\beta_3 = 14.07°$ and face width $b_4 = 120$ mm. Design the countershaft with an application factor of 1.5 by performing the following tasks:

(1) Propose a shaft layout, including means to locate the gears and bearings;
(2) Perform a force analysis and generate shear and bending moment diagrams;
(3) Determine the fatigue and static strength at the critical locations of the shaft;
(4) Specify diameters and length of each segment of the shaft, sketch the shaft to scale showing all fillet sizes, keyways, shoulders, diameters and so on;
(5) Check the deflection at the gear, and the slopes at the gear and the bearings;
(6) Make appropriate changes if either stresses or deflections exceed recommended limits.

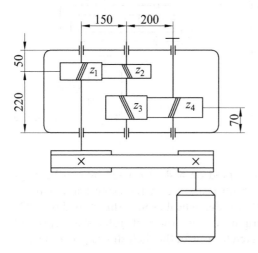

Figure P10.8 Illustration for Design Problem 3.

Structure Design Problems

1   Find the errors in the design in Figure P10.9 and correct them.

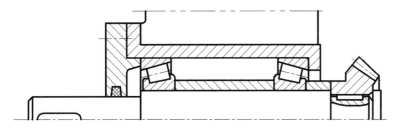

Figure P10.9 Illustration for Structure Design Problem 1.

2   Please mark the errors in Figure P10.10 and correct them.

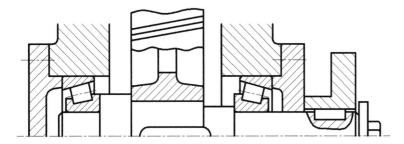

Figure P10.10 Illustration for Structure Design Problem 2.

3   A spur gear is mounted on a shaft that is supported by a pair of deep groove ball bearings. Each bearing retains one directional movement. The gear is lubricated by lubricant oil and bearings by grease. Complete the design of the shaft system.

Structure Design Problems

1. Sketch the details the design for Figure 7.67 and 7.68 assumption and form.

Figure 7.68 Structure for Structure Design Problem 1

2. Determine the force on Figure 7.69, 7.70 and correct them.

Figure 7.70 Structure for Structure Design Problem 2

3. A pump rotor is mounted on a shaft that is supported on a pair of deep groove ball bearings. It on necessary, analyze the thrust load movement. The search is lubricated by forced through oil bearing by grease. Complete the analysis of the shaft system.

11

Rolling Contact Bearings

Nomenclature

A	axial load, N	L_{10}	rated life at reliability of 90%, rev
C	basic dynamic load rating, N	L_{10h}	rated life in hours at reliability of 90%, h
C_0	basic static load rating, N	L_n	bearing life at diffident reliability, h
e	factor	$[L_h]$	allowable life in hours, h
F_a	external axial load, or thrust force, N	N_i	reaction force, N
		n	rotational speed, rpm
F_r	external radial load, or radial force, N	P	equivalent dynamic load, N
		P_0	equivalent static load, N
f_p	load factor	P_i	resolved radial load component, N
f_t	temperature factor	R	radial load on bearing, N
H	total operating hours under variable loads, h	r_1	radius of bore, mm
		r_2	radius of rolling elements, mm
		r_3	radius of outer ring, mm
L_h	bearing life in hours, h	r_4	radius of retainer, mm

Analysis and Design of Machine Elements, First Edition. Wei Jiang.
© 2019 John Wiley & Sons Singapore Pte. Ltd. Published 2019 by John Wiley & Sons Singapore Pte. Ltd.
Companion website: www.wiley.com/go/Jiang/analysis_of_machine_elements

S induced thrust load, N

S' appended thrust load, N

S_0 static safety factor

S_i resolved axial load component, N

t temperature, °C

X dynamic radial load factor

X_0 static radial load factor

Y dynamic axial load factor

Y_0 static axial load factor

α contact angle, deg.

α_1 life-adjustment factor for reliability

ε load-life exponent

θ angular deviation, deg.

ω_1 angular velocity of inner ring, rad s^{-1}

ω_2 angular velocity of rolling elements, rad s^{-1}

ω_3 angular velocity of outer ring, rad s^{-1}

ω_4 angular velocity of retainer, rad s^{-1}

Subscripts

1, 2   bearing number

m average value

11.1 Introduction

11.1.1 Applications, Characteristics and Structures

Rolling contact bearings (or rolling bearings) are standardized machine elements widely used in various machines. They support rotating shafts while permitting relative motion between two elements [1]. Rolling contact bearings have low starting and good operating friction, and are ideal for applications with high starting loads, such as in vehicles, trains, aeroplanes and mobile equipment in general [2].

Compared with sliding bearings, rolling contact bearings require minimum lubrication and less axial space. They have high efficiency and reliability, and minimum maintenance requirements. Besides, rolling contact bearings can be preloaded to eliminate internal clearance, which is crucial for high precision rotation and fatigue life. As standardized products, rolling bearings are interchangeable among manufacturers. Nevertheless, high rotating speeds may lead to rapid accumulation of fatigue cycles and high centrifugal force on rolling elements [2]. The noise level is usually high due to poor damping capability of rolling bearings.

A typical rolling contact bearing composes of an inner ring (race), an outer ring (race), rolling elements and a retainer. The inner ring and outer ring are called cone and cup, respectively, for a tapered roller bearing [2]. Figure 11.1 shows the structure of a single-row, deep-groove ball bearing. The inner ring is pressed onto a shaft with a slight interference fit to ensure that it rotates with the shaft. The outer ring is usually stationary and is held by the housing of machine. The inner ring and outer ring form a narrow, dual, circular track within which rolling elements roll during operation [2]. The presence of rolling elements allows a low friction rotation of shafts. Typical rolling elements include spherical balls, cylindrical, tapered or spherical rollers, needles and so on. The retainer, also called a cage or separator, is used to keep rolling elements separated and evenly spaced around the raceway to prevent them contact with each other during operation.

When a rolling contact bearing carries a load, the load is exerted on a small area where rolling elements contact with the inner or outer rings. The resultant contact stresses are usually quite high, regardless of bearing type. To withstand high stress and to resist wear, rolling elements and races are normally made from hard, high strength steel or ceramic. The most widely used bearing steel is high carbon chromium steels, including GCr15

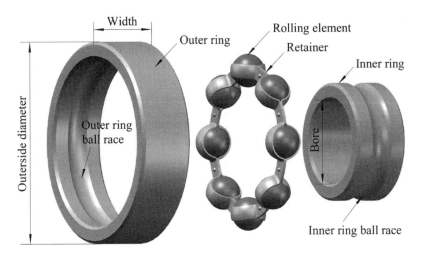

Figure 11.1 The structure of a single-row, deep-groove ball bearing.

steel (or AISI 52100) and carburized, case-hardened G20CrNiMo alloy steel (or AISI 3310, 4620 and 8620) [3]. The bearing steels are through hardened or case hardened to a hardness of 58–65 HRC. Low carbon steel, alloy steel and copper alloy, are usually used for retainers.

11.1.2 Characteristic Factors of Rolling Contact Bearings

11.1.2.1 Internal Clearance

Virtually all rolling bearings are designed with a specific internal clearance, either radial clearance or axial clearance. The radial clearance is measured normally to the bearing axis between the raceway and rolling elements, as shown in Figure 11.2a. This clearance changes with the expansion or contraction of bearing rings. The axial clearance is the distance that one ring can move relative to the other in the axial direction, as shown in Figure 11.2b. The total internal clearance is the amount that one ring can be displaced relative to the other ring, either radially or axially. The internal clearance provides free rotation of rolling elements and compensation for thermal expansion.

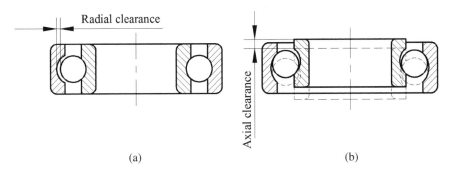

(a)   (b)

Figure 11.2 Internal clearances.

11.1.2.2 Contact Angle α

Bearings are manufactured to support a rotating shaft, taking radial loads, axial (or thrust) loads or their combinations. Correspondingly, bearings include radial bearings, thrust bearings and angular contact bearings, depending on the loads they carry. Radial loads act towards the centre of bearing along a radius, while axial loads act parallel to the axis of bearing. Both radial loads and axial loads are produced by power transmission elements on the shafts, such as gears, worms and wormgears, pulleys and sprockets.

When a bearing carries a load, an angle is formed between the radial direction and the normal line through the contact point between a rolling element and the raceway of the outer ring. This is contact angle. The larger the contact angle, the higher the axial load carrying capacity of the bearing. Figure 11.3a,b shows radial bearings that have contact angles of 0°. Figure 11.3c,d shows thrust bearings that have contact angles in a range of 45°–90°. Figure 11.3e,f illustrates angular contact bearings that have contact angles from 0° to 45°.

11.1.2.3 Angular Deviation θ

Misalignment refers to the angular deviation of the shaft axis from the bearing axis, as shown in Figure 11.4. Misalignment may be caused by shaft deflection or inaccurate assembly and may cause improper meshing of mounted gears. An excellent rating for misalignment indicates that a bearing can accommodate up to 4.0° of angular deviation [1].

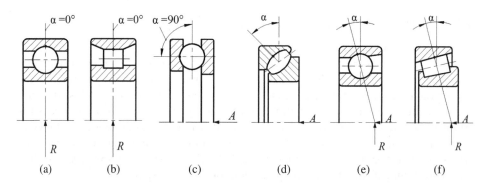

Figure 11.3 Contact angles of different types of bearings.

Figure 11.4 Angular deviation.

11.1.3 Types of Rolling Contact Bearings and Their Selection

11.1.3.1 Classification of Rolling Contact Bearings

Rolling contact bearings can be classified either by the shape of rolling elements or by the load a bearing carries. According to the shape of rolling elements, rolling contact bearings include ball bearings and roller bearings, both of which have many variants within the divisions. Ball bearings are generally nonseparable, that is, the balls, retainer and rings are installed as an assembly [4]. They operate with less friction than roller bearings and are suitable for higher speeds. Roller bearings are distinguished by the design of rollers and raceways. They are usually separable, that is, the rollers and retainer may be not permanently assembled with one of rings [4]. Roller bearings have rolling contact along a line rather than at a point like ball bearings and therefore are suitable for heavier loads.

As described before, according to the loads they carry, rolling contact bearings can be classified as radial bearings, thrust bearings and angular contact bearings.

11.1.3.2 Types of Rolling Contact Bearings

Single-Row, Deep-Groove Ball Bearings, 60000
    Single-row, deep-groove ball bearings (Figure 11.3a) are the most widely used rolling contact bearings because they can support both radial loads and moderate axial loads in either direction. Further, these bearings operate smoothly over a wide speed range and are relatively inexpensive.

    When a deep-groove ball bearing carries a load, the load passes across the side of groove, through the ball, to the opposite side of outer ring and finally to the housing [1]. The contact between a ball and raceway is theoretically a point, but actually a small circular area due to deformation leading to high local contact stresses. Compared with roller bearings, deep grove ball bearings operate at higher speeds yet carry less load.

Cylindrical Roller Bearings, N0000
    Cylindrical roller bearings (Figure 11.3b) are designed for heavy duty applications. The contact between a roller and races is theoretically a line, but practically a small rectangular area because of deformation, resulting in lower contact stresses compared to those for the equivalent-sized ball bearings. Even heavier radial load support can be achieved with double-row roller bearings.

    In contrast with ball bearings, the axial load carrying capacity of a cylindrical roller bearing is usually poor. In a cylindrical roller bearing, rollers are located axially by flanges on the inner, outer or both raceways. The inner ring and outer ring are usually made axially separable. Therefore, a roller bearing can readily accommodate small axial displacements of a shaft relative to the housing due to differential thermal expansion.

Needle Bearings (NAs), NA0000
    Needle bearings can be regarded as a special case of cylindrical roller bearings in which rollers have a far smaller diameter. They are often used without an inner ring, running directly on hardened shafts. Needle bearings are very useful in limited radial space and at low operating speeds. As with other roller bearings, axial load carrying capacity and misalignment capabilities of needle bearings are poor.

Angular Contact Bearings, 70000, 30000

Angular contact ball bearings (70000, Figure 11.3e) carry both radial loads and unidirectional axial loads, while tapered roller bearings (30000, Figure 11.3f) are designed to take substantial radial loads, axial loads and their combinations compared with angular contact ball bearings. Both bearings generate an induced thrust load even though no external axial load is present. To counteract axial loads, these bearings are usually employed in pairs, mounted in opposition either face to face or back to back. Tapered roller bearings are separable and highly suited for combined radial and axial loads.

Thrust Ball Bearings, 50000

Thrust ball bearings carry axial loads acting in one direction (51000, Figure 11.3c) or in both directions (52000). Several types of thrust bearings are commercially available. Shaft speeds must be low, otherwise the centrifugal force on the bearing becomes unacceptable.

Self-aligning bearings, 10000, 20000, 29000

Self-aligning bearings are made to compensate for an appreciable misalignment between shaft and housing due to shaft deflections, mounting inaccuracies or other causes commonly encountered in assembly or in service. A double-row of self-aligning ball bearings (10000) consists of a double row of balls running on a circumferential inner raceway. The outer raceway is spherically ground to tolerate substantial angular misalignment of shaft. A double row of self-aligning roller bearings (20000, Figure 11.4) has a similar structure, but with a greater radial load carrying capacity compared to a self-aligning ball bearing. Self-aligning thrust bearings (29000, Figure 11.3d) are a better choice when both self-aligning capability and combined radial and axial load carrying ability are required.

Extensive descriptions of many types of bearings are available in manufacturers' catalogues or references [6, 7] (http://www.machinedesign.com).

11.1.3.3 Bearing Type Selection

While selecting bearing types, load carrying capacity, speed limitation, misalignment capability, space allowance and cost are the factors that need to be considered and weighted.

The load magnitude and direction a bearing will take is the main consideration in bearing selection. Generally, roller bearings have a higher load carrying capacity than comparably sized ball bearings. If loads are mainly radial loads, deep-groove ball bearings (60000), cylindrical roller bearings (N) and needle bearings (NA) can be selected, with an increasing radial load carrying capacity. If loads are mainly axial loads, thrust bearings are selected. Thrust ball bearings (50000) are used for small axial loads, while thrust roller bearings (29000) are used for great axial loads. If both radial and axial loads are to be taken, deep-groove ball bearings (60000), angular contact ball bearings with small contact angles (70000A, 70000AC), or tapered roller bearings (30000) can be selected for relatively small axial loads, while angular contact ball bearings with a large contact angle (70000B), tapered roller bearings with a large contact angle (30000B) or the combination of radial and thrust bearings are preferred for greater axial loads.

Most catalogues list limiting speeds for each bearing. The limitation is for linear surface speed rather than rotating speed; hence, small bearings can operate at higher rotating speeds than large bearings [4]. Exceeding these limits may result in excessively high operating temperatures due to friction. Generally, point contact in ball bearings leads to lower friction compared to line contact in roller bearings. Thus, ball bearings can operate at higher speeds. Besides, a bearing has a lower limiting speed as loads increase.

Self-aligning ball or roller bearings are selected if excessive shaft deflection or misalignment between shaft and housing needs to be accommodated. Needle bearings or light series bearings are used in cases where space is limited. As for the cost, ball bearings are generally less expensive than roller bearings.

11.1.4 Designation of Rolling Contact Bearings

Rolling contact bearings are standardized products ready for immediate installation. They are standardized by American Bearing Manufacturers Association (ABMA), American National Standards Institute (ANSI), Standardization Administration of the People's Republic of China (SAC), International Organization for Standardization (ISO) and other organizations to facilitate manufacturing and selection for the bearing industry.

Organizations or manufacturers adopt different identification codes to represent rolling contact bearings [6]. Designers need to consult different standards and bearing manufacturer catalogues for corresponding identification codes and other detailed information [6, 7] (http://www.machinedesign.com). Due to limited space, this book only introduces the basic and the most commonly used part of bearing codes specified by the SAC.

The code specified by SAC in the standard GB/T272-1993 [8] is comprised of a pre-code, basic code and post-code, which are represented by letters and numbers. The pre-code indicates bearing components by letters. For example, L represents separable rings of separable bearings.

Basic code indicates the type and size of bearings. It is composed of 5 digits, representing the type code, width series code, diameter series code and bore dimension series code, respectively. The first digit is type code which specifies the bearing types. The second is width code, indicating a series of different widths for the same bore diameter, with 0 for the narrow series, 1 for normal series, 2 for wide series and 3 for the extra wide series. It is usually omitted if a bearing is of a normal width. Following that, diameter code designates a series of different outside diameters for the same bore diameter, with 100 to indicate the extra-light series, 200 for light, 300 for medium and 400 for the heavy series. Figure 11.5a,b shows an example of width and diameter series, respectively, of 100 mm bore of deep-groove ball bearings from GB/T 276-2013 [9]. Most bearings are manufactured with nominal dimensions in metric units, and the

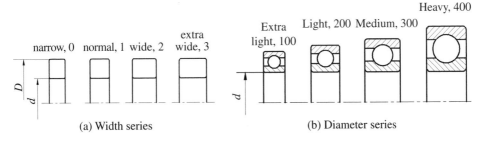

Figure 11.5 Diameter and width series for rolling contact bearings.

last two digits is the bore dimension code indicating the size of bore. For bore sizes of 04 and above, the nominal bore dimension is five times the bore dimension code in millimetres. For bore sizes of 00, 01, 02 and 03, the nominal bore dimensions are 10, 12, 15 and 17 in millimetres, respectively.

Post-code uses letters and numbers to designate structure, tolerance and material of bearings. Items in post-codes are in the sequence of internal structure code, seal and shield structure code, retainer and its material code, special bearing material code, tolerance class code, clearance code and so on. Here, we just introduce some of the most commonly used codes.

Internal structure code indicates the same type of bearings with different structures. For example, angular contact ball bearings with contact angles of 15°, 25° and 40° use C, AC and B to indicate the differences and a tapered roller bearing with contact angles of 27–30° uses B to indicate the differences.

Manufacturing tolerances are extremely crucial for precise operation and have been standardized. Six tolerance classes are commonly used in the bearing industry to accommodate the needs of a wide variety of equipment using rolling contact bearings. From fine to coarse tolerances, tolerance class codes are $P2$, $P4$, $P5$, $P6X$, $P6$ and $P0$, respectively. $P0$ is the ordinary tolerance, adequate for most normal applications, and could be omitted in bearing code.

Generally, internal clearances are designated from the tightest clearance $C1$ through to the loosest or largest clearance $C5$. The normal clearance is 0, which sits between $C2$ and $C3$. It is worth noting that if a bearing clearance is not explicitly stated, it is assumed to have a normal clearance.

As an example, bearing code 6308 represents a medium series deep-groove ball bearing with the bore of 40 mm, ordinary tolerance and normal clearance. As another example, bearing code 7211C/P5 indicates a light series angular contact ball bearing with the bore of 55 mm, contact angle of 15°, class 5 tolerance and normal clearance.

11.2 Working Condition Analysis

11.2.1 Kinematic Analysis

Assume the rotational speed of the inner ring is n_1, from Figure 11.6, the angular velocity of the inner ring ω_1 is

$$\omega_1 = \frac{2\pi n_1}{60} \tag{11.1}$$

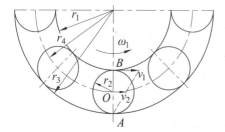

Figure 11.6 Velocity analysis.

Assume the outer ring is stationary, that is, the angular velocity of the outer ring ω_3 is

$$\omega_3 = 0$$

Rolling elements roll purely between the inner ring and outer ring. Point A represents the absolute velocity centre and point B is the relative instantaneous velocity centre. Thus, the linear velocities at point B and point O are expressed as

$$v_1 = \omega_1 \cdot r_1 = \omega_2 \cdot 2r_2 \tag{11.2}$$

and

$$v_2 = \omega_2 \cdot r_2 = \omega_4 \cdot r_4 \tag{11.3}$$

respectively. Therefore, the angular velocity of rolling elements ω_2 is

$$\omega_2 = \frac{r_1}{2r_2}\omega_1 \tag{11.4}$$

and the angular velocity of retainer ω_4 is

$$\omega_4 = \frac{r_2}{r_4}\omega_2 \tag{11.5}$$

Therefore, when a bearing is in operation, the inner ring, outer ring, rolling elements and retainer rotate at different speeds.

11.2.2 Force Analysis

Rolling contact bearings include thrust bearings, radial bearings and angular contact bearings. When these bearings take loads, the load carried by each rolling element is determined by the type of bearing and the characteristic of load.

11.2.2.1 Thrust Bearings (50000, 290000)
When a thrust bearing carries an axial load, the load is shared by each rolling element.

11.2.2.2 Radial Bearings (60000, 10000, 20000, N, NA)
In a radial bearing, a radial load R acts on the inner ring through the shaft. The rolling elements in the upper half circle of the bearing will not take any load while the rolling elements in the lower half circle transmit the load unevenly to the outer ring. The load on the lowest rolling element will be the largest and gradually decrease on both sides, as shown in Figure 11.7. According to the force equilibrium principle, the total reaction on the inner ring by rolling elements will be balanced with the radial load.

11.2.2.3 Angular Contact Bearings (70000, 30000)
Angular contact bearings can carry radial loads, axial loads or their combinations. The following presents detailed analyses under different loading conditions.

(1) A bearing carries a pure radial load R only

When a pure radial load R is applied at the effective force centre O on an angular contact bearing, as shown in Figure 11.8, the reaction force N_i taken by each rolling element is not in the radial direction because of the existence of contact angle α. It can be resolved into a radial component P_i and an axial component S_i. The relationship between

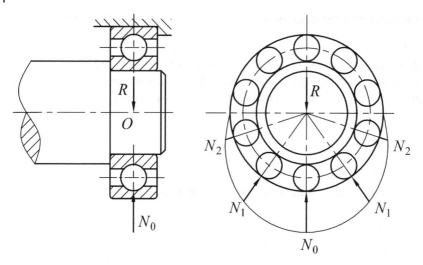

Figure 11.7 Radial load distribution in a radial bearing.

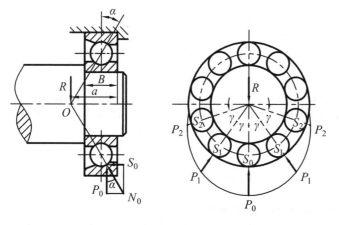

Figure 11.8 Force analysis of an angular contact bearing.

them is $S_i = P_i \tan \alpha$. The vector summary of total radial components P_i is in equilibrium with the applied radial load R. The sum of total axial component S_i is the induced thrust load S.

In other words, the induced thrust load S is created by internal reaction force N_i in the bearing. It is in the direction that forces the separation of the inner and outer rings. To avoid such separation, these bearings usually work in pairs so that the induced thrust load can be counteracted. That is, angular contact bearings should be installed face to face or back to back, as illustrated in Figure 11.9.

(2) A bearing carries both radial load R and axial load A

As shown in Figure 11.8, the bearing carries a radial load R creates an induced thrust load S, which forces the shaft, the inner ring and rolling elements to move left, assuming they move a short distance to the left and are finally balanced with an axial load A (not

Figure 11.9 Axial load analysis of angular contact ball bearings. (a) Face to face. (b) Back to back.

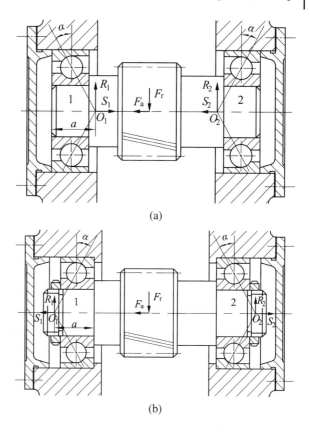

(a)

(b)

shown in the figure). When the radial load R keeps constant, increasing the axial load A implies increasing the induced thrust load S. That is, the number of rolling elements taking the load increases or the loading area increases [5]. In normal operation, usually half the rolling elements take the load. Correspondingly, the magnitude of induced thrust load of an angular contact ball bearing is calculated by [4]

$$S = eR \tag{11.6}$$

The induced thrust load of a tapered roller bearing is calculated by

$$S = R/2Y \tag{11.7}$$

where the values of Y and e can be found in Table 11.1.

(3) Axial load calculation of angular contact bearings

Figure 11.9 shows a pair of angular contact ball bearings carrying both radial and axial loads. The radial reaction R_1 or R_2 on the bearings act through the effective force centres O_1 or O_2, which are at the intersection of the line normal to the contact at the outer race and the shaft centreline. The distance a can be found in bearing catalogues. Effective force centres shorten the span when bearings are mounted face to face and lengthen it when bearings are mounted back to back. If the span is sufficiently large, the radial load

Table 11.1 Dynamic radial and axial load factors.

Bearing type		Relative axial load	e	$A/R > e$	
Name	Code	A/C_0		X	Y
Tapered roller bearings	30000	–	$1.5 \tan \alpha$	0.40	$0.4 \cot \alpha$
Deep-groove ball bearings	60000	0.014	0.19	0.56	2.30
		0.028	0.22		1.99
		0.056	0.26		1.71
		0.084	0.28		1.56
		0.11	0.30		1.45
		0.17	0.34		1.31
		0.28	0.38		1.15
		0.42	0.42		1.04
		0.560	0.44		1.00
Angular contact ball bearings	70000C $\alpha = 15°$	0.015	0.38	0.44	1.47
		0.029	0.40		1.40
		0.058	0.43		1.30
		0.087	0.46		1.23
		0.12	0.47		1.19
		0.17	0.50		1.12
		0.29	0.55		1.02
		0.44	0.56		1.00
		0.58	0.56		1.00
	70000AC $\alpha = 25°$	–	0.68	0.41	0.87
	70000B $\alpha = 40°$	–	1.14	0.35	0.57

Note: Tapered roller bearings 302XX have $e = 0.42$, $Y = 1.4$; and 303XX bearings have $e = 0.35$, $Y = 1.7$ [10].
For detailed data for other tapered roller bearings, readers need to refer catalogues or Standards.
Source: Adapted from Harris 1984, Table 14.3, p. 472. Reproduced with permission of John Wiley & Sons, Inc.

can be assumed as acting at the midpoint of bearings to simplify calculation. Additionally, there may be an externally applied axial load F_a on the shaft, which is from power transmission elements such as helical, bevel or wormgears.

The radial load R induces a thrust load S, which is equivalent to the normal installation, that is, roughly half rolling elements carry radial loads. The magnitude can be obtained from Eq. (11.6) or Eq. (11.7), and the direction is the tendency to separate the inner ring and outer ring. The induced thrust load in each bearing combined with the externally applied axial load F_a establishes equilibrium in the shaft.

Considering Figure 11.9a, if $F_a + S_2 = S_1$, the shaft is in equilibrium. The axial load that each bearing is subjected to will be

$$\begin{cases} A_1 = S_1 \\ A_2 = S_2 \end{cases} \tag{11.8}$$

If $F_a + S_2 > S_1$, the shaft has tendency to move to the left. The left bearing will be squeezed while the right bearing will be relaxed. That is, for the left bearing, the shaft together with the inner ring will press against the rolling elements and the outer ring, while for the right bearing, the shaft and the inner ring will be pulled apart from the outer ring. This will cause the separation of the inner and outer ring and reduces load carrying area, which is not allowed. An appended thrust load S_1' will generated in the left bearing to establish a force equilibrium condition. Therefore, we have

$$S_1' + S_1 = F_a + S_2$$

The axial load each bearing carries will then be

$$\begin{cases} A_1 = S_1' + S_1 = F_a + S_2 \\ A_2 = S_1' + S_1 - F_a = S_2 \end{cases} \tag{11.9}$$

Similarly, if $F_a + S_2 < S_1$, the shaft has tendency to move right. There will be an additional thrust load S_2' on the right bearing, which makes

$$S_1 = F_a + S_2 + S_2'$$

The axial load of each bearing will then be

$$\begin{cases} A_1 = S_1 \\ A_2 = S_2' + S_2 = S_1 - F_a \end{cases} \tag{11.10}$$

In summary, angular contact bearings often operate under the combination of radial and axial loads. The axial load that an angular contact bearing can carry depends on the bearing mounting direction, the magnitudes of induced thrust loads and the direction and magnitude of external axial load. The procedure for determining axial loads on a bearing is summarized as follows:

1. Determine the magnitude and direction of external axial load F_a and induced thrust loads S_1, S_2;
2. According to the force equilibrium principle, judge the moving tendency of the shaft, decide which bearing is relaxed and which one is squeezed
3. The axial load on the relaxed bearing is the induced thrust load itself, while the axial load on the squeezed bearing is equal to the algebraic sum of the induced thrust load at the relaxed bearing and the external axial load.

Readers can use a similar procedure to analyse axial load in the back to back mounted bearings in Figure 11.9b.

11.2.3   Stress Analysis

Figure 11.10a shows selected points of *a*, *b*, *c* on the inner ring, outer ring and a rolling element, respectively. As the rolling contact bearing is in operation, contact stresses generate between the contact surfaces of rolling elements and raceways each time a rolling element passes through the loading zone. Considering the angular velocity of bearing components discussed in Section 11.2.1, as well as load distributions in Figures 11.7 and 11.8, the stress variation at points of *a*, *b* and *c* is illustrated in Figure 11.10b. When the number of repeated stress cycles accumulates and eventually exceeds the fatigue limit of material, fatigue ensues.

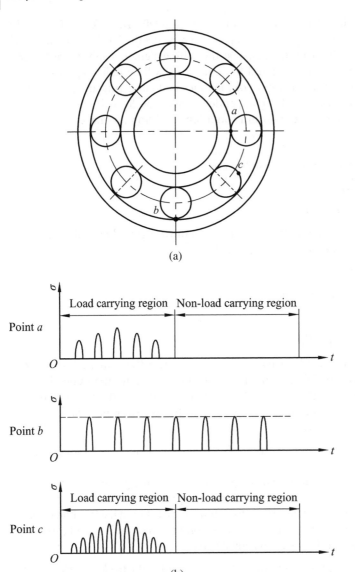

(a)

(b)

Figure 11.10 Stress variation in bearing components.

11.2.4 Potential Failure Modes

Since the inner ring, outer ring and rolling elements experience variable stresses during bearing operation, surface fatigue failure is a predominant failure mode in bearings. As discussed before, the cyclic subsurface Hertzian contact stresses, combined with manufacturing defects, may initiate minute cracks. These minute cracks may propagate during cyclic loading and ultimately cause surface pitting. Failure is considered to occur when either raceway or rolling elements exhibit first pit. Typically, the raceways pit first, resulting in noise, vibration and heat.

Another typical failure mode in rolling bearings is excessive plastic deformation. Under large static loads, balls or rollers will slightly indent the rings and cause permanent deformation. The indentations, permanent deformation or local geometrical discontinuities, will cause subsequent noise, vibration and heat as rolling elements pass by.

When rolling bearings work in dusty environment and/or are not properly lubricated, wear is unavoidable. Insufficient lubrication may also cause scuffing, heat and even burning in high speed bearings. Fretting wear happens even at light loads when bearings operate with small-amplitude reversing cyclic motion. Wear results in unacceptable noise, vibration or heat, as well as contamination of lubricant from accumulated wear debris [11]. Besides, incorrect mounting may cause cage breakage.

11.3 Life Expectancy and Load Carrying Capacities

The size of bearing is selected according to the load carrying capacity, required design life and reliability. The load carrying capacity of rolling contact bearing depends on the total contact area and contact stress, or Hertz stress, which is determined by the magnitude of applied load, the modulus of elasticity of materials in contact and the radii of contacting bodies. It refers to the basic dynamic load rating and basic static load rating, which are quoted in the manufacturer's catalogue.

11.3.1 Life Prediction under Constant Loads

The direct consequence of rolling bearings operating at high cyclic contact stress is fatigue. The sign of fatigue may be minute spalls or pit appearing on any components of a bearing. Whenever the first visible evidences of fatigue occur, the number of revolutions or hours that the bearing can operate at a uniform speed is termed the bearing life.

The life of a rolling bearing is a stochastic variable due to differences in material, manufacturing and assembly. Therefore, reliability is introduced in evaluating bearing life. The rated life is defined as the number of revolutions or hours that 90% of a random sample of apparently identical bearings will reach or exceed at a rated load and a design speed before the first evidence of fatigue appears. For a single bearing, the rated life L_{10} refers to the life associate with 90% reliability [7].

To measure load carrying capacity of a bearing against pitting fatigue, basic dynamic load rating C or sometimes just basic load rating, is introduced. It is the constant load that a bearing can carry for a rated life of one million revolutions with a 90% probability against surface pitting. For radial bearings, the basic dynamic load ratings are radial loads, and for thrust bearings they are axial loads.

11.3.1.1 Relations Between Bearing Load and Bearing Life

Despite using high strength and high hardness steels, all bearings have a finite life and will eventually fail due to fatigue because of high cyclic contact stresses. Obviously, the life of a bearing will increase with the decrease of load. A load-life curve of a group of bearings is obtained by recording the applied loads and corresponding revolution time when pitting starts at a temperature less than 120° under various stable loads, as shown in Figure 11.11. Based on extensive experimental data, the relationship between

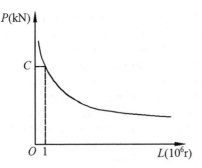

Figure 11.11 Load-life relationship of bearings.

the constant applied load, P and the fatigue life expressed in millions of revolutions, L_{10}, can be stated as

$$P^{\varepsilon} L_{10} = const \tag{11.11}$$

where
ε – load-life exponent, $\varepsilon = 3.0$ for ball bearings and $\varepsilon = 10/3$ for roller bearings.

Although a rated life is one million revolutions at basic dynamic load rating C, where the basic dynamic load rating C is defined by manufacturers and published for each bearing in bearing catalogues, bearing applications usually require lives different from that listed in the catalogues. Therefore, from the relationship between load and life expressed by Eq. (11.11), we have

$$P^{\varepsilon} \cdot L_{10} = C^{\varepsilon} \cdot 1$$

The rated life of bearing in revolutions at load P is

$$L_{10} = \left(\frac{C}{P}\right)^{\varepsilon} \tag{11.12}$$

For bearings operating at a constant speed, it may be more convenient to express rated life in hours using the equation

$$L_{10h} = \frac{10^6}{60n} \left(\frac{C}{P}\right)^{\varepsilon} \tag{11.13}$$

11.3.1.2 Modification of Life Prediction

When at operating temperature, load features are different from those reported in the manufacturer's catalogues, influence factors can be introduced to modify Eq. (11.13) to predict the expected life of bearings.

Bearing ratings shown in manufacturers' catalogues are usually for radial loads, except in the special case of pure thrust bearings. When both radial and axial loads are exerted on a bearing, an equivalent dynamic load is used to predict the life of bearings. The equivalent dynamic load is an imaginary load that would produce the same rated life as it will attain under the actual combined radial and axial loads. The following equation is commonly used to calculate an equivalent dynamic load [12]

$$P = XR + YA \tag{11.14}$$

X – dynamic radial load factor (see Table 11.1);
Y – dynamic axial load factor (see Table 11.1).

The values of X and Y vary with specific bearings and with the ratio of axial load to radial load. To evaluate the effect of axial loads on the load carrying capacity of a bearing, manufacturers specify a factor e for different types of bearing and define a ratio of axial and radial loads for comparison [1]. That is, for relatively small axial loads or a ratio $A/R \le e$, select $X = 1$ and $Y = 0$, that is, the equivalent load is pure radial load and the effect of axial load can be ignored. If $A/R > e$, the effect of axial load on the load carrying capacity of bearing must be considered and Eq. (11.14) is used to compute equivalent dynamic load P. For deep-groove ball bearings, both e and Y depend on the ratio of A/C_0, where C_0 is the static load rating of a particularly selected bearing. Therefore, a simple trial-and-error method is used [1]. Table 11.1 gives dynamic radial and axial load factors for selected bearing types. Values of X, Y and e for loads or contact angles other than those shown are obtained by linear interpolation. Designers can refer to manufacturer's catalogue for detailed data for different types of bearings.

In summary, for bearings such as cylindrical roller bearings, which are subject to radial loads only, the equivalent load is a radial load, that is, $P = R$. For bearings such as thrust bearings, which are subject to axial loads only, the equivalent load is an axial load, that is, $P = A$. For other bearings that are subject to the combination of radial and axial loads, the equivalent load follows $P = XR + YA$. For a pair of bearings supporting a shaft, values for both bearings X_1, Y_1 and X_2, Y_2 should be calculated and whichever combination gives a larger equivalent dynamic load P is used for life calculation.

The bearing rated capacity is for stable loading. Shock or impact loads may reduce bearing life and load factor f_p is introduced to consider the severity of shock. The load factor f_p varies from light impact ($f_p = 1.0$–1.2), through moderate impact ($f_p = 1.2$–1.8) to heavy impact ($f_p = 1.8$–3.0) loading. The equivalent dynamic load is thus modified as $f_p P$.

When a bearing operates at a temperature higher than 120°C, the basic dynamic load rating will decrease. The effect of temperature on load carrying capacity is considered by multiplying the basic dynamic load rating C by a temperature factor of f_t. The temperature factor f_t is selected as 1.0 for temperatures less than 120°C, and as 0.9, 0.8, 0.7, 0.6 and 0.5 for temperatures at 150°C, 200°C, 250°C, 300°C and 350°C, respectively [5]. Temperature factors at other temperatures can be decided by linear interpolation. The design life can be computed by a modified formula as,

$$L_{10h} = \frac{10^6}{60n} \left(\frac{f_t C}{f_p P} \right)^\varepsilon \ge [L_h] \tag{11.15}$$

If the design life L_{10h} and design load P are known, we use the following design formula to compute the required basic dynamic load rating C for bearing size selection,

$$C = \frac{f_p P}{f_t} \left(\frac{60nL_{10h}}{10^6} \right)^{\frac{1}{\varepsilon}} \tag{11.16}$$

11.3.1.3 Rated Life at Different Reliability

The basic dynamic load rating listed in the manufacturer's catalogue has a reliability of 90%. When a higher reliability is desired by designers, a life-adjustment factor for reliability α_1 is incorporated in bearing life calculation. Thus, the rated bearing life, L_{10}, may be adjusted to a higher reliability using

$$L_n = \alpha_1 L_{10} \tag{11.17}$$

where life-adjustment factors α_1 for reliability of 90%, 95%, 96%, 97%, 98% and 99% are selected as 1, 0.62, 0.53, 0.44, 0.33 and 0.21, respectively [6, 7].

Then the rated life at a different reliability is

$$L_n = \frac{10^6 \alpha_1}{60n} \left(\frac{f_t C}{f_p P} \right)^\varepsilon \geq [L_h] \tag{11.18}$$

The basic dynamic load rating with a rated life of L_n corresponding to a different reliability is

$$C = \frac{f_p P}{f_t} \left(\frac{60 n L_n}{10^6 \alpha_1} \right)^{\frac{1}{\varepsilon}} \leq [C] \tag{11.19}$$

11.3.2 Life Prediction under Variable Loads

The analysis discussed so far has assumed that bearings operate at a constant speed under a constant load throughout service life. More often than not, in many applications bearings are subjected to a spectrum of different loads at different speeds during each duty cycle. In such cases, the Palmgren linear cumulative damage rule, or Miner's rule for short introduced in Chapter 2, may be utilized for bearing life prediction.

According to Miner's rule, if a bearing is subjected to a load spectrum, each load contributes to the eventual failure of the bearing. Assume equivalent dynamic loads P_1, $P_2, \ldots P_k$ are applied to a bearing. The rotational speed corresponding to each load is n_1, $n_2, \ldots n_k$. The percentage of operating time under each load is $a_1, a_2 \ldots a_k$.

Assume the bearing reaches life limit after operation under the load of $P_1, P_2, \ldots P_k$ for a total of H hours. The actual number of cycle under the load P_i is

$$L'_i = a_i n_i H \tag{11.20}$$

And the number of cycles when failure happens under the load of P_i is L_i. Therefore, according to Miner's rule, when the bearing fails under the variable loads, we have

$$\sum_{i=1}^{k} \frac{L'_i}{L_i} = 1 \tag{11.21}$$

The total number of cycles under all the load is

$$L_m = a_1 n_1 H + a_2 n_2 H + \ldots + a_k n_k H = (a_1 n_1 + a_2 n_2 + \ldots + a_k n_k) H = n_m H \tag{11.22}$$

Therefore, the average rotational speed n_m is

$$n_m = a_1 n_1 + a_2 n_2 + \ldots + a_k n_k \tag{11.23}$$

Assuming the load spectrum is substituted by an average equivalent dynamic load P_m, under the load of which the bearing fails after operating L_m number of cycles, that is,

$$P_m^\varepsilon L_m = P_i^\varepsilon L_i \tag{11.24}$$

Therefore,

$$L_i = \left(\frac{P_m}{P_i} \right)^\varepsilon L_m \tag{11.25}$$

Substitute Eq. (11.20), (11.22) and (11.25) to Eq. (11.21), we have

$$\sum_{i=1}^{k} \frac{a_i n_i H}{\left(\dfrac{P_m}{P_i}\right)^{\varepsilon} L_m} = \sum_{i=1}^{k} \frac{P_i^{\varepsilon} a_i n_i}{P_m^{\varepsilon} n_m} = 1 \qquad (11.26)$$

The average equivalent dynamic load P_m is

$$P_m = \left(\frac{\displaystyle\sum_{i=1}^{k} a_i n_i P_i^{\varepsilon}}{n_m} \right)^{\frac{1}{\varepsilon}} = \sqrt[\varepsilon]{\frac{a_1 n_1 P_1^{\varepsilon} + a_2 n_2 P_2^{\varepsilon} + \cdots + a_k n_k P_k^{\varepsilon}}{n_m}} \qquad (11.27)$$

Then the expected life of the bearing in hours would be [13]

$$L_h = \frac{10^6}{60 n_m} \left(\frac{C}{P_m} \right)^{\varepsilon} = \frac{10^6}{60} \times \frac{C^{\varepsilon}}{a_1 n_1 P_1^{\varepsilon} + a_2 n_2 P_2^{\varepsilon} + \cdots + a_k n_k P_k^{\varepsilon}} \qquad (11.28)$$

11.3.3 Static Strength Analysis

When a bearing carries a static load exceeding its load capacity, rolling elements will slightly indent the raceway and cause plastic deformation. Large deformation will easily cause increased vibration and noise, leading to premature fatigue failure. The allowable permanent deformation decides the permissible static load or basic static load rating.

Basic static load rating C_0 is the maximum static load that will produce a permanent deformation of approximately 0.0001 times the diameter of the rolling element at the most heavily loaded element contact site [11, 14]. For radial bearings, the basic static load ratings are radial loads, and for thrust bearings these are axial loads.

Similar to the definition of equivalent dynamic load, when a bearing rotates at extremely low speed carrying a combined static radial and axial load, under an imaginary load the maximum contact stress between rolling elements and raceway groove is the same as that under the actual load. The imaginary load is called the equivalent static load, given by

$$P_0 = X_0 R + Y_0 A \qquad (11.29)$$

The static radial and axial load factor X_0 and Y_0 can be found in bearing catalogues or design handbooks [6].

The basic static load rating C_0 is used for selecting a bearing when bearings rotate at slow speeds, or when heavy shock loads act on bearings, evaluated by

$$C_0 \geq S_0 P_0 \qquad (11.30)$$

where S_0 is a static safety factor, which depends on the requirements of application and loading conditions. For high requirements of operation precision and stability, select a large static safety factor within 1.2–2.5; otherwise, select a static safety factor less than 1.0. More detailed data can be found in the design handbook [4].

11.4 Design of Bearing Support Systems

11.4.1 Introduction

Shafts in a machine are usually supported by precisely produced and statistically tested standardized rolling contact bearings. Apart from selecting appropriate bearings from a commercially available inventory, the structural design of a bearing support system is even more crucial for the successful operation of shafts and bearings. This section discusses issues relevant to the mounting and application of bearings, covering the selection, installation, positioning, retaining, preloading, tolerance fit, lubrication and sealing of bearings.

11.4.2 Bearing Selection

The selection of rolling contact bearings from a manufacturer's catalogue involves the determination of bearing types, dimensions and load carrying capacity for a given application. The bearing type is selected according to the load amplitude and direction, speed limitation, alignment requirement and cost, as discussed in Section 11.1.3.

The selection of bearing dimension considers load carrying capacity. To prevent failure, bearings must have sufficient resistance to surface fatigue failure, measured by basic dynamic load rating C; and sufficient resistance to plastic deformation, measured by basic static load rating C_0. These design criteria are universally used in characterizing load carrying capacity for all types of rolling bearings. The values of basic dynamic load rating C and basic static load rating C_0 are tabulated in bearing catalogues. The catalogues identify bearings by codes, basic sizes, mounting dimensions, capacity ratings and so on, which help designers to select bearings that most efficiently and economically fit the particular application. Table 11.2 tabulates basic dynamic load rating and basic static load rating of selected bearing sizes of deep-groove ball bearings, angular contact ball bearings and tapered roller bearings. More data can be found in manufacturers' catalogues for bearings other than those listed in Table 11.2.

Table 11.2 Capacity ratings of selected bearings, C, kN [9, 15, 16].

Bearing bore		Deep-groove ball bearings				Angular contact ball bearings				Tapered roller bearings			
Bore No.	$d\,mm^{-1}$	62XX		63XX		72XXC		72XXAC		302XX		303XX	
		Dynamic C	Static C_0	Dynamic C	Static C_0	Dynamic C	Static C_0	Dynamic C	Static C_0	Dynamic C	Static C_0	Dynamic C	Static C_0
10	50	35.0	23.2	61.8	38.0	42.8	32.0	40.8	30.5	73.2	92.0	130	158
12	60	47.8	32.8	81.8	51.8	61.0	48.5	58.5	46.2	102	130	170	210
14	70	60.8	45.0	105	68.0	70.2	60.0	69.2	57.5	132	175	218	272
15	75	66.0	49.5	113	76.8	79.2	65.8	75.2	63.0	138	185	252	318
16	80	71.5	54.2	123	86.5	89.5	78.2	85.0	74.5	160	212	278	352
18	90	95.8	71.5	145	108	122	105	118	100	200	270	342	440
20	100	122	92.8	173	140	148	128	142	122	255	350	405	525

11.4.3 Design Procedures and Guidelines

Bearings are normally selected after the minimum diameter of mating shaft has been determined from stress or rigidity analyses. Also, the radial and axial forces acting on the power transmission elements, along with the location of bearings with respect to other elements have been decided.

The procedures for making appropriate selections and design of bearing system are:

1. Tentatively select the type of bearing according to the specific bearing application;
2. Decide the bore size of bearing according to the diameter of mating shaft;
3. Specify radial load R and axial load A on each bearing, determine the equivalent load by $P = XR + YA$;
4. Specify the life expectancy L_h and reliability appropriate for the application;
5. Calculate the required basic dynamic load rating C;
6. Select the smallest bearing with the required basic dynamic load rating C and a bore that closely matches the shaft diameter by consulting bearing catalogues;
7. Calculate the equivalent static load by $P_0 = X_0R + Y_0A$ if needed;
8. Determine mounting conditions, such as the diameter and tolerance of shaft seat and housing bore, means of axially locating bearings, and other special needs, such as seals or shields, to ensure bearings can be installed conveniently in a machine. Designers should consult manufacturers' catalogues for more details.

11.4.4 Practical Considerations in the Application of Bearings

After bearing type, bore size and load carrying capacity have been decided, some practical considerations including assembly and disassembly, positioning, retaining, adjustment, tolerance, preloading, lubrication and sealing become crucial in determining successful performance of a bearing.

11.4.4.1 Assembly and Disassembly

Bearings are installed with a light interference fit between the bore of bearing and shaft. Therefore, a rather heavy load is required to apply axially to the bearing inner ring. To reduce extremely high installation load in large bearing assemblies, it may be necessary either to thermally expand the inner ring or to contract the shaft in a mixture of dry ice and alcohol. Heating a bearing should not damage bearing materials or any preinstalled lubricant. Removal of bearings can be realized by bearing pullers, as shown in Figure 11.12.

11.4.4.2 Axial Positioning

A shaft usually requires two supports. Each support may be composed of one or more bearings. Radially, bearings are confined by a housing. Axially, bearings are held in position by either a shaft shoulder, a bolted cover or other retaining elements. When installing bearings, it is important to prevent both axial movement and over constrain. If both bearings are held tightly, any dimensional changes due to thermal expansion could lead to dangerous unexpected loads on bearings.

Three methods are usually employed for axially positioning bearings. Figure 11.13 shows a common practice of retaining two bearings each in one direction. When using deep-groove ball bearings in Figure 11.13a, the right bearing has its outer ring fit loosely

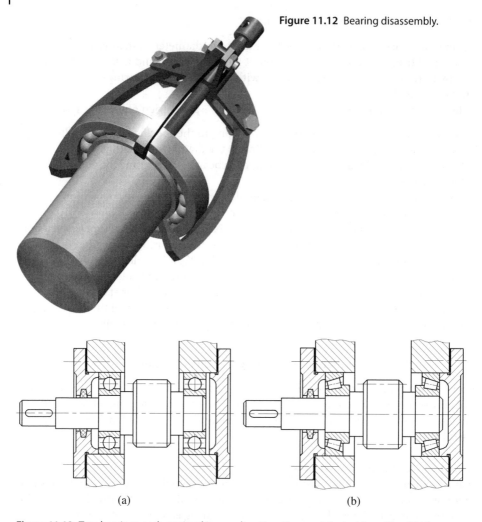

Figure 11.12 Bearing disassembly.

(a)   (b)

Figure 11.13 Two bearings each retained in one direction. Source: Adapted from Wen 2015.

with the bore of housing so that the shaft can move freely when it expands. A clearance is left to accommodate thermal expansion. This method is for short shafts with small thermal expansion. Figure 11.13b has a similar design to Figure 11.13a, except using separable tapered roller bearings to retain bearings and to accommodate thermal expansion. The load is transferred from the shaft, to the inner ring, to the rolling elements, to the outer ring and finally reach the housing.

Figure 11.14 gives examples where one bearing is retained in both directions and the other bearing is free to move in the housing. Such an arrangement will accommodate relatively large thermal elongation for a long shaft. Figure 11.14a shows a method of retaining both the inner ring and outer ring of left bearing and freeing the right bearing. When mounting the free bearing, it is important to ensure sufficient axial clearance.

It is often necessary to use two or more bearings at one end of the shaft to increase load carrying capacity or rigidity of bearings. Figure 11.14b gives an example of using

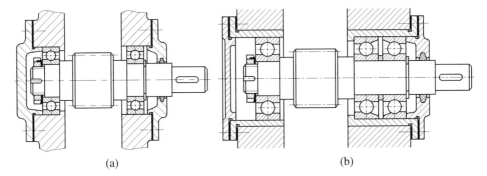

(a)   (b)

Figure 11.14 One bearing retained in two directions, one bearing floats.

Figure 11.15 Two bearings float.

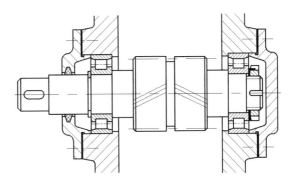

two bearings in a shoulder cartridge to retain both directions at the right end and floating the left bearing.

Figure 11.15 shows that both bearings can float freely while supporting a herringbone gear. With both the inner and outer ring located, the rollers and inner rings could move with the shaft along the raceway of outer rings. This design can accommodate thermal expansion in both directions.

In fact, many alternatives in mounting are available, depending on bearing types, axial retaining approaches and application requirements.

11.4.4.3 Axial Retaining

The inner ring and outer ring of bearing can be retained by various methods. The inner ring of left bearing in Figure 11.14a is held against the shaft shoulder by a locknut threaded at the end of shaft. The internal tab on the lockwasher engages a groove in the shaft, and one of the external tabs is bent into a groove on the nut to prevent the nut from rotating [1]. In Figure 11.14a, the inner ring of right bearing is backed up against the shaft shoulder and held in position by a retaining ring. In the same figure, the outer ring of left bearing is backed up against a housing shoulder and held in position by a bearing cover.

To effectively retain a bearing axially by either shaft shoulders or housing shoulders, detailed specifications for the shoulder diameters to seat the inner ring or to locate the outer ring and connecting fillet radii (see Figure 10.7b) should be consulted in bearing catalogues. The shoulder diameter should be around the specified value to facilitate disassembly of bearings.

11.4.4.4 Axial Adjustment

Bearings require axial adjustment to obtain necessary running clearance without impairing design life. The amount of clearance to be maintained is determined by experience. In most cases, a small axial clearance must be left to allow shaft expansion during operation.

Figure 11.13a illustrates typical axial adjustment methods by shims between bearing cover and the housing. Usually stainless-steel shims are preferred, as brass shims wear more easily and abrasive particles produced may contaminate bearings. Shims should be used only against a nonrotating ring. Figure 11.14a illustrates axial adjustment by a locknut against a rotating ring. When it is required to fine-tune the position of power transmission elements on a shaft, the shims between the shaft cartridge and housing can be adjusted, as shown in Figure 11.14b.

11.4.4.5 Rolling Bearing Fits

Tolerances and fits of shafts and housings are critical for rolling bearing operation. Since bearings are manufactured either by national standards, such as the ANSI/ABMA in USA, DIN in Germany, JIS in Japan, GB in China, or by the international standard ISO [3], proper fits are obtained only by selecting proper tolerances for the mating shaft and housing bore. A suitable fit must be carefully selected as too loose a fit may abrade shafts and/or housing bores, while too tight a fit can cause internal interference and shorten bearing life [2].

In normal practice, bearings are usually mounted on the rotating ring with a light interference fit to prevent relative motion during operation, whether it is an inner ring or outer ring [14]. The stationary ring is mounted with a close clearance fit to accommodate axial sliding and to avoid undesirable thermally induced thrust loads. To ensure proper operation and bearing life, mounting dimensions must be controlled to a total tolerance of only a few thousandths of a millimetre [1]. Recommended fits depend on bearing type, size, tolerance grade, as well as loads, speeds and applications. Since bearings are standard products, shafts usually adopt the basic hole system and the tolerance of shaft diameter selects $k6$, $m6$, $n6$ and so on. Tight interference fits are for high speed, heavy load, high temperature and vibration applications. Housing bores take up the basic shaft system and the tolerance of housing bore selects $H7$, $Js7$ and so on [17]. Most catalogues specify limit dimensions for both shaft diameters and housing bore diameters.

11.4.4.6 Preloading

When high rotational precision, long life, small deflection and vibration, are required for a bearing system, it is common practice to preload bearings to remove excessive internal radial and axial clearances. Preloading are realized by producing an axial displacement of one race relative to the other either by reducing the width of inner ring or outer ring, or by inserting a shim between them [13]. For example, the outer rings in Figure 11.16a are ground to provide the required gap, the bearings are preloaded automatically when the pair of bearings is tightly clamped together. In Figure 11.16b, a pair of angular contact ball bearings is preloaded by inserting a shim between inner rings and clamping the inner ring with a nut on the shaft, and the outer ring with bolts in the housing.

11.4.4.7 Lubrication

Lubrication is important for proper functioning of bearings, especially in high speed applications. Lubrication not only provides a low frictional film between contact surfaces of rolling elements and raceways, but also helps dissipate heat and dispel

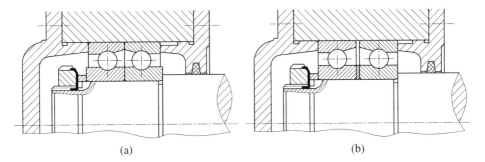

Figure 11.16 Bearing preloading.

contaminants, protects bearing components from corrosion and prolongs the expected life of bearings. Lower friction and longer life are obtained by adding a lubricant of either grease or oil. Operating temperature, load and rotational speed are the main factors considered while selecting lubricants.

Grease is used at relatively slow speeds under normal ambient temperature. Since grease is easy to retain in a bearing housing over a long period of time and acts as a seal against contaminants, most rolling bearings are grease lubricated. Grease can be delivered to bearings by either hand-packing, or manually lubricated with a grease gun.

When bearings operate at high speeds or at high ambient temperatures, oil lubricants with a steady supply of a continuous oil flow are required. In oil lubrication, the maintenance of oil film thickness at the contact surface is extremely important, as oil starvation of the contact area is a main reason for bearing failure. To maintain sufficient oil film thickness at the contact area between rolling elements and raceways, lubricant oil must have a minimum viscosity at the operating temperature.

Lubricant oil can be supplied by oil bath, recirculation splash, oil jet or oil-air mist. In bath lubrication, the height of oil level should not exceed the centre of the lowest rolling elements to avoid energy loss due to oil churning. In splash lubrication, lubricant oil is thrown to the housing cover and is directed in its draining to a bearing [14]. As rotational speed and load increase, oil jet lubrication can be used. Pressurized oil spray on the rolling elements and recirculate for future use. In even high-speed devices, such as jet engines, lubricating oil is pumped under pressure to form a fine oil mist on rolling elements.

11.4.4.8 Sealing

Seals are used to retain lubricant in bearings and also to prevent the entry of contaminants and moisture. They are especially helpful when bearings operate in dirty or moist environments. The selection of a suitable sealing arrangement depends upon lubricant types, rotational speeds, operating temperature, available space and so on. Two common types of sealing, that is contact and non-contact sealing, are used for low speeds and high speeds, respectively.

In contact sealing, seals directly contact rotating shafts. The presence of seals increases friction and wear; therefore, the rubbing surfaces should have a high polish. Figure 11.17a shows a felt seal placed in a machined groove for grease lubrication at low speeds. The commercial seal is an assembly consisting of elastomer and a spring backing [14], encased in a counterbored hole in a bearing cover. The opening of a commercial seal facing a bearing is used to retain lubricant, as shown in Figure 11.17b,

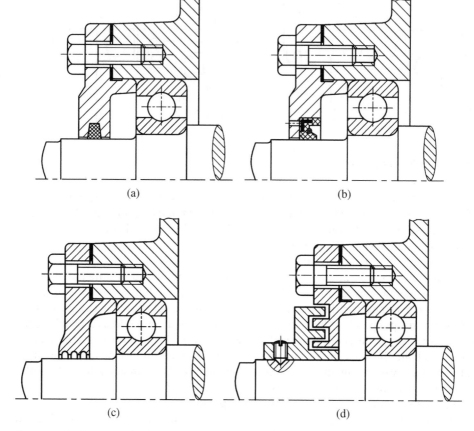

Figure 11.17 Contact and non-contact sealing. (a) Felt seal, (b) commercial seal, (c) groove seal and (d) labyrinth seal.

while the opposite opening direction is for preventing the entry of contaminants and moisture. Sealing function is achieved by the spring and elastomeric seal.

To reduce friction on high speed shafts, non-contact sealing like a groove seal or a labyrinth seal can be used. In Figure 11.17c, grooves in the form of a thread are cut in the bore of the bearing cover. A labyrinth seal in Figure 11.17d consists of an intricate narrow passage with a few hundredths of a millimetre radial clearance [1]. In both seals, the relative motion of shaft with respect to the stationary parts creates a sealing action.

11.4.5 Design Cases

Example Problem 11.1
Select a pair of single-row, deep-groove ball bearings to support a shaft that rotates at 1500 rpm for continuous one-shift (i.e. 8 h a day) operation. The radial load on the bearing is $R = 5000\,\text{N}$, and the axial load is $A = 2880\,\text{N}$, with light-to-moderate impact. The diameter of the shaft is 75 mm. The design life is to be 5000 hours with a reliability of 90%.

Solution

Since the bearing has not been selected, bearing parameters C_0, e, X, Y are unknown. Here a simple trial-and-error method is used. We initially selected bearings 6215 and 6315.

Steps	Computation	Results	
		6215	6315
1. Decide the basic dynamic and static load rating of selected bearings	From a manufacturer's catalogue or Table 11.2, the basic dynamic and static load rating of selected bearings can be obtained.	$C = 66\,000\,\text{N}$ $C_0 = 49\,500\,\text{N}$	$C = 113\,000\,\text{N}$ $C_0 = 76\,800\,\text{N}$
2. Decide the ratio of A/C_0	Bearing 6215 $\dfrac{A}{C_0} = \dfrac{2880}{49500} = 0.058$ Bearing 6315 $\dfrac{A}{C_0} = \dfrac{2880}{76800} = 0.0375$	$A/C_0 = 0.058$	$A/C_0 = 0.0375$
3. The value of e	The value of e can be obtained from Table 11.1 by interpolating.	$e = 0.261$	$e = 0.234$
4. Calculate the ratio of A/R	$\dfrac{A}{R} = \dfrac{2880}{5000} = 0.576$	$0.576 > e$	$0.576 > e$
5. Decide X and Y	X and Y can be selected from Table 11.1 by interpolating.	$X = 0.56$ $Y = 1.699$	$X = 0.56$ $Y = 1.895$
6. Calculate equivalent load P	From Eq. (11.14) $P = XR + YA$ Bearing 6215 $P = XR + YA = 0.56 \times 5000 +$ $1.699 \times 2880 = 7693\,\text{N}$ Bearing 6315 $P = XR + YA = 0.56 \times 5000 +$ $1.895 \times 2880 = 8258\,\text{N}$	$P = 7693\,\text{N}$	$P = 8258\,\text{N}$
7. Compute the bearing life	For light-to-moderate impact, select $f_p = 1.2$. From Eq. (11.15), the design life of bearing 6215 is $L_{10h} = \dfrac{10^6}{60n}\left(\dfrac{f_t C}{f_p P}\right)^{\varepsilon} =$ $\dfrac{10^6}{60 \times 1500}\left(\dfrac{1 \times 66000}{1.2 \times 7693}\right)^3 = 4060\,\text{h}$	$4060\,\text{h}$ $< 5000\,\text{h}$	$16\,475\,\text{h}$ $> 5000\,\text{h}$

Steps	Computation	Results	
		6215	**6315**

The design life of the 6315 bearing is

$$L_{10h} = \frac{10^6}{60n}\left(\frac{f_t C}{f_p P}\right)^\varepsilon =$$

$$\frac{10^6}{60 \times 1500}\left(\frac{1 \times 113000}{1.2 \times 8258}\right)^3 = 16475 \, h$$

| 8. Conclusion | Select bearing 6315 | | |

Example Problem 11.2

A shaft carrying a gear as part of a power transmission system rotates at 1000 rpm. The shaft was designed to be supported on two bearings mounting back to back at points 1 and 2, as shown in Figure E11.1. The moderate impact forces on the gear with a pitch diameter of 360 mm are $F_t = 4500\,N$, $F_r = 2000\,N$ and $F_a = 1000\,N$. Bearings are expected to have a life over 25 000 hours. Estimate the life expectancy if the initially selected bearing is 30210.

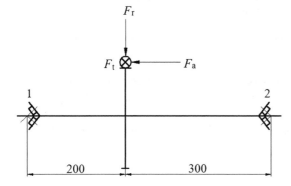

Figure E11.1 Illustration for Example Problem 11.2.

Solution

Steps	Computation	Results	Units
1. Basic data of bearing 30210	From the manufacturer's catalogue or Table 11.2, for bearing 30210, $C = 73\,200\,N$, $C_0 = 92\,000\,N$.	$C = 73\,200$ $C_0 = 92\,000$	N N
2. Radial reaction forces on the pair of bearings	Compute the reaction forces on the bearings $$R_{1v} = \frac{F_r \times 300 + F_a \times \frac{d}{2}}{200 + 300} = \frac{2000 \times 300 + 1000 \times \frac{360}{2}}{500} N = 1560N$$	$R_1 = 3118$ $R_2 = 1853$	N N

Steps	Computation	Results	Units

2. Radial reaction forces on the pair of bearings

$$R_{2v} = \frac{F_r \times 200 - F_a \times \frac{d}{2}}{200 + 300} =$$

$$\frac{2000 \times 200 - 1000 \times \frac{360}{2}}{500} N = 440N$$

$$R_{1H} = \frac{300}{200 + 300} \quad F_t = \frac{300}{500} \times 4500N = 2700N$$

$$R_{2H} = \frac{200}{200 + 300} \quad F_t = \frac{200}{500} \times 4500N = 1800N$$

$$R_1 = \sqrt{R_{1v}^2 + R_{1H}^2} = \sqrt{1560^2 + 2700^2}$$

$$= 3118N$$

$$R_2 = \sqrt{R_{2v}^2 + R_{2H}^2} = \sqrt{440^2 + 1800^2}$$

$$= 1853N$$

3. Compute the axial loads on each bearing

For tapered roller bearings, the magnitude of induced thrust loads is calculated by

$$S = R/2Y$$

The value of Y and e can be found in Table 11.1 as $Y = 1.4$, $e = 0.42$.

$$S_1 = \frac{R_1}{2Y} = \frac{3118}{2 \times 1.4} = 1114N$$

$$S_2 = \frac{R_2}{2Y} = \frac{1853}{2 \times 1.4} = 662N$$

Since

$$S_1 + F_a > S_2$$

Therefore

$$A_1 = S_1 = 1114N$$

$$A_2 = S_1 + F_a = 1000N + 1114N$$

$$= 2114N$$

Results: $A_1 = 1114$ N, $A_2 = 2114$ N

4. Compute the equivalent load P_1 and P_2

$$\frac{A_1}{R_1} = \frac{1114}{3118} = 0.36 < e$$

$$\frac{A_2}{R_2} = \frac{2114}{1853} = 1.14 > e$$

Dynamic radial load factor X and axial load factor Y can be obtained from Table 11.1.

Results: $P_1 = 3118$ N, $P_2 = 3701$ N

Steps	Computation	Results	Units
	For bearing 1 $X_1 = 1.0$, $Y_1 = 0$		
	For bearing 2 $X_2 = 0.4$, $Y_2 = 1.4$		
	$P_1 = X_1 R_1 + Y_1 A_1 = 1.0 \times 3118$		
	$+0 \times 2062 = 3118$ N		
	$P_2 = X_2 R_2 + Y_2 A_2 = 0.4 \times 1853$		
	$+1.4 \times 2114 = 3701$ N		
5. Estimate the life of bearing	For moderate shock, select $f_p = 1.5$.	$L_h = 90\,291$	h
	Since $P_2 > P_1$, check the life of bearing by P_2		
	$L_h = \dfrac{10^6}{60n}\left(\dfrac{f_t C}{f_p P_1}\right)^\varepsilon =$		
	$\dfrac{10^6}{60 \times 1000} \times \left(\dfrac{1 \times 73200}{1.5 \times 3701}\right)^{\frac{10}{3}} =$		
	$90291h > 25000$ h		
	The bearings are satisfactory.		

Example Problem 11.3

A single-row, deep-groove ball bearing is required to carry a stable radial load of 10 kN at rotational speed of 1000 rpm for 1500 hours with a reliability of 99%, determine

1. the rated bearing life at the reliability of 90%;
2. the required basic dynamic load rating of the bearing.

Solution

Steps	Computation	Results	Units
1. The rated life at the reliability of 90%	Find the reliability life-adjustment factor as $\alpha_1 = 0.21$. Calculate the corresponding rated life by Eq. (11.17) $$L_{10} = \frac{L_n}{\alpha_1} = \frac{1500}{0.21} = 7142 \text{ h}$$	$L_{10} = 7142$	h
2. The basic dynamic load rating of the bearing	From Eq. (11.19), compute the basic dynamic load capacity Assume $f_t = 1.0, f_p = 1.0$ $$C = \frac{f_p P}{f_t}\left(\frac{60nL_n}{10^6 \alpha_1}\right)^{\frac{1}{\varepsilon}} =$$ $$10 \times \left(\frac{60 \times 1000 \times 1500}{10^6 \times 0.21}\right)^{\frac{1}{3}} = 75.4 \text{ kN}$$	$C = 75.4$	kN

Example Problem 11.4

A single-row, deep-groove ball bearing 6210 is subjected to the following set of loads and operates at the listed rotational speeds for the given percentage of time:

Condition	Applied load P_i (N)	Rotational speed n_i (rpm)	Time percentage a_i (%)
1	5400	150	30
2	2000	200	50
3	1000	750	20

This load cycle is repeated continuously throughout the life of bearing. Estimate the total life of the bearing.

Solution

Steps	Computation	Results	Units
1. Basic dynamic load rating C	From bearing catalogues or Table 11.2, the basic dynamic load rating C of the bearing 6210 is 35 000 N.	$C = 35\,000$	N
2.	From Eq. (11.28) $$L_h = \frac{10^6}{60} \times \frac{C^\varepsilon}{a_1 n_1 P_1^\varepsilon + a_2 n_2 P_2^\varepsilon + \cdots + a_n n_n P_n^\varepsilon}$$ $$= \frac{10^6}{60} \times$$ $$\frac{35000^3}{0.3 \times 150 \times 5400^3 + 0.5 \times 200 \times 2000^3 + 0.2 \times 750 \times 1000^3} =$$ $$88924$$	$L_h = 88\,924$	h

References

1 Mott, R.L. (2003). *Machine Elements in Mechanical Design*, 4e. Prentice Hall.
2 Hindhede, U., Zimmerman, J.R., Hopkins, R.B. et al. (1983). *Machine Design Fundamentals: A Practical Approach*. New York: Wiley.
3 Juvinall, R.C. and Marshek, K.M. (2011). *Fundamentals of Machine Component Design*, 5e. New York: Wiley.
4 Wen, B.C. (2015). *Machine Design Handbook*, 5e, vol. 3. Beijing: China Machine Press.
5 Pu, L.G. and Ji, M.G. (2006). *Mechanical Design*, 8e. Beijing: Higher Education Press.
6 Harris, T.A. and Kotzalas, M.N. (2007). *Essential Concepts of Bearing Technology*, 5e. New York: CRC Press.
7 Oberg, E. (2012). *Machinery's Handbook*, 29e. New York: Industrial Press.
8 Li, F. and Ma, Y. (1993). *GB/T 272-1993 Rolling Bearing--Identification Code*. Beijing: Standardization Administration of the People's Republic of China.
9 Li, F. and Ma, Y. (2007). *GB/T 276-2013 Rolling Bearings – Deep Groove Ball Bearings – Boundary Dimensions*. Beijing: Standardization Administration of the People's Republic of China.

10 Gong, Y.P., Tian, W.L., Zhang, W.H., and Huang, Q.B. (2008). *Project Design of Mechanical Design.* Beijing: Science Press.

11 Collins, J.A. (2002). *Mechanical Design of Machine Elements and Machines: A Failure Prevention Perspective*, 1e. New York: Wiley.

12 Harris, T.A. (1984). *Rolling Bearing Analysis*, 2e. New York: Wiley.

13 Xu, Z.Y. and Qiu, X.H. (1986). *Machine Elements*, 2e. Beijing: Higher Education Press.

14 Budynas, R.G. and Nisbett, J.K. (2010). *Shigley's Mechanical Engineering Design*, 9e. New York: McGraw-Hill.

15 Li, F. (2007). *GB/T 292-2007 Rolling Bearings - Angular Contact Ball Bearings - Boundary Dimensions.* Beijing: Standardization Administration of the People's Republic of China.

16 Li, F. (1994). *GB/T 297-1994 Rolling Bearings – Tapered Roller Bearings-Boundary Dimensions.* Beijing: Standardization Administration of the People's Republic of China.

17 Li, F., Zhang, X., Zheng, Z. et al. (2015). *GB/T 275-2015 Rolling Bearings-Fits.* Beijing: Standardization Administration of the People's Republic of China.

Problems

Review Questions

1 Why are rolling contact bearings used more widely than sliding bearings in mechanical equipment?

2 What factors are usually considered while selecting a bearing type?

3 What factors are usually considered while designing assembly details of a bearing?

4 Why is lubrication important for the proper operation of rolling contact bearings?

5 When should two bearings be used to support a shaft with each bearing retained in one direction?

Objective Questions

1 The sealing of contact bearings does not prevent _____.
   (a) outside dust and waste getting in
   (b) outside water vapour getting in
   (c) oil inside flowing out
   (d) operational noise

2 _____ must be used in pairs.
   (a) Deep-groove ball bearings
   (b) Cylindrical roller bearings

(c) Thrust ball bearings

(d) Tapered roller bearings

3 The correct designation for the fit between the shaft and the inner ring of a rolling contact bearing is _____.

(a) $\phi 30\frac{H7}{k6}$

(b) $\varphi 30H7$

(c) $\phi 30k6$

(d) $\phi 30\frac{k6}{H7}$

4 A deep-groove ball bearing carries a constant load. If its rotational speed changes from $960\,\mathrm{r\,min^{-1}}$ to $480\,\mathrm{r\,min^{-1}}$, the life of the bearing changes from _____.

(a) from L_h increases to $2L_h$ (h)

(b) from L_r reduces to $L_r/2$ (r)

(c) from L_r increases to $2L_r$ (r)

(d) from L_h reduce to $L_h/2$ (h)

5 An angular contact ball bearing rotates at a constant speed. If the equivalent dynamic load reduces from $2P$ to P, the life of the bearing changes from L_{10} _____.

(a) decrease to $0.2L_{10}$

(b) increase to $2L_{10}$

(c) increase to $8L_{10}$

(d) unchanged

Calculation Questions

1 A shaft is supported by a pair of angular contact ball bearings 7212 AC, as shown in Figure P11.1. The bearings operate at a speed of $n = 1200\,\mathrm{r\,min^{-1}}$ under constant radial loads of $R_1 = 2500\,\mathrm{N}$, $R_2 = 5000\,\mathrm{N}$ on bearing 1 and bearing 2, respectively. The external axial load acting on the gears are $F_{a1} = 400\,\mathrm{N}$, $F_{a2} = 2400\,\mathrm{N}$. If the load factor is $f_p = 1.0$, and the bearings work at room temperature. Estimate the life of bearing. ($e = 0.68$, $X = 0.41$, $Y = 0.87$, $S = 0.68R$, $C = 61.0\,\mathrm{kN}$).

Figure P11.1 Illustration for Calculation Question 1.

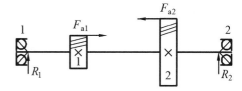

2 A bevel gear is supported by a pair of taper roller bearings 30 206 with a basic dynamic load rating of $C = 41\,200\,\mathrm{N}$, as shown in Figure P11.2. The radial loads act on the bearings are $R_1 = 1200\,\mathrm{N}$ and $R_2 = 2000\,\mathrm{N}$, respectively. The shaft rotates at

$n = 1200 \, \text{r min}^{-1}$ at room temperature. The external axial load acting on the gear is $F_a = 1000 \, \text{N}$. Assume the load factor is $f_p = 1.2$. For designs (a) and (b), calculate the life of bearings and the ratio of bearing lives in the two designs.

e	A/R ≤ e	A/R > e	S
0.37	X = 1, Y = 0	X = 0.4, Y = 1.6	R/2Y

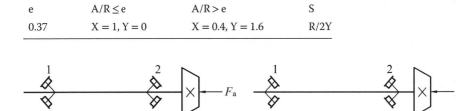

(a) (b)

Figure P11.2 Illustration for Calculation Question 2.

3   A shaft is supported by a pair of 7212C bearings with the basic dynamic load rating of $C = 61.0 \, \text{kN}$, as shown in Figure P11.3. The radial loads on the bearings are $R_1 = 640 \, \text{N}$ and $R_2 = 1965 \, \text{N}$ and the external axial load is $F_a = 230 \, \text{N}$. The rotational speed of the shaft is 960 rpm. The bearings work at a stable loading condition and normal temperature. Determine the life of the bearings.

Figure P11.3 Illustration for Calculation Question 3.

4   A deep-groove ball bearing 6310 with a dynamic load rating of $C = 61.8 \, \text{kN}$ supports a shaft that rotates at 1000 rpm. A uniform radial load spectrum of 3000 N, 5000 N and 7000 N operates for 50%, 30% and 20% of the time, respectively. Estimate the life of the bearing for a reliability of 97%.

Design Problems

1   A pair of deep-groove ball bearings carry stable radial loads of $R_1 = 5000 \, \text{N}$ and $R_2 = 6000 \, \text{N}$ and an axial load of $F_a = 2300 \, \text{N}$, as indicated in Figure P11.4. The 60 mm diameter shaft supporting the bearings operates at a speed of 800 rpm below 100°C. The design life is 20 000 hours. Select proper bearings for the application.

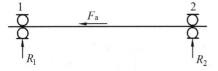

Figure P11.4 Illustration for Design Problem 1.

2   A pair of tapered roller bearings is to support a shaft rotating at 1500 rpm for continuous operation with an expected life of 20 000 h, as shown in Figure P11.5. The radial loads on the bearings are $R_1 = 3000\,\text{N}$ and $R_2 = 5000\,\text{N}$ and the axial load is $F_a = 1600\,\text{N}$ with light-to-moderate impact. The diameter of the shaft is 50 mm. Specify suitable tapered roller bearings for the shaft.

Figure P11.5 Illustration for Design Problem 2.

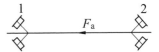

3   A shaft is supported by a pair of single-row, deep-groove ball bearings 6210 with a reliability of 90%. It is now required to increase the reliability to 99% with the same life expectancy. Decide on the bearings that can be used as a replacement.

Structure Design Problems

1   Design a radial and an axial labyrinth seal structure.

2   A gear mounted on a shaft is supported by a pair of rolling contact bearings. Which of the following designs in Figure P11.6 is better?

Figure P11.6 Illustration for Structure Design Problem 2.

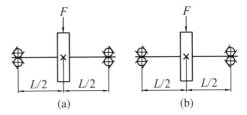

3   Two bearing assemblies are shown in Figure P11.7. Which design is better?

4   In Figure P11.8,
   1. Describe of assembly process of the bearing shaft system;
   2. How could one adjust clearances in the bearings?
   3. How could one adjust the axial position of the shaft to ensure the cone tips of a pair of bevel gears coincide?
   4. How is the axial load transferred to the housing?

5   A wormgear is mounted on a shaft supported by a pair of tapered roller bearings. Each bearing retains one direction of movement. The wormgear is lubricated by lubricant oil and the bearings by grease. Partial design is shown in Figure P11.9, please complete the design. If the wormgear is changed to a spur gear, make modifications to the design if required.

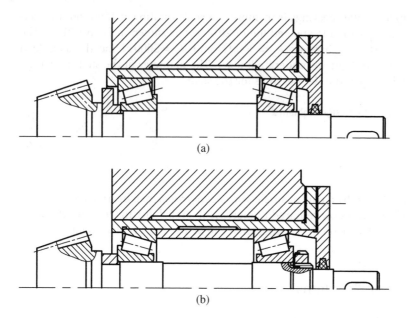

(a)

(b)

Figure P11.7 Illustration for Structure Design Problem 3.

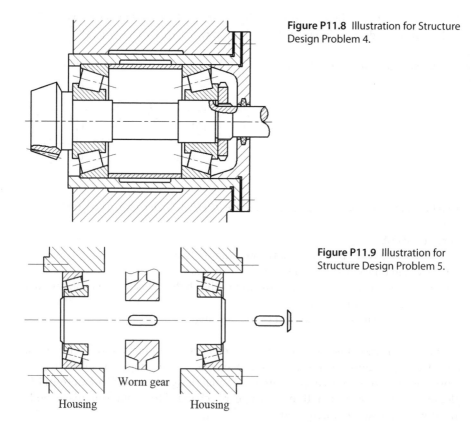

Figure P11.8 Illustration for Structure Design Problem 4.

Figure P11.9 Illustration for Structure Design Problem 5.

Worm gear

Housing   Housing

CAD Problems

1 Write a flow chart for the bearing design process to complete the Example Problem 11.1.

2 Develop a program to implement a user interface similar to Figure P11.10 and complete the Example Problem 11.1.

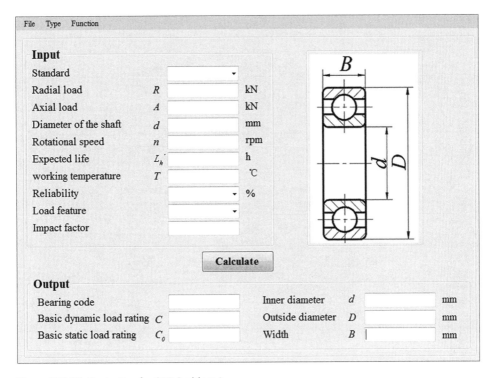

Figure P11.10 Illustration for CAD Problem 2.

12

Sliding Bearings

Nomenclature

B	bearing width, m or mm	c	specific heat, J/(kg °C)
C_f	coefficient of friction variable	D	bearing diameter, mm
C_p	coefficient of load carrying capacity	d	journal diameter, mm
C_Q	coefficient of oil flow	e	eccentricity, mm
		F	force, N
		f	coefficient of friction

Analysis and Design of Machine Elements, First Edition. Wei Jiang.
© 2019 John Wiley & Sons Singapore Pte. Ltd. Published 2019 by John Wiley & Sons Singapore Pte. Ltd.
Companion website: www.wiley.com/go/Jiang/analysis_of_machine_elements

H	frictional power rate, W	t_o	outlet oil temperature, °C
H_1	heat removed by oil flow per second, W	Δt	temperature rise from inlet to outlet, °C
H_2	heat dissipated by housing per second, W	u,v	velocity, m/s
h	lubricant film thickness, μm	α	wrap angle, °
h_{min}	minimum film thickness, μm	α_s	heat transfer coefficient, W/(m² °C)
h_0	oil film thickness at the highest pressure, μm	Δ	diametrical clearance, mm
$[h]$	allowable minimum film thickness, μm	δ	radial clearance, mm
		ε	eccentricity radio
n	rotational speed, rpm	η	dynamic viscosity, Pa s, N s m^{-2}
p	unit bearing load, MPa	ρ	lubricant density, kg m^{-3}
$[p]$	allowable pressure, MPa	τ	shear stress, MPa
Q	volumetric flow rate of lubricant, m³ s^{-1}	υ	kinematic viscosity, m² s^{-1} or cSt
R	radius of bearing, mm	φ	angular coordinate in circumferential direction, °
R_z	surface roughness, μm	φ_1	angular coordinate of pressure leading edge, °
r	radius of journal, mm	φ_2	angular coordinate of pressure trailing edge, °
S	factor of safety	φ_a	attitude angle, °
t_i	oil temperature, °C	ψ	relative clearance
t_m	mean temperature, °C	ω	angular velocity, rad s^{-1}

12.1 Introduction

12.1.1 Applications, Characteristics and Structures

A sliding bearing (or plain bearing) is characterized by direct sliding of a journal on a bearing. The journal is usually a part of rotating shaft carrying radial loads and the bearing is the stationary element that mates the journal. Loads are transferred from the journal through the bearing to the housing of machine.

Sliding bearings are better suited to high speed, heavy load applications. They can also realize high precision position and quite operation. The application for sliding bearings is immense, from bearings working in extreme conditions in an internal combustion engine or in a steam turbine of power-generating station, to bearings working in extreme high accuracy in a radio antenna or a huge telescope [1–3].

Compared with rolling contact bearings, sliding bearings require less radial space, yet greater axial space. Although sliding bearings may have less running friction, their start friction can be much higher [4]. Besides, they need a lubrication system and may become instable at high speeds.

A sliding bearing normally consists of a housing, a liner or insert supporting shafts, and lubricating and protective devices [5]. Typical structures of sliding bearings are shown in Figure 12.1.

A housing may be a separate casting or weldment attached to a machine. They support sliding bearings by either solid design or split design, as shown in Figure 12.1a and b, respectively. Shafts are supported by bushings press-fitted into a bore in the housing.

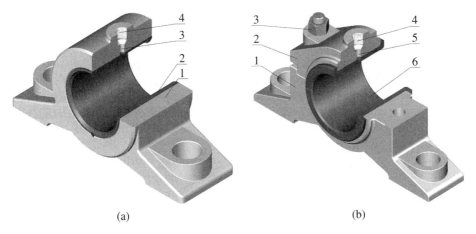

(a)   (b)

Figure 12.1 Structures of sliding bearings: (a) 1 housing, 2 bushing, 3 oil hole and 4 screw; (b) 1 housing, 2 housing cover, 3 stud, 4 screw, 5 oil hole and 6 liner or bushing.

To reduce wear, costs and also to facilitate repair of worn bushings, liners or inserts are fabricated by casting molten expensive antifriction lining materials continuously on split bushings. Locating pins are used to position the bushing in the housing [1].

Lubricating and protective devices include elements pertinent to lubrication. Proper functioning of a sliding bearing requires sufficient lubricant supply. Lubricant oil is usually delivered through the oil hole, and distributed over the width of bearing by grooves cut in the bushings. Oil holes normally locate at the centre of bushing in the low-pressure region (see Figure 12.1). Grooves of various patterns, as illustrated in Figure 12.2, should not be cut in the load carrying region and should terminate before the end of bearing to avoid pressure drop and side leakage.

12.1.2 Types of Sliding Bearings

According to the configuration, sliding bearings include solid bearings and split bearings. Solid bearings (Figure 12.1a) have simple structure, high rigidity, and are easy to manufacture at low cost. However, they present difficulties for axial assembly of heavy

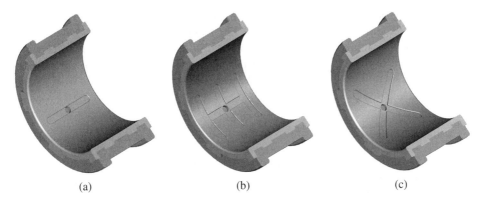

(a)   (b)   (c)

Figure 12.2 Typical groove patterns.

or large diameter shafts. Split bearings (Figure 12.1b), on the other hand, facilitate such assembly and allow adjustment of bearing clearance. They are especially suitable for supporting heavy elements like crankshafts.

Sliding bearings include radial bearings and thrust bearings according to loading directions. Radial bearings, more commonly called journal bearings or sleeve bearings, are cylindrical and carry radial loads perpendicular to the shaft axis. Thrust bearings are generally flat and carry axial loads parallel to the shaft axis. Although the design and analysis of thrust bearings is also important, a detailed study of thrust bearings is not covered here due to limited space.

Lubrication is an important factor in classifying sliding bearings. Accordingly, we have boundary and hydrodynamically lubricated bearings. In a boundary-lubricated bearing, the journal and bearing have intermittent contact, that is, only a partial lubricant film exits between contact surfaces. Boundary-lubricated bearings are used successfully in light service, low speed or unimportant applications, such as electric fans, office machinery and home appliances. In a hydrodynamically lubricated bearing, continuously supplied lubricant is fed by the rotating journal into a wedged-shaped region, lifting the rotating journal off the stationary bearing. As the journal and bearing are completely separated, there is no metal-to-metal contact, resulting comparably low friction. Hydrodynamically lubricated bearings are usually for high rotating speeds with impact and momentary overloads.

Finally, according to the load carrying mechanism, we have hydrodynamic and hydrostatic bearings. Unlike hydrodynamically lubricated bearings, lubrication in a hydrostatic bearing is obtained by continuous supply of externally pressurized lubricant to the bearing interface at a pressure high enough to separate the surfaces with a relatively thick film of lubricant. The hydrostatic lubrication can achieve extremely low friction at all times, including at startup and low-speed operation. However, it requires an expensive and complicated external fluid pressurization system. Hydrostatic bearings are commonly used in machine tools, huge telescopes and antenna systems, where slow, smooth and accurate operation is required.

12.2 Working Condition Analysis

The analysis and design of sliding bearings involves substantial knowledge of tribology, which studies friction, wear and lubrication. Just like strength theory, tribology plays a significant role in machine element life. For a machine element to function successfully, design must not only satisfy strength and rigidity requirements, but also tribological demands. This section will introduce some basic tribological principles that will be used in sliding bearing design.

12.2.1 Friction

Friction is regarded as a force resisting relative movement between surfaces in contact. Although friction generates frictional heat and causes power losses, not all friction is undesirable. For instance, belt drives rely on friction to transmit torque and motion.

Friction between stationary surfaces is called static friction, while between moving surfaces it is called kinetic friction. The two main kinds of kinetic friction are rolling friction and sliding friction, appearing in contact bearings and sliding bearings, respectively. The sliding friction between solid objects is dry friction, and between a solid and a gas or liquid is fluid friction [6].

Fluid friction occurs under proper lubrication conditions. Consequently, lubrication regimes are classified as boundary lubrication, mixed-film lubrication and hydrodynamic lubrication, depending on the degree to which the lubricant separates sliding surfaces.

In boundary lubrication, there is a very thin lubricant film adhering to the contact surfaces, and the highest asperities may be separated by lubricant films only several molecular dimensions in thickness [1]. Local metal-to-metal surface contact is continuous and extensive, which leads to a high rate of friction and wear. The surface films vary in thickness from 1 to 10 nm. Typical values of friction coefficient are from 0.05 to 0.20 [2].

In mixed-film lubrication, surface peaks are intermittently in contact and there is partial hydrodynamic support. The average film thickness in a mixed lubrication is less than 1 μm and greater than 0.01 μm. The coefficients of friction commonly range from 0.004 to 0.1 [2].

The most desirable type of lubrication is obviously hydrodynamic lubrication. In hydrodynamic lubrication, film thickness is significantly greater than the height of surface roughness. Surfaces are completely separated by relatively thick lubricant film. Both metal-to-metal contact and surface wear are prevented. The frictional losses originate only within the lubricant film. Typical film thicknesses are from 0.008 to 0.02 mm. Coefficients of friction are within the range of 0.002–0.01 [2]. Hydrodynamic lubrication is also called full-film, or fluid lubrication.

Lubrication regimes determine the performance of a sliding bearing. The performance of bearing is evaluated by the bearing characteristic $\eta n/p$, that is, the combined effect of lubricant viscosity η, the journal rotational speed n and the unit bearing load p. Figure 12.3 illustrates the influence of bearing characteristic, $\eta n/p$, on the type

Figure 12.3 The variation of coefficient of friction and lubrication regime with bearing characteristic $\eta n/p$. Source: Juvinall and Marshek 2001, Figure 13.4, p.549. Reproduced with permission of John Wiley & Sons, Inc.

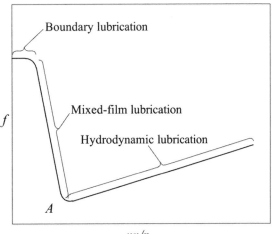

of lubrication and resulting coefficient of friction, f, which was initially obtained by the McKee brothers in an actual test of friction.

At low values of $\eta n/p$, sliding surfaces rub together and lubricant film is extremely thin. There is a great possibility of metal-to-metal contact. Boundary lubrication occurs with high coefficient of friction. In general, boundary lubrication is expected to occurs in hydrodynamically lubricated bearings during starting, stopping, overloading or lubricant deficiency.

At high values of $\eta n/p$, hydrodynamic lubrication produces a complete physical separation between sliding surfaces and greatly reduces friction and wear. To promote hydrodynamic lubrication, light loads, high relative speeds between moving surfaces and a copious supply of high viscosity lubricant are required. Hydrodynamic lubrication implies minimum power losses and maximum life expectancy for bearings.

Between boundary and hydrodynamic lubrication within the range of point A and B in Figure 12.3 is mixed-film lubrication. Mixed-film lubrication occurs due to scarcity of lubricant, low viscosity, low bearing speeds, overload, tight clearance and so on. The mixed-film lubrication zone should be avoided, as a small change in any of three factors of η, n or p produces a sharp change in friction coefficient f, resulting in unpredictable performance of machine [3].

From starting-up and accelerating to its operating speed, a sliding bearing may experience through all three modes of lubrication, which will be discussed in detail in Section 12.2.4.

12.2.2 Wear

Wear is undesirable cumulative profile change due to progressive loss of materials from the contact surfaces of mating elements as the result of loads and relative motion. It has long been recognized as a most important detrimental process in machine elements. Wear is classified as abrasion, adhesion, fatigue, fretting and corrosion wear by the physical nature of the underlying process. In practical engineering, two or more types of wear often occur simultaneously.

Abrasive wear arises either due to direct contact of a hard and a soft surface, or due to rubbing of abrasive particles on a surface. For the former, the asperities of the harder surface penetrate the softer surface under a normal load, producing plastic deformation; while for the latter, small, hard or sharp-edged particles, like grains of sand or of metal or metal oxide, rub off metal surface. Usually, the harder the surface, the more resistant it is to abrasive wear.

Adhesive wear occurs because of local welding at surface asperity junctions. The extremely high local pressure and temperature, combined with continuous relative motion of contact surfaces, cause metal removal from one surface with a lower yield strength to another by solid-phase welding. Such adhesive wear or surface damage is called scoring. A mild adhesive wear is often called scuffing, while a severe adhesive wear is called galling. Seizure is an extension of the galling process when two surfaces are virtually welded together and relative motion is no longer possible [1–3]. As before, harder surfaces are more resistant to adhesive wear.

Fatigue wear is caused by propagation of subsurface microcracks under repeated cyclic loads. Stresses develop on and below the contact surfaces of mating elements. As cyclic loads are applied, microcracks may initiate and propagate and, eventually, coalesce at and near the surface. Materials at the surface then easily spall out, degrading the element surface by either small deep pits or large shallow pits, releasing work-hardened particulate contaminates [7]. Fatigue wear is a prevalent form of failure in such machine elements as rolling contact bearings, gears and so on.

Fretting wear occurs at oscillatory sliding contact surfaces under large loads. It causes damage at the asperities of contact surfaces, producing increased surface roughness and micropits. Typical sites of fretting damage include interference fits, bolts, keyed and splined joints.

Corrosive wear is undesired oxidation or rusting of metal surface caused by the combined mechanical, chemical or electrochemical interaction with the environment. A typical example is the deterioration of lubricated surfaces in sliding bearings.

Wear process includes three stages. The first stage is break-in wear or a running-in process, a beneficial process that allows wear to occur so that mating surfaces can adjust to each other to provide smooth running. After running-in, the tips of asperities in contact become flattened. During the second stage, wear increases with the operating time. The time span at this stage decides product life. The last stage is destructive and is characterized by a rapidly increased wear rate that leads to failure.

The severity of wear can be alleviated by reducing contact forces, relative speeds and mating surface properties. And the most important is to maintain continuous lubrication between rubbing surfaces to prolong element life and postpone the advent of the destructive wear stage.

12.2.3   Lubrication

The objective of lubrication is to reduce friction, wear and friction generated heat by introducing a lubricant, usually petroleum-based mineral oil, between contact surfaces. Lubricant films are usually assumed to be Newtonian fluids, following Newton's law of viscous flow.

12.2.3.1   Newton's Law of Viscous Flow
In Figure 12.4, plate A moves at a velocity v on a film of a lubricant with thickness h. As force F is applied on the plate, shear stress τ will generate within the lubricant. Assume

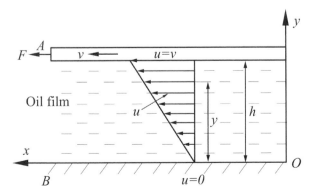

Figure 12.4 Laminar flow between parallel plates.

the lubricant film is composed of a series of horizontal layers that can slide relative to each other, that is, laminar flow. The lubricant layers adjacent to the surfaces A and B have identical velocities to these surfaces. Intermediate layers have velocities following a linear velocity gradient, as shown in Figure 12.4.

Neglecting inertial force and gravity, experimental results indicate that the shear stress in the fluid is proportional to the rate of shearing strain $\partial u / \partial y$ [4, 5]. This is Newton's law of viscous flow and is expressed as

$$\tau = \eta \frac{\partial u}{\partial y} \tag{12.1}$$

where
η –dynamic viscosity, also called absolute viscosity, measured in Pa s;
$\dfrac{\partial u}{\partial y}$ –the rate of shearing strain, or velocity gradient.

12.2.3.2 Viscosity of Lubricants

The viscosity of lubricants is a measure of internal frictional resistance of fluid. Commonly used viscosities are dynamic viscosity and kinematic viscosity.

(1) Dynamic viscosity, η

Three physical units are used to describe dynamic viscosity, that is, SI metric units, C.G.S. units and imperial (FPS) units. The dynamic viscosity is measured by Newton-second per square metre ($\mathrm{N\,sm^{-2}}$), or Pascal-second (Pa s) in SI units. To give a direct feel for dynamic viscosity, the viscosity of water at room temperature is about 1 mPa s, while honey has a viscosity of 1500 mPa s [4]. The unit of dynamic viscosity in C.G.S system is dyne-second per square centimetre ($\mathrm{dyn\,s\,cm^{-2}}$) or poise (P). It has been customary to use centipoise (cP) in analysis for convenience, where $1\mathrm{cP} = 10^{-2}$ P. In imperial units, dynamic viscosity has units of pound force-second per square inch ($\mathrm{lbf\,s/in^2}$), called a reyn in honour of Osborne Reynolds. The conversions of these units are listed in Table 12.1.

(2) Kinematic viscosity, υ

In many situations, it is convenient to use kinematic viscosity rather than dynamic viscosity. Kinematic viscosity is defined as

$$\upsilon = \frac{\eta}{\rho} \tag{12.2}$$

Table 12.1 Conversion factors for dynamic viscosity.

To convert from	To		
	SI units ($\mathrm{N\,s\,m^{-2}}$)	C.G.S units (cP)	Imperial units ($\mathrm{lbf\,s\,in^{-2}}$)
	Multiply by		
SI metric units ($\mathrm{N\,s\,m^{-2}}$)	1	10^3	1.45×10^{-4}
C.G.S units (cP)	10^{-3}	1	1.45×10^{-7}
Imperial units ($\mathrm{lbf\,s\,in^{-2}}$)	6.9×10^3	6.9×10^6	1

where ρ is lubricant density. The densities for mineral oils are between 850 and 900 kg m^{-3}. The SI metric unit for kinematic viscosity is square metres per second (m^2 s^{-1}); while C.G.S unit is square centimetres per second (cm^2 s^{-1}), called a stoke (St), where 1 St = 100 cSt.

The mean value of kinematic viscosity in centistokes (cSt) at 40°C is commonly used to specifies oil viscosity grade number, which appear in ISO Standard 3448 and the Chinese National Standard GB/T3141-1994. Twenty equivalent ISO/GB grades are specified, with kinematic viscosity values at 40°C of 2, 3, 5, 7, 10, 15, 22, 32, 46, 68, 100, 150, 220, 320, 460, 680, 1000, 1500, 2200 and 3200 cSt [8]. The viscosities defined by the American Gear Manufacturers Association (AGMA) correlate its grades of 0~8 with ISO viscosity grades of 32~680 [3]. The Society of Automotive Engineers (SAE) viscosity grades specify oil kinematic viscosity in centistokes at 100°C (212°F). Common grades are SAE 10, 20, 30, 40, 50, 60, 85, 90, 140 and 250 [3].

ISO VG 32 grade industrial oil, or an equivalent L-AN 32 or SAE 10 grade oils are commonly used for general lubrication and for gear-type power transmissions in many types of machinery, like automobiles, turbines, compressors, electric motors and various other equipment [9].

(3) Temperature and pressure effects on viscosity

The viscosities of mineral and synthetic oils decrease substantially with an increase of temperature. Figure 12.5 shows the kinematic viscosity as a function of temperature for GB standard L-AN oils. The dynamic viscosity as a function of temperature for ISO VG oil [10] and SAE oils [11] shows similar trends. Lubricating oils experience an increase in viscosity with pressure only at pressures greater than 200 MPa. The magnitude of viscosity will affect friction, the formation of lubrication film and the load carrying capacity of sliding bearings, which is of great importance in hydrodynamically lubricated bearing design.

12.2.4 Formation of Hydrodynamic Lubrication in a Journal Bearing

12.2.4.1 Formation of Hydrodynamic Lubrication in Plates

In hydrodynamic lubrication, load carrying surfaces are separated by a relatively thick film of lubricant to prevent metal-to-metal contact. Figure 12.6 explains explicitly the required geometry and motion conditions for developing hydrodynamic lubrication in plates.

Assume the plates are extremely wide so that side-leakage flow can be neglected. In Figure 12.6a, two parallel plates move relatively, separated by lubricant film with constant thickness. The velocity within the lubricant film varies uniformly from zero at bottom surface B_1B_2 to v at top surface A_1A_2, form a shear flow. The volume of fluid flowing across section A_2B_2 in a unit time is equal to that flowing across section A_1B_1. Therefore, no pressure is established within the film. In Figure 12.6b, both parallel plates are stationary and a force is applied to the top surface A_1A_2, the lubricant will be squeezed out, form a pressure flow [12].

Now consider a case of two nonparallel plates moving relatively, with a force acting on the top surface A_1A_2, as shown in Figure 12.6c. If velocity profiles are like those indicated by dashed lines, the volume of lubricant into the wedge space through inlet section A_2B_2 during a unit time is obviously greater than the volume discharged through the

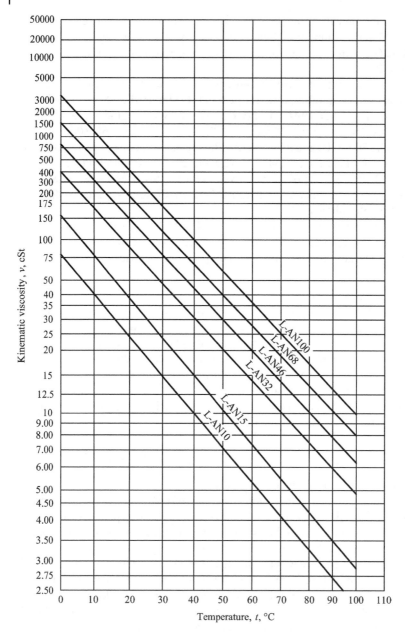

Figure 12.5 Viscosity-temperature curves for industrial mineral oils [5].

outlet section A_1B_1. Since the lubricant is assumed to be incompressible, the velocity distribution profiles must satisfy the condition that the volumetric flow rate through inlet section A_2B_2 equals that through outlet section A_1B_1. Combining the shear flow in Figure 12.6a and pressure flow in Figure 12.6b, the resulting inlet velocity profile should be concave and the outlet velocity profile convex, as shown in Figure 12.6c. Detailed mathematical derivation can be found in Section 12.3.2. The external load is thus supported by fluid pressure generated by relative motion of surfaces.

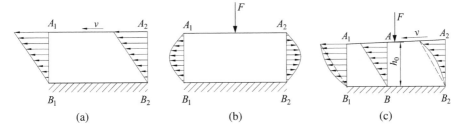

Figure 12.6 Formation of hydrodynamic lubrication in plates.

Therefore, to establish film pressure to separate the surfaces against the applied load, the relative moving surface is required to move at a sufficiently high velocity to draw lubricant of certain viscosity into a narrow wedge-shaped zone.

12.2.4.2 Formation of Hydrodynamic Lubrication in a Journal Bearing

The explanation of hydrodynamic lubrication formation in plates is applicable to journal bearings. Imagine the nonparallel plates are wrapped into a cylindrical geometry for which the moving plate becomes a rotating journal, and the stationary plate becomes a fixed bearing. Figure 12.7 illustrates the formation of hydrodynamic lubrication in a journal bearing from startup to a steady-state hydrodynamic operation.

When a journal is at rest (Figure 12.7a), the applied radial load F forces the journal to move down in the loading direction, squeeze out the oil film at bottom and contact directly with the bearing.

At startup, the journal rotates clockwise slowly. The friction between the journal and bearing cause the journal to roll to the right and climb up the wall of bearing surface. Boundary lubrication and mixed-film lubrication precede the establishment of hydrodynamic lubrication will be observed in Figure 12.7b.

When the rotating speed is progressively increased, the journal draws lubricant oil into the wedge-shaped space above the contact region continuously, producing a pressure high enough in the oil film to lift the journal off the bearing. The gradually built up pressure forces the journal to move to the left, and eventually balanced with the applied radial load F. Thus hydrodynamic lubrication is established. The journal stays at

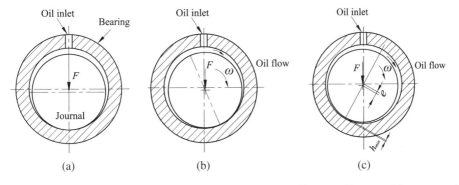

Figure 12.7 Formation of hydrodynamic lubrication in a journal bearing. (a) At rest, (b) startup and (c) steady-state operation.

steady-state position, offset from the loading direction, with an eccentricity e, between the centre of bearing and the centre of journal, as shown in Figure 12.7c.

To sum up, a journal bearing experiences boundary, mixed-film and hydrodynamic lubrication from startup to normal operation. At a normal operating speed, the rotating journal draws a sufficient amount of lubricant of certain viscosity into a wedge-shaped zone between the bearing at a velocity high enough to create a pressure large enough to lift the journal off the bearing. Thus, the external load applied to the journal bearing is supported by a continuous film of oil lubricant.

12.2.5 Potential Failure Modes

Typical failure modes of a sliding bearing include abrasive wear, adhesive wear, scuffing, galling and seizure, fatigue and corrosion.

Foreign particles from operating environment, oxidized wear particles or grit may cause scraping or scratching on the mating element surfaces. In particular, during startup, shutdown or when a journal contacts with a bearing, abrasive wear will accelerate, causing an impaired profile and increased clearance, leading to malfunction of the bearing. Abrasive wear can be reduced by using filters in lubrication systems.

When a sliding bearing carries a heavy load or the supplied lubricant is insufficient, the lubricant film may be so thin that metal-to-metal contact occurs at the bearing interface. Combined with continuous relative motion of the journal and bearing, adhesive wear may occur, leading to the damage of bearing surfaces. In serious cases, adhesive wear results in scoring, scuffing, galling or even seizure.

When a sliding bearing is subject to a cyclic load or a cyclic small amplitude sliding, cracks may generate and fatigue or fretting wear may occur. Finally, acid formation during oxidation of a lubricant may induce unacceptable corrosion on bearing surfaces.

12.3 Load Carrying Capacities

12.3.1 Boundary-Lubricated Bearings

The design of boundary-lubricated bearing is largely an empirical process based on documented user experiences. The primary design variables of unit bearing load p, sliding velocity v and their product pv should be within allowable values.

Unit bearing load p is the ratio of the applied bearing load F to the projected area of a bearing, that is, the product of journal diameter d and bearing width B, that is,

$$p = \frac{F}{dB} \leq [p] \tag{12.3}$$

Because excessive velocity may speed up the wear process, the linear velocity of journal must be within allowable values and is expressed as

$$v = \frac{\pi dn}{60 \times 1000} \leq [v] \tag{12.4}$$

In addition to the individual consideration of load capacity, p, and sliding velocity, v, the pv factor is an important variable for the evaluation of temperature rise and wear rate at sliding interfaces. Combined with Eqs. (12.3) and (12.4), the *pv* factor is calculated by

$$pv = \frac{F}{Bd} \cdot \frac{\pi dn}{60 \times 1000} \le [pv] \tag{12.5}$$

When designing a boundary-lubricated bearing, these design criteria must be satisfied. The allowable values of $[p]$, $[v]$, $[pv]$ of commonly used bearing materials can be found in Table 12.5 later.

12.3.2 Hydrodynamically Lubricated Bearings

To simplify the analysis, it is assumed that [1–5]:

- the lubricant is an incompressible Newtonian fluid with constant viscosity throughout the film;
- the lubricant experiences laminar flow, with no slip at boundary surfaces;
- the lubricant has no inertial or gravitational forces;
- the bushing and journal extend infinitely in the axial direction (z-coordinate), that is, no lubricant flow in the axial direction;
- the film is so thin that it experiences negligible pressure variation over its thickness, thus the pressure depends only on coordinate x;
- the velocity of any particle of lubricant in the film depends only on coordinates x and y.

12.3.2.1 Reynolds Equation

Figure 12.8a shows the forces acting on an elemental cube M with dimensions of dx, dy and dz selected from the lubricant film in a journal bearing (Figure 12.8b). The right and left sides of the element are subjected to normal forces due to pressure, while the top and bottom sides of the element are subjected to shear forces due to viscosity and velocity. Establish the force equilibrium equation in the x direction, which gives

$$\sum F_x = pdydz - \left(p + \frac{\partial p}{\partial x} dx \right) dydz - \tau dxdz + \left(\tau + \frac{\partial \tau}{\partial y} dy \right) dxdz = 0 \tag{a}$$

This reduces to

$$\frac{\partial p}{\partial x} = \frac{\partial \tau}{\partial y} \tag{b}$$

According to Newton's law of viscous flow in Eq. (12.1), the shearing stress is simply proportional to the rate of shearing strain, that is,

$$\tau = \eta \frac{\partial u}{\partial y} \tag{c}$$

Substituting Eq. (c) in Eq. (b), and convert it to

$$\frac{\partial^2 u}{\partial y^2} = \frac{1}{\eta} \frac{\partial p}{\partial x} \tag{d}$$

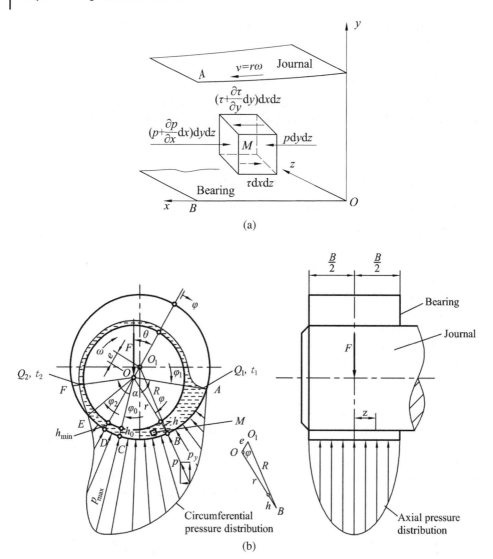

Figure 12.8 Hydrodynamic lubrication in a journal bearing. Source: Adapted from Pu and Ji, 2006.

Holding x constant, integrating twice with respect to y and evaluating constants of integration by imposing boundary conditions as $y = 0$, $u = 0$; and $y = h$, $u = v$ yields the expression

$$u = \frac{v}{h}y + \frac{y(y - h)}{2\eta} \cdot \frac{\partial p}{\partial x} \tag{12.6}$$

This equation gives the velocity distribution of lubricant film as the function of coordinate y and pressure gradient $\partial p/\partial x$. The velocity distribution across the film consists of two terms: a linear distribution due to shear flow and a parabolic distribution due to pressure flow.

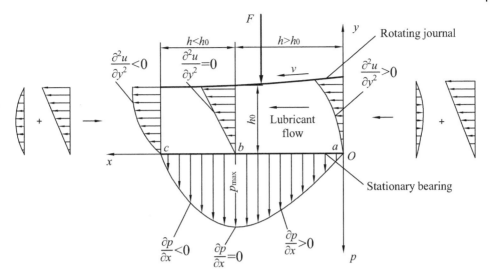

Figure 12.9 Velocity and pressure distributions in a converged wedge.

Figure 12.9 illustrates the superposition of these two terms to obtain the velocity at the inlet, the maximum pressure point and the outlet. The parabolic term may be additive or subtractive to the linear term, depending on the sign of pressure gradient at the inlet and outlet. The figure also presents the variation of pressure gradient and pressure distribution. At the section where pressure reaches the maximum value, that is, $\partial p/\partial x = 0$, and the velocity gradient is linear.

Define Q as the volume of lubricant flowing across a section in the x direction per unit of time. For unit width in the z direction,

$$Q = \int_0^h u\, dy \tag{12.7}$$

Substituting the value of u from Eq. (12.6) and integrating, gives

$$Q = \frac{vh}{2} - \frac{h^3}{12\eta}\frac{\partial p}{\partial x} \tag{12.8}$$

When $\partial p/\partial x = 0$, the pressure reaches the maximum value p_{max}, and $h = h_0$, the volume flow rate is

$$Q = \frac{vh_0}{2} \tag{12.9}$$

Because it is assumed that the lubricant is incompressible, the flow rate is constant for all sections. Thus

$$-\frac{h^3}{12\eta}\frac{\partial p}{\partial x} + \frac{vh}{2} = \frac{vh_0}{2}$$

Convert it to

$$\frac{\partial p}{\partial x} = \frac{6\eta v}{h^3}(h - h_0) \tag{12.10}$$

This is the classical one-dimensional Reynolds equation. It neglects side leakage, that is, the flow in the z direction.

From the derivation in (12.10), similar conditions for establishing hydrodynamic lubrication to the analysis in Section 12.2.4 can be summarized as; an abundant supply of lubricant with sufficient viscosity, a wedge-shaped space between two relatively moving plates and lubricant flow from big entrance to small exit.

12.3.2.2 Hydrodynamic Lubrication in a Journal Bearing

Figure 12.8b shows a journal rotating at a constant speed in the clockwise direction, supported by a film of lubricant of variable thickness h on a fixed bearing. The clearance between the bearing and the journal is highly exaggerated. The nomenclature and main design variables of the journal bearing are also illustrated in the figure.

(1) Geometrical relationship of design variables

In Figure 12.8b, the line OO_1 connects journal centre O and bearing centre O_1. A radial clearance δ is the difference in the radii of bearing and journal, that is,

$$\delta = R - r \tag{12.11}$$

Similarly, a diametrical clearance Δ is

$$\Delta = D - d \tag{12.12}$$

The ratio of diametrical clearance to the journal diameter is defined as relative clearance,

$$\psi = \frac{\Delta}{d} = \frac{\delta}{r} \tag{12.13}$$

The distance between bearing centre and journal centre is eccentricity e. The dimensionless eccentricity radio ε is defined as

$$\varepsilon = \frac{e}{\delta} \tag{12.14}$$

The minimum film thickness is designated by h_{min}, and it occurs at point D on the line of centres.

$$h_{min} = \delta - e = \delta(1 - \varepsilon) = r\psi(1 - \varepsilon) \tag{12.15}$$

In journal bearing design, one of principal objectives is to ensure the minimum film thickness is sufficiently large to separate the surfaces of journal and bearing completely.

The film thickness at any other point is designated by h; from the triangle OO_1B in Figure 12.8, we have

$$R^2 = e^2 + (r + h)^2 - 2e(r + h) \cos \varphi$$

Therefore,

$$r + h = e \cos \varphi \pm R\sqrt{1 - \left(\frac{e}{R}\right)^2 \sin^2 \varphi}$$

Neglecting $\left(\frac{e}{R}\right)^2 \sin^2 \varphi$, the oil film thickness at any position is given by

$$h = \delta(1 + \varepsilon \cos \varphi) = r\psi(1 + \varepsilon \cos \varphi) \tag{12.16}$$

The film thickness at the highest pressure on point C is

$$h_0 = \delta(1 + \varepsilon \cos \varphi_0) \tag{12.17}$$

(2) Load carrying capacity coefficient

Convert variables in an xyz rectangular coordinate system to a polar-cylindrical coordinate system with z axis along the axis of journal, we have

$$dx = rd\varphi$$

$$v = r\omega$$

Substitute these relations and Eq. (12.16), Eq. (12.17) into Eq. (12.10) and the Reynolds equation in a cylindrical coordinate system is expressed as

$$\frac{dp}{d\varphi} = 6\eta \cdot \frac{\omega}{\psi^2} \cdot \frac{\varepsilon(\cos\varphi - \cos\varphi_0)}{(1 + \varepsilon\cos\varphi)^3} \tag{12.18}$$

Integrating from φ_1 to φ where pressure is created in the bearing, we have

$$P_\varphi = 6\eta\frac{\omega}{\psi^2}\int_{\varphi_1}^{\varphi}\frac{\varepsilon(\cos\varphi - \cos\varphi_0)}{(1 + \varepsilon\cos\varphi)^3}d\varphi \tag{12.19}$$

Integrating from φ_1 to φ_2, the total load in the radial load direction for unit width is

$$\begin{aligned}
P_y &= \int_{\varphi_1}^{\varphi_2} P_\varphi \cos[180^\circ - (\varphi_a + \varphi)]rd\varphi \\
&= 6\eta\frac{\omega r}{\psi^2}\int_{\varphi_1}^{\varphi_2}\left[\int_{\varphi_1}^{\varphi}\frac{\varepsilon(\cos\varphi - \cos\varphi_0)}{(1 + \varepsilon\cos\varphi)^3}d\varphi\right]\cos[180^\circ - (\varphi_a + \varphi)]\,d\varphi
\end{aligned} \tag{12.20}$$

The circumferential pressure distribution in a hydrodynamically lubricated bearing is illustrated in Figure 12.8b. The pressure increases gradually as the rotating journal draws viscous oil into the converging wedge, approaching the maximum value and falls rapidly as the space between the journal and bearing diverges again. Equation (12.20) expresses the pressure in the external load direction in one cross section of the journal bearing.

Neglecting side leakage, the integrated effect of the pressure distribution is a force sufficient to balance the applied load and support the shaft on the oil film without metal-to-metal contact [3]. Therefore, for a limited width sliding bearing, the total load capacity is

$$F = p_y B = 6\eta\frac{\omega r}{\psi^2}B\int_{\varphi_1}^{\varphi_2}\left[\int_{\varphi_1}^{\varphi}\frac{\varepsilon(\cos\varphi - \cos\varphi_0)}{(1 + \varepsilon\cos\varphi)^3}d\varphi\right]\cos[180^\circ - (\varphi_a + \varphi)]\,d\varphi \tag{12.21}$$

Therefore

$$\frac{F\psi^2}{Bd\eta\omega} = 3\varepsilon\int_{\varphi_1}^{\varphi_2}\left[\int_{\varphi_1}^{\varphi}\frac{(\cos\varphi - \cos\varphi_0)}{(1 + \varepsilon\cos\varphi)^3}d\varphi\right]\cos[180^\circ - (\varphi_a + \varphi)]\,d\varphi \tag{12.22}$$

Define the load carrying capacity coefficient C_p, or Sommerfeld number [18], as

$$C_p = \frac{F\psi^2}{Bd\eta\omega} = \frac{F\psi^2}{2\eta vB} \tag{12.23}$$

The unit for bearing width B is selected as m. The unit of dynamic viscosity η is N·s m^{-2} and the journal velocity is m s^{-1}. Load carrying capacity coefficient and relative clearance are dimensionless.

In reality, due to side leakage, the highest pressure occurs at the middle of bearing width and falls rapidly close to both ends. Figure 12.8c shows the axial pressure distributions along the journal through the line of maximum pressure. In journal bearing design, this effect is considered by using load carrying capacity coefficient C_p from two-dimensional numerical analysis results.

Load carrying capacity coefficient C_p is a function of eccentricity radio ε, the ratio of bearing width to diameter B/d and wrap angle α (see Figure 12.8). Physically, it implies the position of journal in a bearing. Computerized solutions for C_p are presented in charts or tables in standards or design handbooks. Table 12.2 lists values of C_p when a journal bearing is supplied by a non-pressure oil opposite the load carrying area, and pressure generates within an 180° arc between the journal and liner. It can be noticed that with the increase of ε and B/d, the load carrying capacity coefficient C_p also increases.

As discussed before, the moving journal and stationary bearing are separated by lubricant film that supports load. Reducing the minimum film thickness will increase eccentricity ratio ε, and consequently increase the load carrying capacity coefficient C_p. However, the minimum film thickness is limited by the asperities of both journal and bearing. The minimum film thickness must meet

$$h_{\min} = \delta(1 - \varepsilon) = r\psi(1 - \varepsilon) \geq [h] = S(R_{z1} + R_{z2}) \tag{12.24}$$

where

S – factor of safety. Safety factor is usually selected as $S = 2\sim3$ to provide a margin of safety to prevent metal-to-metal contact and to pass particles that escape the oil filter.

R_{z1}, R_{z2} – surface roughness of the journal and bearing, respectively. For common sliding bearings, R_{z1} of a ground journal varies from 1.6 to 3.2 μm, while R_{z2} of a bearing is from 3.2 to 6.3 μm. For high precision sliding bearings, polishing or lapping can produce a surface finish in the order of 0.2 to 0.8 μm for a journal and 0.4 to 1.6 μm for a bearing [5, 16].

$[h]$ – the allowable minimum film thickness, which depends on the surface roughness of the journal and bearing.

Table 12.2 Load carrying capacity coefficient, C_p [15].

B/d	ε							
	0.2	0.4	0.6	0.8	0.9	0.925	0.95	0.975
0.25	0.0190	0.0537	0.1465	0.6054	2.1725	3.4617	7.3485	23.4266
0.50	0.0747	0.2002	0.4900	1.7222	5.1676	7.8436	13.852	38.055
0.75	0.1513	0.381	0.8883	2.6987	7.1375	10.4252	16.7465	43.5128
1.00	0.2328	0.5601	1.2448	3.4514	8.4037	11.9405	18.5775	45.771
1.25	0.3106	0.7326	1.5312	3.9818	9.2279	12.8584	19.9471	46.7535
1.50	0.3781	0.8711	1.7528	4.3531	9.6987	13.4451	20.5259	47.2761

12.3.3 Heat Balance Analysis

The viscosity of lubricant is a crucial parameter for the performance of sliding bearing. In the previous analysis, the viscosity is assumed to be constant as the lubricant flows through the bearing. However, oil temperature is actually higher when the lubricant leaves the loading zone than it was on entry [1], and the viscosity drops off significantly with the rise of temperature. It is thus important to estimate oil film temperature at equilibrium operating conditions so as to use proper viscosity at operating temperature in the analysis.

Under equilibrium operating conditions, the rate at which heat is generated by friction must be equal to the rate at which heat is removed from the bearing so as to prevent continued temperature rise to an unsatisfactory level. The removed heat includes the heat taken away by lubricant flow and the heat dissipated from the exposed metal surface area of bearing housing to the ambient environment. Therefore

$$H = H_1 + H_2 \tag{12.25}$$

H –heat generated by friction per second, or frictional power rate, calculated by
      $H = fFv$

H_1 –heat removed by lubricant flow per second $H_1 = cQ\rho(t_o - t_i)$

H_2 –heat dissipated from the bearing housing per second $H_2 = \alpha_s \pi dB(t_o - t_i)$

   Therefore, we have

$$fFv = cQ\rho(t_0 - t_i) + \alpha_s \pi dB(t_0 - t_i)$$

Then

$$\Delta t = t_o - t_i = \frac{fFv}{Qc\rho + \pi \alpha_s dB} = \frac{\frac{fFv}{\psi vBd}}{\frac{Q}{\psi vBd}c\rho + \frac{\pi \alpha_s dB}{\psi vBd}} = \frac{\frac{f}{\psi}p}{2c\rho \frac{d}{B}C_Q + \frac{\pi \alpha_s}{\psi v}} \tag{12.26}$$

where

Q –volumetric flow rate of lubricant, m^3 s^{-1};

ρ –lubricant density, for mineral oils $\rho = 850\sim900$ kg m^{-3};

c –specific heat of lubricant, for mineral oils $c = 1675\sim2090$ J (kg °C)$^{-1}$;

t_o –outlet oil temperature, °C;

t_i –inlet oil temperature, $t_i = 35\sim45$ °C;

α_s –heat transfer coefficient. It depends on many factors, such as bearing material, surface geometry, the temperature difference between the housing and surrounding, and ventilation. For light loaded bearings, or high ambient temperature, select $\alpha_s = 50$ W (m^2 °C)$^{-1}$; for medium-sized bearings and general ventilation, select $\alpha_s = 80$ W (m^2 °C)$^{-1}$ and for heavy bearings and good ventilation, select $\alpha_s = 140$ W (m^2 °C)$^{-1}$ [5].

$C_Q = \frac{Q}{\omega \psi d^3}$ –coefficient of oil flow. Table 12.3 lists the values of C_Q for various values of eccentricity radio ε and width to diameter ratio B/d when pressure generates within 180° arc between the journal and liner.

$C_f = \frac{f}{\psi}$ –coefficient of friction variable. Table 12.4 lists the value of C_f for various values of eccentricity radio ε and width to diameter ratio B/d when pressure generates within 180° arc between the journal and liner.

Table 12.3 Coefficient of oil flow, C_Q [15].

B/d	ε							
	0.2	0.4	0.6	0.8	0.9	0.925	0.95	0.975
0.25	0.0110	0.0214	0.0295	0.0339	0.0334	0.0332	0.0321	0.0299
0.50	0.0213	0.0394	0.0522	0.0573	0.0547	0.0530	0.0508	0.0451
0.75	0.0289	0.0513	0.0660	0.0698	0.0649	0.0617	0.0580	0.0513
1.00	0.0338	0.0575	0.0729	0.0735	0.0679	0.0640	0.0601	0.0513
1.25	0.0369	0.0615	0.0748	0.0747	0.0682	0.0642	0.0601	0.0507
1.50	0.0385	0.0628	0.0755	0.0750	0.0675	0.0632	0.0589	0.0501

Table 12.4 Coefficient of friction variable, C_f [15].

B/d	ε							
	0.2	0.4	0.6	0.8	0.9	0.925	0.95	0.975
0.25	87.6215	36.5135	16.9891	6.0472	2.7036	2.0119	1.2603	0.5668
0.50	22.4028	9.9610	5.4518	2.5502	1.3352	1.0390	0.7419	0.4120
0.75	11.1513	5.3207	3.1078	1.7050	1.0322	0.8278	0.6390	0.3801
1.00	7.3068	3.6873	2.2797	1.3799	0.9061	0.7446	0.5894	0.3754
1.25	5.5146	2.8666	1.8931	1.2248	0.8220	0.7075	0.5589	0.3706
1.50	4.5565	2.4424	1.6803	1.1391	0.796	0.6782	0.5495	0.3701

Since temperatures are different from point to point within bearing, the average of inlet and outlet temperature t_m is used to determine the viscosity to be used in the analysis. The average temperature is

$$t_m = t_i + \frac{\Delta t}{2} \tag{12.27}$$

Δt —temperature rise of lubricant from inlet to outlet, usually $\Delta t \leq 30°C$;

t_m —average or mean temperature. Average temperatures within the range of $t_m = 50–75°C$ are commonly used. The viscosity used in the analysis must correspond to t_m.

t_i —inlet oil temperature. Inlet oil temperature is usually slightly higher than ambient temperature, normally select $t_i = 35–45°C$.

Finding the equilibrium temperature is an iterative process. To find out temperature rise from Eq. (12.26), a viscosity η at average temperature is required; while to find out average temperature from Eq. (12.27), a temperature rise is required. To solve this logical circle during design process, we firstly estimate an average oil film temperature and select the corresponding viscosity, then use Eq. (12.26) to calculate temperature rise, followed by average temperature calculation by Eq. (12.27). This process is continued until the estimated and computed temperatures agree.

12.4   Design of Sliding Bearings

12.4.1   Introduction

The design of a sliding bearing is a considerably complicated procedure, involving variable selections and design decisions. With properly selected variables, the rotation of journal could develop sufficient film thickness and necessary oil pressure to carry external loads, while an improper selection may result in direct metal-to-metal contact, rapid heating and, eventually, failure. Therefore, the minimum film thickness and temperature rise must be within the specified limits to ensure satisfactory bearing performance.

While designing a sliding bearing, the information provided includes the magnitude and direction of bearing load F, the journal diameters d and the rotational speed of journal n. The design decisions to be made are the material of bearing, the type of lubricant and its viscosity η, the width of bearing B, the relative clearance ψ, as well as operating temperature and so on.

The combined effect of these variables decides the performance of a sliding bearing, which is evaluated by the minimum film thickness h_{min} and the temperature rise Δt according to the detailed lubrication analyses. Limitations on these values are determined by the surface finish of journal and bearing, the characteristics of bearing material and lubricant. It is important to ensure that the design is satisfactory for all reasonably anticipated combinations of design variables.

Due to limited space, data presented in this book are for journal bearings with a wrap angle of 180°. When designing journal bearings with wrap angles other than 180°, readers can obtain relevant data in design handbooks or standards using the design method introduced here.

12.4.2   Materials for Sliding Bearings

12.4.2.1   Property Requirements for Sliding Bearing Materials

The bearing and journal should be made of different materials. Usually, bearing materials are softer than journal surfaces so that hard abrasive particulates can be embedded completely in the bearing without protruding above the surface and causing damage to the rotating journal. That is, bearing materials should provide embeddability. On the other hand, bearing materials should have sufficient compressive and fatigue strength to resist externally applied cyclic loads, even at elevated temperatures. They also need to have conformability to permit the journal and bearing contours to conform with each other or to relieve local high pressures caused by misalignment or shaft deflection. Compatibility with the journal is equally important for bearing materials to resist local welding, scoring and seizing, especially at startup or stopping when direct metal contact occurs. Besides, the coefficient of friction between journal and bearing must be as low as possible to reduce frictional wear when sliding bearings operate at boundary lubrication during startup or at low speeds [2].

Bearing materials should have high thermal conductivity to dissipate away the heat generated during operation. The thermal expansion coefficient of bearing housing and journal material should be similar to maintain suitable clearance. Bearing materials should also have appreciable corrosion resistance to acids that may form during lubricant oxidation and considerable wear resistance to outside contamination [2]. Additional considerations in the selection of bearing material are durability, machinability and, of course, cost.

The properties required for desirable bearing materials are comprehensive and sometimes conflicting. As with many machine elements, tradeoff decisions must frequently be made to conciliate contradictory requirements while choosing bearing materials.

12.4.2.2 Commonly Used Bearing Materials

The combination of large loads and high sliding velocities encountered in sliding bearings is somewhat analogous to wormgear drive. Since wear is inevitable after a period of operation, an ideal design is to confine most wear to the element that can be easily and economically replaced.

In a sliding bearing, the journal is part of a shaft, frequently made of steel, heat treated by carburizing, nitriding, flame or induction hardening. Hard surfaces on rotating shafts usually mate with soft, wear compatible bearing surfaces. Commonly used bearing materials are:

(1) Babbitts

Babbitts may be lead-based (as 75%Pb, 15%Sb, 10%Sn) or tin-based (as 89%Sn, 8%Pb, 3%Cu) [2, 3]. They use lead or tin as a soft matrix metal and hard particles of Sb-Sn or Cu-Sn to resist wear [5]. Because of their softness, babbitts are unrivalled in conformability and embeddability. However, their compressive and fatigue strengths are rather low, particularly at elevated temperature. Because of low strength and high cost, a thin babbitt overlay, usually about 0.025 mm [2], is often deposited as liners over steel bushings to combine the great load carrying capacity of steel with conformability and corrosion resistant of babbitts. Babbitts are usually used for high speed, heavy duty applications such as crankshaft bearings in internal combustion engines.

(2) Bronze alloys

Bronze refers to copper alloyed with lead, tin or aluminium, either singly or in combination. Consequently, copper alloys used in bearings include lead bronze, tin bronze and aluminium bronze. Lead bronzes have good embeddability and resistance to seizing, yet relative low strength. Tin bronzes and aluminium bronzes have higher strength and hardness, yet poor embeddability [3]. They are used in demanding applications in engines, machine tools and aircrafts.

(3) Grey cast iron and steel

Grey cast iron and steel are inexpensive bearing materials suitable for light loads, low speed applications. The free graphite in cast iron adds lubricity but liquid lubricant is needed as well. Hardened steel can run against any material with proper lubrication.

(4) Sintered porous metals

Porous metal bearings are produced by sintering powders of bronze, tin, aluminium or mixed with lead or copper into a desired shape. The sintered porous structure has a large number of voids into which lubricating oil penetrates and is held by capillary action. During operation, the porous structure releases the oil out of pores to the bearing surface to provide lubricant film. When the shaft stops rotation, the lubricant flows back into pores [4]. The self-contained lubrication makes porous metal bearings ideal for applications where lubrication supply is difficult, inadequate or infrequent.

(5) Nonmetallic materials

Nonmetallic bearing materials mainly refer to various plastic materials, including, fiuoropolymers, phenolics, polycarbonates, acetals, nylons, Teflon and many others [3]. They have inherently low coefficient of friction against metal surfaces, and offer the possibility of dry running or with a very limited supply of lubricant. Many of them have low strength, low melting point and poor thermal conduction, which severely limits the load carrying capacity and operating speed. Therefore, plastic bearings are widely used for low to moderate loads and speeds. They are lightweight, easy to machine and inexpensive. Other nonmetallic materials for bearings include graphite, rubber, woods and so on [13].

Commonly used materials for boundary-lubricated bearings, their operating limits and performance ratings, are listed in Table 12.5. Material properties of porous materials and nonmetallic materials can be referred to in references [13, 14].

12.4.3 Lubricants, Their Properties and Supply

12.4.3.1 Lubricants

Lubricants are applied to the interface between moving elements to reduce friction, wear and heat. They enhance smooth operation and prolong lifetime of machine elements. Oils, greases and solids are three main types of lubricants used in sliding bearings.

Table 12.5 Common bearing materials, their operating limits and performance ratings [3, 5, 13, 14].

Materials		$[p]$	$[v]$	$[pv]$	T_{max}	Friction and compatibility	Conformability and embedability	Corrosion resistance	Fatigue strength	Applications
		MPa	m s^{-1}	MPa m s^{-1}	°C					
Babbitt	Tin-base Babbitt	20–25	60–80	15–20	150	5	5	5	1	High speed, heavy loads
	Lead-base Babbitt	5–15	8–12	5–15	150	5	5	3	1	Medium speed and loads
Bronze	Tin bronze	8–15	3–10	15	260	3	1	4	5	Medium speed, medium to heavy loads
	Lead bronze	5–25	8–12	30	260	3	2	2	4	High speed, variable, impact or heavy loads
	Aluminium bronze	15–20	5	15	260	1	1	4	4	Low speed, heavy loads
Iron	Grey cast iron	1–4	2	1		2	1	5	5	Low speed, light loads

Notes: 5 is excellent; 1 is poor.

Lubricating oils include natural oils and synthetic oils. Natural oils cost less and can meet general lubrication demands. Chemical additives are frequently added to enhance viscosity, to reduce friction, wear and corrosion or to retard oxidation. Synthetic oils are specially designed chemical formulations with better performance yet higher cost than natural oils [3].

Greases are essentially a mixture of natural or synthetic oils and thickeners. Thickeners act as carriers for the oil, which are metallic soaps formed by the reaction of animal or vegetable fats with alkaline substances such as lithium, calcium, an aluminium complex and others [3]. Unlike oils that circulate and serve a cooling and cleaning function, after being applied to the interface between moving elements, the grease remains in place to provide lubrication and to prevent harmful contaminants from entering contact surfaces. Therefore, periodically replenishing greases is required to discharge contaminated or oxidized grease [2, 3].

A solid lubricant is a thin solid film that adheres to mating surfaces to reduce friction and wear. Solid lubricants blended with binders are applied to critical surfaces by brushing, spraying or dipping [3]. Molybdenum disulfide (MoS_2), graphite, powdered metal and Teflon coating are commonly used solid lubricants. They are usually used in applications at excessively high or low temperatures or operation in a vacuum.

12.4.3.2 Lubricant Properties and Their Selection

Lubricant properties contribute to satisfactory performance of sliding bearings. Lubricants should have adequate viscosity and good lubricity at operation temperature and pressure. Their chemical and thermal properties should be stable within operating temperature for a reasonable service period. Besides, lubricant should be compatible with surrounding elements, such as bearings and seals, with regard to corrosion resistance and degradation [3].

The selection of lubricant depends on a number of factors, such as lubrication type, load, speed, operating temperature and working environment. Viscosity grade number, viscosity index (VI), and corrosion protection are the main properties need to be considered [3]. Viscosity grade number discussed in Section 12.2.3 affects oil film forming capability. High viscosity facilitates film formation and is recommended for heavy load, low speed and high temperature application. However, too high a viscosity may result in a waste of power in overcoming the internal friction of oil. Lower viscosities may be used, on the other hand, for light load, high speed or low temperature operation.

Viscosity index (VI) is a measure of temperature effect on viscosity. Generally, a high viscosity index is desirable as it exhibit a more uniform performance as temperature varies [3]. Corrosion protection can be improved by adding additives to base oils. For example, corrosion inhibitors can form a protective shield against water on ferrous metals and bronze bearing metals. Oxidation inhibitors are used to prolong the life of oils. The selection of lubricant with suitable additives are determined by specific application demands and can be referred to in relevant references [17].

12.4.3.3 Lubricant Supply

To maintain hydrodynamic lubrication, sufficient lubricant must be supplied to sliding bearing surfaces. Continuous lubricant can be supplied by oil holes and grooves, oil rings, splash, oil bath and oil pump [2]. In addition, appropriate oil filters or shaft seals are required to keep lubricant free from contaminants to avoid abrasive wear.

An abundant lubricating oil can be fed through an oil supply hole to the distributing grooves, which are cut into the bearing internal surface axially or circumferentially, as shown in Figure 12.2. Oil flows either by gravity or under pressure to the bearing surface. It is an effective and efficient method for supplying oil and for removing heat.

Oil rings, discs or rotating elements are effective in providing oil for low to medium speed bearings. The oil ring, usually about 1.5–2 times the diameter of journal, hangs loosely from the journal, with the lower part dips into an oil reservoir [2]. As the shaft rotates, the ring lifts oil to the bearing by a combination of inertia and surface tension effects.

Splash feeding supplies oil to the bearing by rotating machine elements. For example, in a gear drive, one of gears is designed to dip into an oil sump. The rapidly rotating gear carries oil up to the gear mesh and splashes oil to be channelled to flow into the bearing.

In oil bath lubrication, oil is supplied by a partially submerged journal in the oil reservoir. The housing should be made oil tight. The shaft speed should be controlled to avoid excessive churning of a large volume of oil and substantial viscous friction losses [2, 10].

Oil supplied by a pumping or oil mist lubrication system distributes oil flow, or oil mist to the bearing surface for demanding applications. It is commonly used for high speed, heavy duty bearings or gearings.

For boundary-lubricated bearings, the required lubricant is in a small quantity. Lubricants may be supplied by grease fittings, hand oiling, oil cups or wick feed, which rely on mechanical feed, gravity or capillary action.

12.4.4 Design Criteria

For a boundary-lubricated bearing, it is important to guarantee the formation of boundary oil film to avoid wear and galling. Therefore, design criteria are to ensure unit bearing load p, sliding velocity v and their product pv within allowable limits, as expressed in Eqs. (12.3)–(12.5).

In a hydrodynamically lubricated bearing, film pressure is created by relatively moving surface drawing lubricant into a narrowing, wedge-shaped space at a velocity sufficiently high to create a film pressure necessary to separate the journal and bearing. The acceptable value of minimum film thickness depends on the surface roughness of the journal and bearing because the film must be thick enough to eliminate any solid contact during expected operating conditions. Therefore, a minimum film thickness must be guaranteed by

$$h_{min} \geq S(R_{Z1} + R_{Z2}) \tag{12.28}$$

Frictional energy in sliding bearings increases oil temperature, which will reduce viscosity and cause oxidation of lubricant. Therefore, the temperature rise should be kept within the range of 10–30°C and the average temperature should not exceed 75°C. Since the maximum oil film temperature can be substantially higher than average temperature, which may even cause damage to some bearing materials, such as babbitt alloys; a widely used rule of thumb is to keep the maximum oil film temperature under 93°C for babbitt bearings [13]. These limitations are specified by the characteristics of bearing materials and lubricants and can be found in design handbooks [13, 14].

12.4.5 Design Procedures and Guidelines

Sliding bearing design starts once the mating journal diameter has been determined. Like many other machine elements, the design of sliding bearings involves variable selections and design decisions and, therefore, several acceptable solutions may be possible. The following only presents the basic design procedures and empirical guidelines for the design of boundary and hydrodynamically lubricated journal bearings. The design cases in the following section show the principle steps in journal bearing design. More realistic and detailed calculations can be referred to in design handbooks or standards [14, 18].

12.4.5.1 Design of Boundary-Lubricated Bearings

The design process for boundary-lubricated bearings is relatively simple. A ratio of bearing width to diameter, B/d, typically within the range of 0.5–2.0, is first selected to obtain the bearing width. Then the unit bearing load p, the linear velocity of journal v and the value of pv are calculated and evaluated against the corresponding allowable values of the selected material. These calculations are also required for hydrodynamic bearings, as they may experience boundary and mixed-film lubrication during startup or at low speeds.

12.4.5.2 Design of Hydrodynamically Lubricated Bearings

The design of hydrodynamically lubricated bearings is more complicated than the design of boundary-lubricated bearings. Although the data provided are the same, in hydrodynamic bearing design, designers need to select bearing width B, relative clearance ψ and lubricant viscosity η, and perform detailed lubrication analyses to ensure the minimum film thickness h_{min} and temperature rise Δt do not exceed limitations. If not, the initially selected variables have to be changed to reiterate analyses. Suggestions on design procedures and important parameter selections are given here:

(1) Select suitable bearing materials capable of providing sufficient strength, conformability, embeddability and corrosion resistance;
(2) Select the ratio of width to diameter B/d, specify the actual design value of bearing width B;

   The bearing and journal have an identical nominal diameter. Since the journal is a part of shaft, its diameter is determined by strength and rigidity requirements of the shaft. The bearing width is specified to provide adequate bearing capacity. Great ratios of B/d will increase load carrying capacity, frictional energy and temperature rise, while small ratios increase side leakage and facilitate heat dissipation. The typical range of B/d is from 0.25 to 2.0 [3], with smaller values for high speeds, light duty applications and larger values for low speed, heavy duty applications.
(3) Ensure the calculated value of p,v, pv is within the limits of selected materials;
(4) Assume an average temperature and specify lubricant viscosity η;

   The viscosity of lubricant is a critical parameter for the performance of a journal bearing. Small viscosity lubricants are usually for high speed, light duty bearings, while large viscosity lubricants are for low speed, heavy load applications, as they can increase load carrying capacity yet cause large power loss and high temperature. Since large clearance usually makes it difficult to form lubrication film, a large viscosity value is chosen. With the increase of B/d, side leakage will be reduced, therefore, small viscosity is preferred. To start design, the initially selected viscosity can be estimated by [14]

$$\eta = \frac{0.068}{n^{1/3}} \qquad\qquad (12.29)$$

where n is rotational speed of journal in rad s^{-1}.

(5) Specify a relative clearance ψ and diametrical clearance Δ;

Many factors influence the selection of relative clearance, such as bearing diameter, operation precision requirement, rotational speed, load variation, shaft deflection, as well as the surface roughness and thermal expansion coefficient of the journal and bearing.

The relative clearance greatly influences the operational behaviour of a journal bearing. Normally, the load carrying capacity and rotation precision increases with the decrease of relative clearance. However, if the clearance is too small, the minimum film thickness will be too thin to form hydrodynamic lubrication. Small clearance results in high operating temperature, excessive wear and friction and possible scoring and seizing. On the other hand, a large clearance permits a greater oil flow for cooling, allows dirt particles to pass through and is usually for high speed bearings. Yet, too large a clearance will cause a bearing becomes noisy [1].

An overall guideline for relative clearance is within the range of 0.001–0.002. High speeds and small loads require a large clearance. The initial relative clearance can be estimated by the empirical formula as [19]

$$\psi = 0.8\sqrt[4]{v} \times 10^{-3} \qquad\qquad (12.30)$$

where v is journal linear velocity in m s^{-1}.

(6) Specify the surface roughness specification for the journal and bearing based on the application.

(7) Calculate temperature rise, inlet and outlet temperature and average temperature. Ensure they do not exceed limits.

(8) Compute the minimum and maximum diametrical clearance, iterate the previous steps to ensure that the minimum film thickness h_{min} and temperature rise Δt satisfy the specified limitation at both the minimum and maximum possible clearance.

(9) Complete the design of hydrodynamically lubricated bearing by summing up previous results including lubricant selection, diametrical clearance, surface roughness and thermal control. Besides, an adequate supply of clean and sufficiently cooled lubricant at the bearing inlet is also needed.

12.4.6 Design Cases

Example Problem 12.1
A bearing is to be designed to carry a radial load of 10.0 kN from a shaft with a minimum diameter of 100 mm and rotating at 750 rpm. Design the bearing to operate under boundary-lubrication conditions.

Solution

Steps	Computation	Results	Units
1. Select diameter	$d = 100\,\text{mm}$	$d = 100$	mm
2. Select the ratio of B/d	Select $B/d = 1.2$ Therefore $B = \dfrac{B}{d} \times d = 1.2 \times 100 = 120\,\text{mm}$	$B = 120$	mm
3. Calculate unit bearing load	From Eq. (12.3) $p = \dfrac{F}{dB} = \dfrac{10000}{100 \times 120} = 0.833\,\text{MPa}$	$p = 0.833$	MPa
4. Calculate journal velocity	From Eq. (12.4) $v = \dfrac{\pi dn}{60 \times 1000} = \dfrac{\pi \times 100 \times 750}{60 \times 1000} = 3.925\,\text{m/s}$	$v = 3.925$	m s^{-1}
5. Calculate pv factor	From Eq. (12.5) $pv = \dfrac{F}{Bd} \cdot \dfrac{\pi dn}{60 \times 1000} = 0.833 \times 3.925$ $= 3.27\,\text{MPa} \cdot \text{m/s}$	$pv = 3.27$	MPa m s^{-1}
6. Select materials for the bearings	Select aluminium bronze, from Table 12.5, the allowable values are $[p] = 15\text{–}20\,\text{MPa}$, $[v] = 5\,\text{m s}^{-1}$ and $[pv] = 15\,\text{MPa m s}^{-1}$. Therefore, the design values are less than the limits.	Aluminium bronze	

Example Problem 12.2

Design a sliding journal bearing on a 1000 rpm steam turbine rotor supports a constant radial load of 17 kN. The shaft stress analysis determines that the minimum acceptable diameter at the journal is 150 mm. The shaft is part of a machine requiring good precision. The lubricant is supplied to the bearing at atmospheric pressure.

Solution

Steps	Computation	Results	Units
Assumptions	Viscosity is constant and corresponds to the average temperature of the oil flowing to and from the bearing. The influence on flow rate of any oil holes or grooves is negligible. The entire heat generated in the bearing is carried away by oil		
1. Select the ratio of bearing width to diameter	$B/d = 0.5$	$B/d = 0.5$	
2. Compute the width of bearing	$B = (B/d) \times d = 0.5 \times 0.15 = 0.075\,\text{m}$	$B = 0.075$	m
3. Compute the journal speed	$v = \dfrac{\pi dn}{60 \times 1000} = \dfrac{\pi \times 150 \times 1000}{60 \times 1000} = 7.85\,\text{m/s}$	$v = 7.85$	m s^{-1}

(continued)

Steps	Computation	Results	Units
4. Compute Unit bearing load	During start-up, the unit load will be the static load. From Eq. (12.3) $p = \dfrac{F}{dB} = \dfrac{17000}{150 \times 75} = 1.511$ MPa	$p = 1.511$	MPa
5. Select bearing material	Select tin bronze ZCuSn10P1 to ensure $p \leq [p]$, $v \leq [v]$, $pv \leq [pv]$	Tin bronze ZCuSn10P1	
6. Estimate the dynamic viscosity of the lubricant	From Eq. (12.29) $\eta = \dfrac{0.068}{n^{1/3}} = \dfrac{0.068}{(1000/60)^{1/3}} = 0.0266$ $Pa \cdot s$		
7. Compute the kinematic viscosity	Assuming the density $\rho = 900 \, kg/m^3$. From Eq. (12.2) $\upsilon = \dfrac{\eta}{\rho} \times 10^6 = \dfrac{0.0266}{900} \times 10^6 = 29.58$ cSt		
8. Select average temperature	$t_m = 50 \,^\circ C$		
9. Select lubricating oil	Select lubricating oil L-AN 46 from Figure 12.5.	lubricating oil L-AN 46	
10. select the kinematic viscosity of lubricant	From Figure 12.5, select the kinematic viscosity of L-AN46 at $t_m = 50 \,^\circ C$ as $v_{50} = 30$ cSt.	$v_{50} = 30$	cSt
11. Compute the dynamic viscosity	Compute the dynamic viscosity of L-AN64 at 50°C $\eta_{50} = \rho v_{50} \times 10^{-6} = 900 \times 30 \times 10^{-6} = 0.027$ Pa s	$\eta = 0.027$	Pa s
12. Estimate the relative clearance	From Eq. (12.30) $\psi = 0.8 \sqrt[4]{v} \times 10^{-3} = 0.8 \times \sqrt[4]{7.85} \times 10^{-3} = 0.001339$	$\psi = 0.001339$	
13. Compute the diametrical clearance	$\Delta = \psi d = 0.001339 \times 150 = 0.20$ mm	$\Delta = 0.20$	mm
14. Compute the load carrying capacity coefficient	From Eq. (12.23) $C_p = \dfrac{F\psi^2}{2\eta v B} = \dfrac{17000 \times (0.001339)^2}{2 \times 0.027 \times 7.85 \times 0.075} = 0.9587$	$C_p = 0.9587$	
15. Compute the eccentricity ratio	According to C_p and B/d, from Table 12.2, by interpolating, we have $\varepsilon = 0.676$	$\varepsilon = 0.676$	
16. Compute the minimum film thickness	From Eq. (12.24), the minimum film thickness is $h_{min} = \dfrac{d}{2}\psi(1 - \varepsilon) = \dfrac{150}{2} \times 0.001339 \times (1 - 0.676)$ $= 0.0325 \, mm$	$h_{min} = 32.5$	μm
17. Determine the surface roughness for the journal and bearing	Select $R_{z1} = 0.0032$ mm for the journal and $R_{z2} = 0.0063$ mm for the bearing	$R_{z1} = 0.0032$ $R_{z2} = 0.0063$	mm mm
18. Compute the allowable film thickness	Assume a safety factor of 2.0, the allowable film thickness is calculated as $[h] = S(R_{z1} + R_{z2}) = 2 \times (0.0032 + 0.0063)$ $= 0.019$ mm Since $h_{min} > [h]$, the design is satisfactory.	$[h] = 19$	μm

(continued)

Steps	Computation	Results	Units
19. Compute the coefficient of friction	According to ε and B/d, from Table 12.4, by interpolating, we have $C_f = 4.349$ $f = C_f\psi = 4.349 \times 0.001339 = 0.00582$	$f = 0.00582$	
20. Select the coefficient of oil flow	According to the ratio of width to diameter B/d, and ratio of eccentricity $\varepsilon = 0.676$, from Table 12.3, we have the coefficient of oil flow $C_Q = 0.0541$.	$C_Q = 0.0541$	
21. Calculate oil flow rate	$Q = C_Q\psi\omega d^3 = 0.0541 \times 0.001339 \times \dfrac{7.85 \times 2}{0.15}$ $\times 0.15^3 = 2.56 \times 10^{-5}$ m^3 s^{-1}	$Q = 2.56 \times 10^{-5}$	m^3 s^{-1}
22. Compute the temperature rise of the lubricant from inlet to outlet	Assuming the density of lubricant $\rho = 900$ kg m^{-3}, specific heat capacity of lubricant is $c = 1800$ J (kg $^\circ$C)$^{-1}$, heat transfer coefficient of lubricant is $\alpha_s = 80$ W (m^2 $^\circ$C)$^{-1}$. From Eq. (12.26), $\Delta t = \dfrac{\frac{f}{\psi}p}{2c\rho\frac{d}{B}C_Q + \frac{\pi\alpha_s}{\psi v}}$ $= \dfrac{\frac{0.00582}{0.001339} \times 1.511 \times 10^6}{2 \times 1800 \times 900 \times \frac{0.15}{0.075} \times 0.0541 + \frac{\pi\times80}{0.001339\times7.85}}$ $= 17.54\,^\circ C$	$\Delta t = 17.54$	$^\circ$C
23. Compute the oil inlet temperature	$t_i = t_m - \dfrac{\Delta t}{2} = 50 - \dfrac{17.54}{2} = 41.23\,^\circ C$ $t_o = t_m + \dfrac{\Delta t}{2} = 50 + \dfrac{17.54}{2} = 58.77\,^\circ C < 80\,^\circ C$ Since normally $t_i = 35\text{–}45^\circ$C, the inlet temperature is acceptable		
24. Select tolerance and fit	According to the diametral clearance $\Delta = 0.20$ mm, the tolerance and fit is selected as F6/d7 according to GB/T 1801-1979. The diameter of the bearing bore is $\phi150^{+0.068}_{+0.043}$ the diameter of the journal is $\phi150^{-0.145}_{-0.185}$		
25. Compute the minimum diametrical clearance and the maximum diametrical clearance	$\Delta_{max} = 0.068 - (-0.185) = 0.253mm$ $\Delta_{min} = 0.043 - (-0.145) = 0.188mm$ Since $\Delta = 0.20$ mm is between Δ_{min} and Δ_{max}, therefore the tolerance is acceptable	$\Delta_{min} = 0.188$ $\Delta_{max} = 0.253$	mm mm
26. Check the minimum film thickness at the maximum clearance	At the maximum clearance $\psi = \dfrac{\Delta_{max}}{d} = \dfrac{0.253}{150} = 0.0016867$ From $C_p = \dfrac{F\psi^2}{2\eta vB} = \dfrac{17000 \times (0.0016867)^2}{2 \times 0.027 \times 7.85 \times 0.075} = 1.5212$ According to C_p and B/d, from Table 12.2, we have $\varepsilon = 0.7674$ From $h_{min} = \dfrac{d}{2}\psi(1 - \varepsilon) = \dfrac{150}{2} \times 0.0016867 \times (1 - 0.7674)$ $= 0.02943\ mm$ Since $h_{min} > [h]$, the design at maximum clearance is satisfactory.	$h_{min} = 0.02943$	mm

(continued)

Steps	Computation	Results	Units
27. Check the minimum film thickness and the temperature rise at the minimum clearance	At the minimum clearance: (1) check the minimum film thickness: $$\psi = \frac{\Delta_{min}}{d} = \frac{0.188}{150} = 0.001253$$ From $$C_p = \frac{F\psi^2}{2\eta vB} = \frac{17000 \times (0.001253)^2}{2 \times 0.027 \times 7.85 \times 0.075} = 0.8399$$ According to C_p and B/d, from Table 12.2, we have $\varepsilon = 0.6568$ From $$h_{min} = \frac{d}{2}\psi(1-\varepsilon) = \frac{150}{2} \times 0.001253 \times (1-0.6568)$$ $$= 0.03225 \; mm$$ Since $h_{min} > [h]$, the design at minimum clearance is satisfactory. (2) Check the temperature rise According to ε and B/d, from Table 12.4, by interpolating, we have $C_f = 4.6277$ $f = C_f\psi = 4.6277 \times 0.001253 = 0.005799$ According to ε and B/d, from Table 12.3, by interpolating, we have $C_Q = 0.0536.$ The temperature rise is $$\Delta t = \frac{\frac{f}{\psi}p}{2c\rho\frac{d}{B}C_Q + \frac{\pi\alpha_s}{\psi v}} =$$ $$\frac{\frac{0.005799}{0.001253} \times 1.511 \times 10^6}{2 \times 1800 \times 900 \times \frac{0.15}{0.075} \times 0.0536 + \frac{\pi \times 80}{0.001253 \times 7.85}} = 18.75$$ The input temperature is $$t_i = t_m - \frac{\Delta t}{2} = 50 - \frac{18.75}{2} = 40.62°C$$ $$t_o = t_m + \frac{\Delta t}{2} = 50 + \frac{18.75}{2} = 59.38°C < 80°C$$ Both t_i and t_o are acceptable.	$h_{min} = 0.03225$ $\Delta t = 18.75$	mm °C
28. Final design result	Lubricating oil L-AN46	$B = 50$ $t_m = 50$ journal: $\phi 150^{-0.145}_{-0.185}$ $R_{z1} = 0.0032$ Bore: $\phi 150^{+0.068}_{+0.043}$ $R_{z2} = 0.0063$	mm °C mm mm

References

1 Budynas, R.G. and Nisbett, J.K. (2011). *Shigley's Mechanical Engineering Design*, 9e. USA: McGraw-Hill.

2 Juvinall, R.C. and Marshek, K.M. (2011). *Fundamentals of Machine Component Design*, 5e. New York: Wiley.

3 Mott, R.L. (2003). *Machine Elements in Mechanical Design*, 4e. Prentice Hall.

4 Hindhede, U., Zimmerman, J.R., Hopkins, R.B. et al. (1983). *Machine Design Fundamentals: A Practical Approach*. New York: Wiley.

5 Pu, L.G. and Ji, M.G. (2006). *Mechanical Design*, 8e. Beijing: Higher Education Press.

6 Hibbeler, R.C. (2006). *Engineering Mechanics-Statics*, 11e. Prentice Hall.

7 Collins, J.A. (2002). *Mechanical Design of Machine Elements and Machines: A Failure Prevention Perspective*, 1e. New York: Wiley.

8 ISO 3448 Industrial liquid lubricants-ISO viscosity classification. Switzerland: International Organization for Standards, 1992.

9 Khonsari, M.M. and Booser, E.R. (2008). *Applied Tribology Bearing Design and Lubrication*, 2e. Wiley Blackwell.

10 Raimondi, A.A. and Szeri, A.Z. (1984). *'Journal and Thrust Bearings' Handbook of Lubrication*, vol. 2, 413–462. Boca Raton, FL: CRC Press.

11 Boyd, J. and Raimondi, A.A. (1958). A solution for the finite journal bearing and its application to analysis and design. Parts I, II, and III. *Transactions of the American Society of Lubrication Engineers* 1 (1): 159–209.

12 Katz, J. (2010). *Introductory Fluid Mechanics*. New York: Cambridge University Press.

13 Oberg, E. (2012). *Machinery's Handbook*, 29e. New York: Industrial Press.

14 Wen, B.C. (2015). *Machine Design Handbook*, 5e, vol. 3. Beijing: China Machine Press.

15 ISO 7902–2:1998 Hydrodynamic plain journal bearings under steady-state conditions – Circular cylindrical bearings – Part 2: Functions used in the calculation procedure. Switzerland: International Organization for Standards, 1998.

16 Qin, D.T. and Xie, L.Y. (2013). *Modern Handbook of Mechanical Design: Mechanical Drawing and Precision Design*. Beijing: Chemical Industry Press.

17 Mang, T. and Dresel, W. (2007). *Lubricants and Lubrication*, 2e. New York: Wiley.

18 ISO 7902–1:1998 Hydrodynamic plain journal bearings under steady-state conditions – Circular cylindrical bearings – Part 1: Calculation procedure. Switzerland: International Organization for Standards, 1998.

19 ISO 7902–3:1998 Hydrodynamic plain journal bearings under steady-state conditions – Circular cylindrical bearings – Part 3: Permissible operational parameters. Switzerland: International Organization for Standards, 1998.

Problems

Review Questions

1 What are the common failure modes in sliding bearings?

2 How should one select the value of relative bearing clearance ψ? What will be the effect of relative clearance on bearing performance?

3 What should one pay attention to while designing grooves on a bushing?

4 In the hydrodynamic sliding bearing design process, if the calculated minimum oil film thickness h_{min} is too small or the temperature rise Δt is too high, what measures could be taken to increase h_{min} or to reduce Δt?

5 What are the specific requirements for the liner materials of sliding bearings?

Objective Questions

1 A journal operates in a steady hydrodynamic lubrication state in a sliding bearing. Which part of Figure P12.1 shows the right position of the journal?

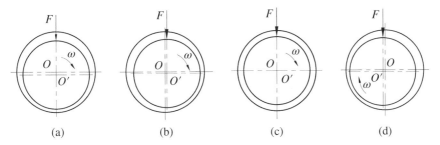

(a) (b) (c) (d)

Figure P12.1 Illustration for Objective question 1.

2 To form hydrodynamic oil film, the inlet and outlet velocity profiles in a wedged lubricated gap should be _____.

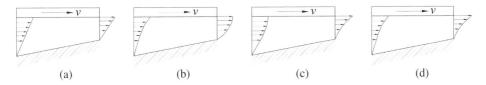

(a) (b) (c) (d)

Figure P12.2 Illustration for Objective question 2.

3 At _____, the viscosity of lubricant in a sliding bearing should *not* be chosen as high.
(a) overload
(b) high operating temperature
(c) high speed
(d) variable loads

4 While designing hydrodynamically lubricated bearings, we usually _____ to reduce the working temperature of sliding bearings.
(a) increase the relative clearance ψ and reduce the ratio of B/d
(b) increase the relative clearance ψ and increase the ratio of B/d
(c) reduce the relative clearance ψ and reduce the ratio of B/d
(d) reduce the relative clearance ψ and increase the ratio of B/d

5 Which of the following is most likely to form hydrodynamic lubrication _____?

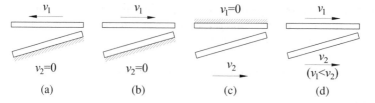

Figure P12.3 Illustration for Objective question 5.

Calculation Questions

1 A boundary-lubricated sliding bearing has a width of $B = 200$ mm and a diameter of $d = 200$ mm. The shaft rotates at $n = 300$ rpm. The material of the bearing is ZCuAl10Fe3. The allowable pressure is $[p] = 15$ MPa, the allowable velocity is $[v] = 4$ m s^{-1}, the allowable value of pv factor is $[pv] = 12$ MPa m s^{-1}. Determine the maximum radial load the bearing can carry.

2 A sliding bearing carries a load of 200 kN, supporting a shaft rotating at a speed of 500 rpm. The ratio of bearing width to diameter is 1.0, where the shaft diameter is 200 mm. The surface roughness of the journal and bearing are 0.0032 mm and 0.0063 mm, respectively. If the diametral clearance is 0.250 mm. The dynamic viscosity is 0.045 Pa s at operating temperature of 50°C. Select the factor of safety as 2. Determine whether the bearing can form fluid hydrodynamic lubrication. If operating temperature is increased to 60°C, the load carrying capacity will increase or decrease? Why?

3 A hydrodynamic sliding bearing, whose journal diameter is 200 mm, rotates at a rotational speed of 750 rpm. The allowable oil film thickness is $[h] = 0.02$ mm. The ratio of the bearing width to diameter is $B/d = 1.0$. Three pairs of variables can be chosen from:

No	Relative bearing clearance ψ	Viscosity of oil η (Pa s)
1	0.0015	0.027
2	0.002	0.027
3	0.002	0.018

Please decide:
(1) By selecting which combination of variables, the sliding bearing can have the largest load carrying capacity?
(2) Calculate the maximum load the sliding bearing can carry according to your selection.

4 A sliding bearing with a journal diameter of 200 mm carries a radial load of 100 kN at 500 r min⁻¹. The B/d ratio is unity. The relative bearing clearance is 0.00125. The average temperature is $t_m = 50°C$, with dynamic viscosity of 0.045 Pa s. Decide:
   (1) the minimum film thickness of the bearing;
   (2) the eccentricity;
   (3) the coefficient of friction of the bearing;
   (4) friction power loss;
   (5) the total oil flow rate through the bearing Q;
   (6) oil temperature rise through the bearing.

5 A sliding bearing has a journal diameter of 50−0.05/−0.075 mm, and a bushing bore diameter of 50 mm, with a unilateral tolerance of +0.025 mm. The ratio of the bearing width to diameter is $B/d = 1.0$. The journal rotates at 1200 r min⁻¹ under a radial steady load of 2000 N. The bearing is lubricated with SAE grade 20 oil (equivalent to ISO viscosity grade 68) at 50°C. If the heat transfer coefficient $α_s = 80$ W $(m^2 \, °C)^{-1}$, decide the inlet and outlet temperature at the maximum and minimum clearance. At both the minimum and maximum clearance assembly estimate the minimum film thickness and coefficient of friction.

Design Problems

1 A journal bearing operates under a steady radial load of 60 kN at a speed of 960 rpm. The journal diameter has been determined as 160 mm from shaft rigidity analysis. Design a suitable hydrodynamic bearing.

2 In an automotive crankshaft application, a hydrodynamic bearing has a diameter of 100 mm with a unit B/d ratio. The bearing must support a radial load of 1000 N, and the journal rotates at 3000 rpm. The lubricant is to be SAE grade 20 oil. Determine a suitable combination of bearing width, radial clearance and lubricant variables suitable for the proposed application.

3 A hydrodynamically lubricated bearing is required for a machine tool. The journal nominal diameter is 25 mm, rotates at 2500 rpm, carrying a radial load of 2250 N. The desired ratio of width to diameter is 1.0. Design the sliding bearing.

4 A partial journal bearing with 127 mm diameter, 63.5 mm width and 0.0635 mm radial clearance supports a radial load of 18 570 N. The shaft rotates 1800 rpm. The lubricant has viscosity of 0.0073 Pa s, supplied at atmospheric pressure. Estimate the minimum oil film thickness, friction power loss and required oil flow rate.

Structure Design Problems

1 Design the structure of a solid radial sliding bearing.

2 Design a typical solid liner and split liner for a radial sliding bearing.

CAD Problems

1 Write a flow chart for bearing design process to complete the Example Problem 12.2.

2 Develop a program to implement a user interface similar to Figure P12.4 and complete the Example Problem 12.2.

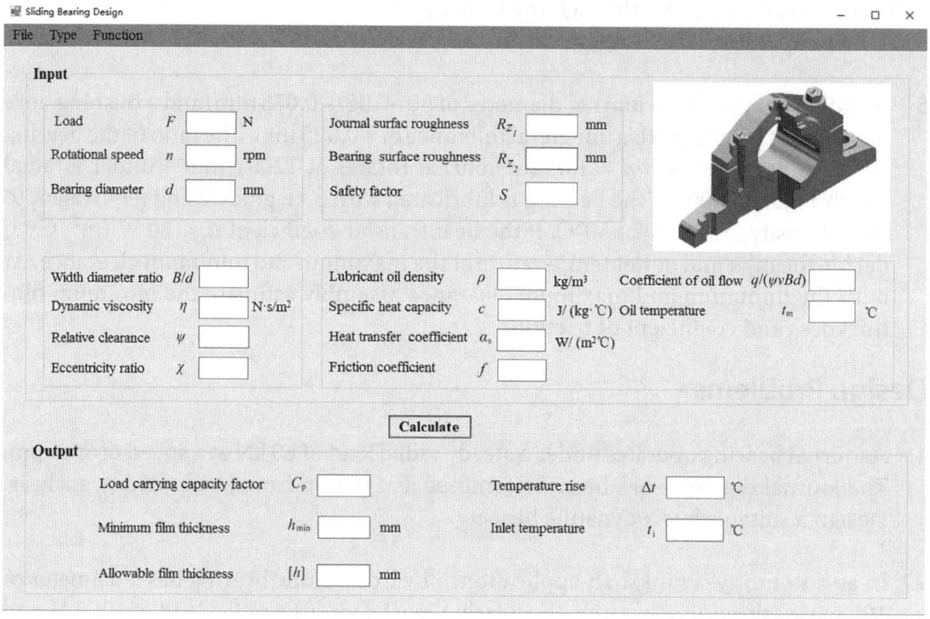

Figure P12.4 Illustration for CAD problem 2.

13

Couplings and Clutches

Nomenclature

D_i inside diameter of clutch surface, mm

D_o outside diameter of clutch surface, mm

d shaft diameter, mm

F normal force at clutch surface, N

f coefficient of friction

K_A service factor

n rotational speed, rpm

$[n]$ limiting rotational speed, rpm

p pressure on frictional surface, MPa

$[p]$ allowable pressure on frictional surface, MPa

T nominal torque, N mm

T_{ca} design torque of a coupling or clutch, N mm

Analysis and Design of Machine Elements, First Edition. Wei Jiang.
© 2019 John Wiley & Sons Singapore Pte. Ltd. Published 2019 by John Wiley & Sons Singapore Pte. Ltd.
Companion website: www.wiley.com/go/Jiang/analysis_of_machine_elements

T_f	friction torque, N mm	y	axial misalignment, or parallel offset, mm
$[T]$	rated torque of a coupling or clutch, N mm		
		z	number of contact surfaces
x	end float, mm	α	angular misalignment, °

13.1 Introduction to Couplings

13.1.1 Applications, Characteristics and Structures

A coupling is a device used to connect two shafts end to end to transmit power and torque. The output shaft of a driver, such as a motor or engine, is connected to the input shaft of a transmission, or a driven machine through a coupling (Figure 6.1) [1]. Since power units, transmission units and driven machines are often manufactured separately, couplings are indispensable in mechanical power transmission system. They are widely used in machines in automotive, petroleum, chemical, gas, electric power and steel and other industries.

Apart from the primary function of connecting shafts to transmit power and torque, couplings can compensate for misalignments between the coupled shafts. Some couplings can also absorb shock and vibration, or act as a safety device to prevent overload on a machine. The shafts connected by couplings normally do not allow disconnection during operation. However, during repair, couplings facilitate temporary disconnection for safety concern.

Couplings differ widely in size, appearance and characteristics. Nevertheless, all couplings consist of three basic parts, that is, two shaft hubs and a connecting element. Most of couplings are highly standardized and commercially available from manufacturers. Couplings can be either selected or designed according to application requirements.

13.1.2 Shaft Misalignments

Although precise alignment is desirable for all machines, perfect alignment is difficult during initial assembly and installation, and almost impossible to achieve and maintain during operation [2]. Shaft misalignments include end float, axial and angular misalignments, as illustrated in Figure 13.1. End float, x, is the displacement when two shafts move relative to each other axially. Axial or parallel offset misalignment, y, appears if the axes of driving and driven shafts are parallel, but not on the same line. Angular misalignment, α, is present when the axes of two shafts are inclined one to the other. In some cases, these misalignments may occur simultaneously.

Shaft misalignments may be caused by initial assembly inaccuracies, or operational misalignments due to thermal expansion, shaft deflection or wear. It is the primary source of premature downtime in high-speed machinery [2]. If misaligned shafts are

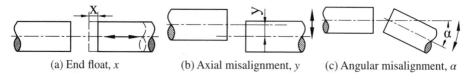

(a) End float, x (b) Axial misalignment, y (c) Angular misalignment, α

Figure 13.1 Shaft misalignments.

connected by a rigid coupling and are forced into alignment, the coupling, shafts and bearings may subject to severe variable stresses, which may lead to rapid fatigue failure. Ideally, a flexible coupling is a better choice as they can produce the greatest flexibility without significant stresses induced in the connected shafts.

13.1.3 Types of Couplings

Mechanical couplings are typically divided into two broad categories: rigid couplings and flexible couplings. Rigid couplings are only for accurate shaft alignment that presents at initial installation and maintains during operation. If significant misalignments occur, flexible couplings are preferred, as they can accommodate misalignments and reduce stresses induced in the connected shafts, which may lead to early fatigue failure. Table 13.1 shows typical types of rigid couplings.

Flanged couplings are probably the most common type of rigid connection. The coupling flanges are mounted on the ends of the driving and driven shaft with keys and are joined together by a group of precision bolted joints or ordinary bolted joints on the bolt circle. A piloted boss is used for precise alignment if ordinary bolted joints are used. An outer protective rim is often added to the flanges to provide a safety shield for the bolt heads and nuts. While transmitting torque, the load is transmitted from the driving shaft to its flange, through the bolts, into the mating flange and out to the driven

Table 13.1 Types of rigid couplings [8].

Types of couplings	Figures	Features and applications
Flanged couplings		Simple structure, low cost; Suitable for heavy-duty, slow speed, stable load applications.
Sleeve couplings		Simple structure, small radial dimension; Suitable for light duty applications.
Ribbed couplings		Convenient for assembly and disassembly; Suitable for small to medium torque, slow speed and stable load transmission, such as in vertical pumps, agitators, and many other types of applications.

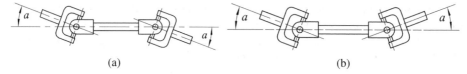

Figure 13.2 Double universal couplings.

shaft [1]. Therefore, the rating of a flanged coupling is limited by the strength of the keys, hubs and bolts. Usually, bolts are assumed to be the weakest link in a coupling. The analysis and design for bolt quantity and size to be used in a coupling is then similar to that for bolted connections introduced in Chapter 3.

Sleeve couplings have either parallel or woodruff keys on the shaft ends to transmit torque. The sleeve is usually locked to the shafts by setscrews or pins. Sleeve couplings are probably the simplest form of couplings. They have small radial dimensions and are mostly used for light duty applications.

Ribbed couplings are used where convenient assembly and disassembly is required. A ribbed coupling involves two similar cast iron halves joined together by bolts at the split. Attachment of coupling hubs to shafts are usually accomplished by keys, tapered sleeves or interference fits [3]. Torque is transmitted by frictional force produced by bolts rather than by direct loading of bolts themselves [4]. The number of bolts varies depending on the size of coupling. Symmetric bolts are installed inversely to improve balance. Ribbed couplings are mainly for small to medium torque, low speed and steady load transmission, such as in vertical pumps, agitators and many other types of applications.

Flexible couplings are desirable when precise alignment of mating shafts is difficult to achieve during installation or operation. They are able to transmit torque smoothly and reliably while at the same time accommodate misalignments in axial, radial and angular directions, reducing stresses induced in the connected shafts. Table 13.2 shows typical types of flexible couplings, their misalignment accommodation capabilities, features and applications.

Misalignments in flexible couplings are compensated for by either mechanical components, resilient materials or a combination of both [2]. Correspondingly, they are classified as mechanical flexible couplings, elastomeric couplings and metallic element couplings. Mechanical flexible couplings include Oldham couplings, slider block couplings, universal couplings, gear couplings and chain couplings. These couplings accommodate misalignments by an interposed element that slides or rolls to introduce small clearance between coupled shafts [3]. However, due to lack of elastomeric elements, these couplings cannot cushion shock or absorb vibration.

An Oldham coupling is composed of two slotted flanges and an intermediate disc. The two facing slotted flanges are coupled by the intermediate disc, with mating cross-keys sliding between flanges [3]. Since the keys of disc are able to slide in the slots of halve couplings, it could compensate relatively large misalignments between two shafts. The disc rotates around its centre at the same speed as the driving and driven shafts. The greater the shaft misalignment, the greater the sliding and consequently, the larger the centrifugal force, dynamic load and wear. Therefore, lubrication and wear must be considered [4]. A slider block coupling is similar to an Oldham coupling, except that the intermediate disc is a square, textolite made slider block.

Table 13.2 Types of flexible couplings [8].

Types of couplings	Figures	Misalignment capabilities			Shock absorption	Features and applications
		End float	Parallel offset	Angular		
Oldham couplings			√	√	N	Simple and compact structure, small radial size, low costs; Suitable for slow to medium speed, high torque drives.
Slider block couplings		√	√	√	N	Simple and compact structure, small radial size, low costs; Suitable for low to medium power, high-speed applications.
Universal couplings	Single universal coupling 			√	N	Allow large angular misalignment; Suitable for transmitting power between shafts that intersect at a constantly varying angle. Typical applications are in the transmission to wheels in automobiles.

Without elastomeric elements

(*continued*)

Table 13.2 (Continued)

Types of couplings	Figures	Misalignment capabilities			Shock absorption	Features and applications
		End float	Parallel offset	Angular		
	Double universal coupling					
Gear couplings		√	√	√	N	Small radial size; large power transmission capacity, yet high cost. Suitable for frequent start/stop cycles, high-speed and heavy-duty applications.
Chain couplings		√	√	√	N	Simple structure, compact size, convenient for maintenance. Suitable for hostile environments, like high-temperature, humid and dusty conditions.

With nonmetallic elastomeric elements						
Pin and bushing couplings	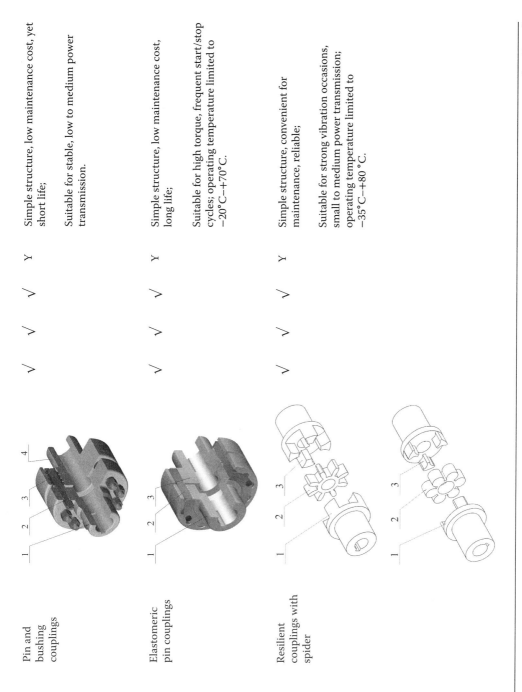	√	√	√	Y	Simple structure, low maintenance cost, yet short life; Suitable for stable, low to medium power transmission.
Elastomeric pin couplings		√	√	√	Y	Simple structure, low maintenance cost, long life; Suitable for high torque, frequent start/stop cycles; operating temperature limited to −20°C−+70°C.
Resilient couplings with spider		√	√	√	Y	Simple structure, convenient for maintenance, reliable; Suitable for strong vibration occasions, small to medium power transmission; operating temperature limited to −35°C−+80°C.

(continued)

Table 13.2 (Continued)

Types of couplings	Figures	Misalignment capabilities			Shock absorption	Features and applications
		End float	Parallel offset	Angular		
Tyre couplings		√	√	√	Y	Simple structure, excellent misalignment compensation and vibration dampening capability, yet require ample radial space; Suitable for frequent start/stop cycles, reversing torque loads, heavy shock, vibration applications.
Metallic grid couplings		√	√	√	Y	Good misalignment compensation capability, complicated structure, require lubrication; Suitable for shock loading, medium to high power transmission.
Diaphragm couplings		√	√	√	Y	Simple structure, no need for lubrication, lightweight; Suitable for high-speed, high-temperature or high-performance applications, like gas turbines, precision machines, servo systems.

With metallic elastic elements

Universal couplings, also known as universal joints or Cardan joints, permit substantial angular deviation between mating shafts, with the maximum value reaching 35–45° [5]. The single universal joint consists of a splined-hub driving yoke 1, and driven yoke 3, connected by a cross link 2 (or a spider) through pivot bearings. The two yokes are oriented at a right angle to each other and are connected to the driving and driven shafts, respectively. The yokes pivot on the arms of cross. Since there are two pivots, the two shafts can be at any angle to one another and can still rotate and transmit power [2].

When the driving yoke rotates at constant angular velocity, the angular velocity ratio between driving and driven shafts varies periodically, although the two shafts make complete revolutions during the same time [2, 5]. The fluctuation of angular output velocity implies acceleration, leading to undesirable dynamic loads and subsequent noise and vibration. Therefore, most practical applications require a double joint, that is, to use an intermediate shaft or Cardan shaft to connect two single universal joints. During assembly, it is important to ensure that the driving shaft, driven shaft, and intermediate shaft lie in the same plane, and the misalignment angles at two joints are equal, that is, the double joint arrangement is adaptable to either parallel shafts (Figure 13.2a) or intersecting shafts (Figure 13.2b). In these arrangements, the second joint tends to compensate for the nonuniform oscillation of the first joint so that the fluctuation in angular velocity ratio is very slight [1–3].

In a gear coupling, two identical, externally geared hubs 1 and 6 mesh internally with flanged sleeves 3 and 5, respectively. The hubs are connected to the driving and driven shafts by keys and are enclosed by bolted flanged sleeves. Torque is transmitted from one hub to the other by the meshing of teeth. The teeth on the sleeves are usually longer than the hub teeth, and there is a gap between two hubs. The hubs are, therefore, able to move axially to ensure that teeth always maintain adequate contact for power transmission. Clearance between the mating teeth in the hub and the flanged sleeve allows gear couplings to accommodate end float, parallel offset and angular misalignments. Crowned gear teeth are often used to accommodate larger angular misalignments without excessive backlash.

A chain coupling is composed of two hubs with sprockets and a detachable, duplex chain. The common duplex roller chain meshes with two sprockets with the same number of teeth on the hub to transmit torque, providing a convenient and rapid method of disconnecting shafts. End float, parallel offset and angular misalignments are accommodated by the built-in clearances between the chain and sprocket, which exist between the rollers and sprocket teeth, the rollers and bushings, the bushings and pins, and the links and sprockets teeth. Chain couplings find wide applications in mechanical equipment and are especially suitable for working in hostile environments such as high-temperature, humid and dusty conditions. In practice, a cover is usually incorporated to maintain lubrication and reduce wear, thereby increasing the life of chain couplings.

Elastomeric couplings are couplings that contain nonmetallic elastomeric elements, including pin and bushing couplings, elastomeric pin couplings, resilient couplings with a spider and tyre couplings. The flexibility of these couplings is obtained from stretching or compressing resilient materials, usually rubber or synthetic rubberlike material. Therefore, they have excellent vibration dampening capabilities and misalignment accommodation capacities, providing extended life for connected equipment.

Pin and bushing couplings are similar to flanged couplings, except that they join two half coupling flanges 1, and 4 by pins 2 with rubber-cushioned bushings 3 instead of bolts. Since they transmit motion and torque through nonmetallic elastomeric bushings, pin and bushing couplings can compensate for misalignments and absorb torsional shock and vibration.

Elastomeric pin couplings use elastomeric material, usually nylon, as pins 2, to connect two halves of coupling flanges 1 and 3. Compared with pin and bushing couplings, elastomeric pin couplings have greater torque transmission capability. Elastomeric pin couplings can absorb shock and vibration, and accommodate axial and minor radial and angular misalignments, so they are usually used on large vehicles or modern industrial machinery. However, because nylon studs are sensitive to temperature, the operating temperature is limited to a range of $-20-+70°C$.

A resilient coupling with spider has an elastomeric element 2 and a pair of axially overlapping rigid jaws 1 and 3. The elastomeric element can be in the form of a star or a plum, with different hardnesses to suit load carrying capacities. The elastomeric spider is placed between the crowning teeth of the two jaws to transmit power through compression and to achieve shaft coupling. Elastomers are generally made from plastic or rubber, which combines resiliency with high damping capacity. Therefore, resilient coupling with spider can cushion and dampen impact and shock, and are usually used in strong vibration occasions. They can compensate for radial, axial and angular misalignments. The flexibility is derived from clearances in mating parts, or from deflection of elastomeric members [2]. However, the operating temperature is limited by the elastomer material, usually from $-35-+80°C$.

A tyre coupling has a moulded, reinforced rubber tyre 2 assembled to the two metal hubs 3 and steel frames by a group of bolts 1. The reinforced rubber tyre is composed of two radially split halves, which facilitates installation and maintenance without the need for moving driving and driven shafts. Tyre couplings have superior misalignment compensation performance. The elastomeric tyre can flex in any direction to accommodate single or combined end float, parallel offset and angular misalignments. It cushions shocks and damps torsional vibrations, providing a resilient shaft connection.

Metallic element couplings refer to the couplings that contains metallic elastic elements, including metallic grid couplings and diaphragm couplings. These couplings compensate for misalignments by the flexure of flexible metallic elements.

A metallic grid coupling consists of a spring steel alloy grid, and two hubs with tapered or contoured parallel slots cut into the peripheries. The two hubs are connected by the serpentlike spring steel alloy grid fitted in the slots. Torque is transmitted through the flexible spring steel alloy grid. Under a light or normal load, the serpentlike springs have a relative long span. As the load increases, the springs deflect and the span is reduced. Greater support is afforded to the springs by the walls of flared slots [5, 6]. The flexible span provides flexible accommodation to varying loads, which is most desirable for damping torsional vibration and reducing shock loads. Metallic grid couplings accommodate end float, parallel offset and angular misalignments by the movement of steel grid in the lubricated slots [7]. The couplings are usually enclosed in oil-tight, dust-proof covers. They are adaptable to nearly all types of industrial equipment.

A typical diaphragm coupling consists of two half couplings 1 and 6, one or more thin diaphragms 2 and 4 and spacer tubes 3. The diaphragms are crisscrossed with the two half couplings and spacer tube by precision bolted joints 5. Axial, angular misalignments and end float of two connected shafts are compensated through flexing of thin metallic diaphragms. The difference between a single diaphragm coupling and a dual diaphragm coupling is the misalignment accommodating capability.

13.2 Design and Selection of Couplings

A wide variety of couplings are highly standardized and commercially available in many configurations. For a coupling, it is required that the rated torque be transmitted, and shaft misalignment be accommodated without premature failure. The design requirement of torque transmission is fulfilled by manufacturer, while the application requirement of misalignment accommodation is accomplished by coupling user. As a matter of fact, most coupling failure occurs as a result of misalignment, misuse, abuse or neglect rather than a design or material deficient [4]. The selection of suitable coupling for a specific application is thus of paramount importance.

The procedure for analysis and selection of coupling is basically similar for all couplings. The type of coupling is usually first selected according to the application requirements. The specification of coupling is then decided from manufacturers' catalogues according to design data, that is, the design torque to be transmitted, the diameters of mating shafts and rotational speeds. Some components may need to be analysed to ensure safety for important applications.

13.2.1 Coupling Type Selection

Couplings are selected based on their capacities and characteristics. Important capacities include the rated torque, the limiting speed and misalignment accommodation capability. The selected coupling should be able to transmit required torque at operating speeds, and accommodate misalignment of connected shafts. Coupling characteristics include torsional stiffness, damping capability, assembly convenience, lubrication and costs. To select a suitable coupling, the following factors need to be considered:

13.2.1.1 The Characteristics of Operation Conditions
Although different types of couplings may be interchangeable, each type is actually restricted to fairly limited applications [2]. Generally, rigid couplings are for machines carrying steady loads at constant speeds and especially with precise shaft alignment, while flexible couplings are for machines carrying variable loads and at variable speeds with shaft misalignments. Flexible couplings with either metallic elastic or nonmetallic elastomeric elements are selected to cushion shock and absorb vibration. For example, tyre couplings are chosen in heavy impact applications to cushion shock and vibration, gear couplings are selected for heavy load transmission and diaphragm couplings are preferred for high-speed transmission.

Since misalignment is the main cause of premature failure in rotating equipment [4], and it is practically impossible for coupled rotating shafts to achieve or maintain perfect alignment, proper coupling selection is extremely important. Different types of couplings have different misalignment accommodation capability in respect of misalignment type and magnitude. For example, slider block couplings can tolerate large radial misalignments while universal couplings are ideal for intersection shafts or large angular misalignments.

13.2.1.2 Reliability and Operating Environments
Since flexible couplings without elastomeric elements rely on sliding or rolling to compensate misalignments, lubrication and wear must be considered. Therefore, they are less reliable than flexible couplings with elastomeric elements that rely on the flexure of elastomeric elements to compensate for misalignments. Besides, couplings with non-metallic elastomeric elements, such as rubber, are sensitive to temperature and corrosive media.

13.2.1.3 Manufacturing, Installation, Maintenance and Cost Considerations
The selected coupling affects both the cost of equipment and the downtime for coupling replacement or failure repair [4]. By careful selection, installation and maintenance of couplings, substantial savings could be made in reducing maintenance costs and downtime.

13.2.2 Coupling Size Selection

While selecting the size of coupling, the design torque T_{ca} is established by multiplying a service factor with the nominal torque T to cope with momentary peaks caused by the fluctuation of either prime mover torque or load torque, that is,

$$T_{ca} = K_A T \leq [T] \tag{13.1}$$

The service factor K_A is affected by the starting inertia of load, the characteristics of prime mover and driven machine. It can be selected by increasing the values in Table 2.1 by 20–30%. While specifying commercially available couplings, the rated torque of selected coupling $[T]$ should be greater than the design torque. Both couplings and clutches can be selected from manufacturers' catalogues according to the design torque obtained from Eq. (13.1).

Another capacity needs to be considered in coupling selection is the limiting speed. Since the power a coupling can transmit is the product of transmitted torque and rotational speed, for a given size of coupling, the transmitted power increases as the speed of rotation increases. And centrifugal effects determine the upper limit of shaft speed. The rotational speed of coupling must satisfy

$$n \leq [n] \tag{13.2}$$

In most cases, each coupling accommodates a range of shaft diameters. The maximum and minimum values of shaft diameter or a series of dimensions, are provided in manufacturers' catalogues or design handbooks [8, 9]. The diameter of mating shafts should be within the specified range of selected couplings. Besides, assembly and installation accuracy must be specified according to the misalignment tolerance of the selected coupling.

Finally, it is important that the selected coupling is compatible with the connected equipment. The introduced coupling should cause little change to the operating conditions of the system and introduce minimum stresses on the components of coupling. When a coupling has an elastomeric element, the operating temperature of the coupling should be lower than the maximum allowable temperature for the elastomeric material.

13.3 Introduction to Clutches

13.3.1 Applications, Characteristics and Structures

A clutch is a device used to connect or disconnect two shafts end to end at any time during operation. The two shafts rotating at different speeds are brought to a common angular velocity after a clutch is actuated. Through a smooth and gradual engagement or disengagement, clutches change the speed or the rotational direction of the driven machine to adjust to different operating conditions. Acting as an on-off switch, clutches control the flow of power to ensure the safe and effective operation of machinery. They can also serve as a variable operating control to automatic react to the predetermined torque, speed or direction [2]. Clutches are indispensable power transmitting components in mechanical power transmission.

A clutch principally consists of two parts that are engaged or disengaged either manually or by power driven device. Each part is connected to a driving shaft and a driven shaft, respectively. Most of clutches are highly standardized and commercially available from manufacturers. Clutches can be either selected or designed according to the application requirements.

13.3.2 Types of Clutches

The classification of clutches depends on actuation methods and torque transmission principles. Clutches can be actuated mechanically, pneumatically, hydraulically, electromagnetically and automatically. Mechanically actuated clutches are controlled manually by a lever or wedges. Pneumatic and hydraulic controlled clutches move the actuating linkages by pneumatic and hydraulic forces. Electrically or electromagnetically operated clutches are engaged electrically and spring-released. Automatic clutches realize engagement and disengagement automatically when predetermined conditions, such as torque, direction or speed, are satisfied.

Clutches work following different physical principles. For example, jaw clutches rely on the interlocking of teeth to transmit torque, while friction clutches depend on friction between contact surfaces. The typical types of clutches, their features and applications are schematically outlined in Table 13.3, followed by a brief introduction to the configuration and operating principles. More specific design information is unique to manufacturers and available through catalogues and design handbooks.

A jaw clutch is usually composed of two halves, connected to a driving shaft and a driven shaft, respectively, either by a key or by a spline. The two halves of clutch have a series of evenly spaced teeth with square, triangular, spiral, or toothed profiles (see Figure 13.3) to provide either one direction, or smoother engagement, or prolonged engagement time [10]. The teeth of mating sets of jaws are brought into engagement by sliding one or both halves axially to form a rigid mechanical junction. Once the teeth are engaged, shafts are effectively joined and torques are positively transmitted [2]. The structure of jaw clutch ensures a strong mechanical coupling between two shafts with no slippage, wear, or heat generation. However, the engagement is constantly accompanied by shock. Therefore, jaw clutches are normally engaged while a system is stopped or is running slowly.

Friction clutches utilize the force between contact surfaces to transmit torque from a driving shaft to a driven shaft. They are the most common type of all clutches. Friction

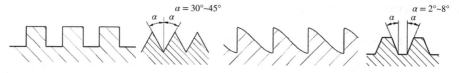

Figure 13.3 Teeth of jaw clutches.

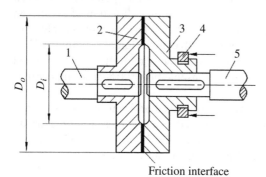

Figure 13.4 A schematic view of a single disc friction clutch.

Friction interface

clutches include cone clutches and disc clutches, with the former being largely displaced by the latter [10]. Disc clutches include single disc friction clutches and multiple disc friction clutches.

The schematic view of a single disc friction clutch is shown in Figure 13.4. The single disc friction clutch is composed of two friction plates 2 and 3, keyed to the driving and driven shafts 1 and 5, respectively, and an actuator 4. During engagement, the operating device 4 actuates the mating friction surfaces into direct contact, producing a normal force between contact surfaces. The generated tangential frictional torque gradually reduces the angular velocity difference between the rotating shafts. The stationary component accelerates from rest to operating speed over a suitable span of time. When the relative sliding velocity is reduced to zero, both shafts rotate at the same rotational speed.

A multiple disc friction clutch increases the number of friction discs rather than the diameter of discs to increase the capacity. In a multiple disc friction clutch, friction discs are alternately attached to the rotating and stationary shafts. As shown in Table 13.3, a group of discs are splined to rotate with the driving shaft while another group of discs are similarly splined to rotate with the driven shaft. When the clutch is engaged, the two groups of friction discs are in contact and the tightly clamped discs provide friction force on the contact surfaces. The friction force is proportional to the applied normal force and acts in the tangential direction, producing a frictional torque. The normal force can be adjusted to achieve maximum capacity and minimum wear.

Figure 13.4 shows friction plates with an outside diameter D_o and an inside diameter D_i. Assume the discs are rigid and wear is uniform. Although there is some variation in pressure over the surface of friction plates, the friction force is still assumed to be acting at the mean radius of annular plate to simplify the analysis. The frictional torque should satisfy [5, 10]

$$T_f = zfF\frac{(D_o + D_i)}{4} \geq K_A T \tag{13.3}$$

The pressure on the friction surface satisfies

$$p = \frac{4F}{\pi(D_o^2 - D_i^2)} \leq [p] \tag{13.4}$$

where K_A is service factor and can be selected from Table 2.1 using the median values. The value of coefficient of friction f and allowable pressure $[p]$ can be found in design handbooks [8, 9].

Friction clutch design demand that the friction torque be produced by an acceptable normal force. In clutch design, the initial value of outside diameter D_o and inside diameter D_i are initially selected as [7]

$$D_i = (1.5 \sim 2)d, D_o = (1.5 \sim 2)D_i \qquad \text{For friction clutches operate dry}$$
$$D_i = (2 \sim 3)d, D_o = (1.5 \sim 2.5)D_i \qquad \text{For friction clutches operate wet}$$

Then the normal force presented to the clutch F can be obtained from Eq. (13.4), and the number of friction surfaces z from Eq. (13.3).

Unlike most machine components, such as bearings, gears and many others, minimizing friction is desirable for reducing energy loss and wear. In contrast, friction clutches rely on friction to transmit torque. One of the aims of friction clutch design is to maximize the friction coefficient and keep it uniform over a wide range of operating conditions, while at the same time minimizing wear [6].

Suitable material for friction surfaces should have a high and stable coefficient of friction to transmit required torque, good wear resistance to guarantee an acceptable period of operation and high thermal conductivity to dissipate friction-induced heat. In a typical friction clutch, one contact surface is metal, the other is a high-friction material called a liner. Contact and friction occurs between a replaceable lining and metallic surface [2]. The metallic surface is commonly made of grey cast iron or steel, while the replaceable lining is usually made of composite materials in which hard particles are embedded in a matrix of thermosetting polymer. The commonly used reinforcing particles are fibreglass and sintered particles that include brass, zinc, ceramic and so on [3].

Friction clutches are designed to operate either dry or wet with oil. Automotive clutches operate dry, using circulating air to dissipate frictional heat. Dry operation implies a high coefficient of friction and high torque capacity, yet high temperature [2]. Nevertheless, they have simple structure, low cost and short starting times. They are widely used in situations requiring instantaneous engagement or overload protection. Most multiple disc clutches can operate wet. Friction discs are soaked in an enclosed box of oil. The oil serves as an effective coolant during clutch engagement, leading to a reduced coefficient of friction and torque capacity, yet smooth engagement and less wear.

Friction clutches can be engaged or disengaged mechanically, electromagnetically, pneumatically and hydraulically. Once the torque capacity of disc friction surfaces is exceeded, the clutch will slip, protecting machines from overload damage. Friction clutches have a large frictional area, resulting in large heat generation and wear.

For special requirements of automatic engagement or detachment, it is demanded that when torque, rotational speed or rotational direction reach the preset value, clutches can engage or disengage automatically. Safety couplings and clutches, centrifugal clutches and overrunning clutches are examples of these devices.

Table 13.3 Types of clutches [8].

Types	Figures		Feature and applications
Clutches operated by mechanical, pneumatic and hydraulic, or electromagnetic forces	Jaw clutches		Simple structure, small size, no slippage; Strong mechanical coupling; suitable for slow speed, small torque engagement. Commonly used in the drive engagement of individual gears in a manual automobile gearbox.
	Friction clutches	Single disc friction clutch	Smooth engagement, large size, heat generated during operation, slippage occurs when overload; The most common type; widely used in transmission system of vehicles, like automobiles and tractors. Multiple disc friction clutches have higher capacity than single disc friction clutches.
		Multiple disc friction clutch	
Clutches operated automatically	Safety clutches		Torque sensitive clutch. Simple structure, yet less precision; Widely used as overload protection devices in machine tools, or in transmission systems in which overload occurs occasionally.

(*continued*)

Table 13.3 (Continued)

Types	Figures	Feature and applications
Centrifugal clutches		Speed sensitive clutch. Suitable for accelerating large masses of inertia to high rotational speeds by a relatively small driving motor.
Overrunning clutches		Direction sensitive clutch. Examples of applications are in bicycle hubs, centrifugal pumps and conveyors, elevators.

Safety clutches, also called overload clutches, are torque sensitive clutches. When the applied torque exceeds the predetermined value, clutches will automatically disengage or slip and thus save driven machines from costly breakdowns. This overload protection is the mechanical equivalent of an electric fuse [2]. Torque limitations can be realized by different mechanical designs. When the torque reaches the specified value, shear pins may break in safety clutches; connected shafts may separate in jaw clutches, and friction discs may slip in friction clutches, thus protest important components from damage.

A safety clutch with shear pins relies on one or more pins to shear at the predetermined torque. As shown in Table 13.3, pins 3 are placed in the holes on flanges 1 and 4 of the two clutch halves. They are usually made of medium carbon steel, heat treated to 50–60 HRC, with precut notches to assist snap. To avoid the sheared pins scratching surfaces, pins are installed in hardened steel sleeves 2. The sheared pins have to be replaced manually before the coupling can operate again.

Centrifugal clutches are speed sensitive clutches. When the rotational speed reaches or exceeds the predetermined value, centrifugal clutches will automatically engage or disengage. A typical structure of a centrifugal clutch in Table 13.3 composes of an input spider 1, an output drum 2 and asbestos blocks 5 connected to the input spider 1 by bolts 3. The input spider deliver power from a prime mover. When the input shaft rotates, asbestos blocks 5 are thrown radially outward by centrifugal force against the inner surface of drum 2. Torque is thus transmitted through friction between asbestos block 5 and drum 2. When the drum 2 reaches operating speed, the clutch acts simply as a coupling. Springs 4 are used to preset engagement speed by providing correct tension during assembly.

An overrunning clutch or directional clutch transmits torque from the driving shaft to drive shaft only in one direction. In the opposite direction, the driven shaft is in overrun condition, with no torque transmitted. The ratchet and pawl is a simple device permitting alternate engagement and disengagement of rotating members. In a more high-precision machine, an overrunning clutch can be used. As shown in Table 13.3, the clutch has an inner ring 1 with cam profiles around the periphery and an outer ring 2. When the driving shaft rotates in the driving direction, or when the driving shaft rotates faster than the driven shaft, the bolts and springs 3 ensure the rollers 4 are wedged tightly between the outer ring 2 and inner ring 1 and thus transmit torque. When the driving shaft rotates in the opposite direction, or when the driven shaft rotates faster than the driving shaft, the rollers 4 move out of engagement, and no torque is transmitted, thus protect equipment from damage due to over speeding [1]. There are usually between three and four rollers and the greater the number of rollers, the higher the torque that can be transmitted.

References

1 Mott, R.L. (2003). *Machine Elements in Mechanical Design*, 4e. Prentice Hall.
2 Hindhede, U., Zimmerman, J.R., Hopkins, R.B. et al. (1983). *Machine Design Fundamentals: A Practical Approach*. New York: Wiley.
3 Collins, J.A. (2002). *Mechanical Design of Machine Elements and Machines: A Failure Prevention Perspective*, 1e. New York: Wiley.
4 Mancuso, J.R. (1986). *Couplings and Joints: Design, Selection, and Application*. New York: Marcel Dekker.
5 Pu, L.G. and Ji, M.G. (2006). *Mechanical Design*, 8e. Beijing: Higher Education Press.
6 Juvinall, R.C. and Marshek, K.M. (2011). *Fundamentals of Machine Component Design*, 5e. New York: Wiley.
7 Xu and, Z.Y. and Qiu, X.H. (1986). *Machine Elements*, 2e. Beijing: Higher Education Press.
8 Wen, B.C. (2015). *Machine Design Handbook*, 5e, vol. 3. Beijing: China Machine Press.
9 Oberg, E. (2012). *Machinery's Handbook*, 29e. New York: Industrial Press.
10 Budynas, R.G. and Nisbett, J.K. (2011). *Shigley's Mechanical Engineering Design*, 9e. New York: McGraw-Hill.

Problems

Review Questions

1 How can one transmit power to or from a rotating shaft?

2 What are the characteristics of a universal coupling? How could one assemble a pair of universal couplings?

3 List measures that could increase the load carrying capacity of a friction clutch.

4 What kinds of couplings have damping capability? Why is damping capability important for a coupling?

Objective Questions

1 The main difference between a coupling and a clutch is _____.
   (a) capability of overload protection
   (b) capability for engagement and disengagement of two shafts at any time
   (c) capability of misalignment compensation

2 The main function of a coupling and a clutch is to _____.
   (a) cushion shock and reduce vibration
   (b) transmit motion and torque
   (c) prevent overload
   (d) compensate for misalignment and thermal expansion

3 The resilient material in flexible couplings can _____.
   (a) accommodate misalignment
   (b) reduce friction
   (c) cushion shock and reduce vibration
   (d) increase load carrying capability

4 When a stiff shaft rotates at high speed and is subject to impact and vibration, a
   _____should be selected.
   (a) flanged coupling
   (b) universal coupling
   (c) Oldham coupling
   (d) pin and bushing coupling

5 When the angular misalignment is 30°, a _____ is the best choice.
   (a) universal coupling
   (b) flanged coupling
   (c) gear coupling
   (d) pin and bushing coupling

Calculation questions

1 A safety clutch with shear pins in Figure P13.1 transmits a torque of $T = 500$ N·m.
   The clutch has two identical pins on the circle of diameter of $D_m = 100$ mm. Pins
   are made of medium carbon steel and normalized, with an allowable stress of
   $[\tau] = 400$ MPa. It is required the safety clutch function at 50% of overload. Decide
   the diameter of the pins.

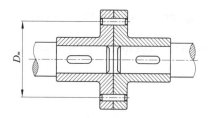

Figure P13.1 Illustration for Calculation Question 1.

2   A multiple friction clutch transmitted a power of 5 kW at $n = 960\,\text{r min}^{-1}$. The friction discs are made of tempered steel, with coefficient of friction of $f = 0.15$ and allowable pressure of $[p] = 0.3\,\text{MPa}$. There are four driving discs and five driven discs. The friction surface has inner diameter $D_i = 80\,\text{mm}$ and outer diameter $D_o = 120\,\text{mm}$. Decide the axial force required to act on the friction disc.

Design Problems

1   Select suitable couplings for the power transmission system in Figure P13.2. The electric motor transmits a power of $P = 7.5\,\text{kW}$ at rotational speed of $n_1 = 720\,\text{rpm}$. The diameter of output shaft of the motor is 42 mm. The speed ratio of the belt drive, gear reducer and chain drive are 2.2, 4.15 and 1.54, respectively. The power transmission efficiency the belt drive, gear reducer and chain drive are 97, 99 and 96%, respectively. The diameter of shaft connecting the driven machine is $d_2 = 70\,\text{mm}$.

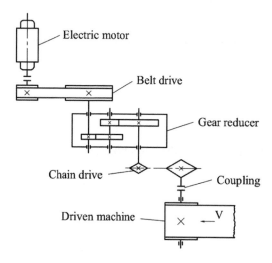

Figure P13.2 Illustration for Design Problem 1.

Electric motor

Belt drive

Gear reducer

Chain drive

Coupling

Driven machine

V

2   Design a flanged coupling using precision bolted joints in Figure P13.3. The coupling transmits a torque of $T = 2000\,\text{N m}$. The allowable stress of bolts is $[\tau] = 150\,\text{MPa}$. Decide the bolt shank diameter, bolt number and the diameter of bolt circle D for bolt assembly.

Bolt diameter d	M6	M8	M10	M12	M16	M20
Shank diameter d_0/mm	7	9	11	13	17	21

Figure P13.3 Illustration for Design Problem 2.

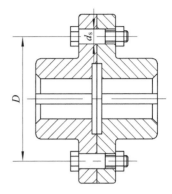

3   A multiple friction clutch transmits a power of 1.7 kW at a speed of $n = 500\,\mathrm{r\,min^{-1}}$. The friction discs are made of tempered steel with coefficient of friction $f = 0.05$, and the allowable pressure is $[p] = 0.6{\sim}0.8\,\mathrm{MPa}$. The friction surface has inner diameter $D_i = 80\,\mathrm{mm}$, and outer diameter $D_o = 120\,\mathrm{mm}$. The axial force on the friction disc is $F = 2000\,\mathrm{N}$. Assume the service factor is 1.5. Decide the number of discs required.

4   Explain the reason for a gear coupling in Figure P13.4 accommodating parallel misalignment and angular misalignment.

Figure P13.4 Illustration for Design Problem 4.

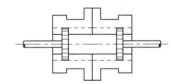

Figure P13.5 Illustration for Structure Design Problem 2.

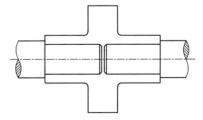

Structure design problems

1 Compare the two designs of flanged couplings in Table 13.1, which are aligned by (a) a piloted boss and (b) precision bolted joints (the joint on the top), respectively.

2 Complete the design of flanged coupling connected by ordinary bolted joints.

3 Design a coupling that can accommodate axial and angular misalignments, and cushion shock and vibration. Produce a drawing.

4 Design a coupling that can accommodate axial and angular misalignments. The coupling has no resilient element. Produce a drawing.

CAD Problems

1 Write a flow chart for the design process of a coupling.

2 Develop a program to implement a user interface similar to Figure P13.6 to complete coupling design.

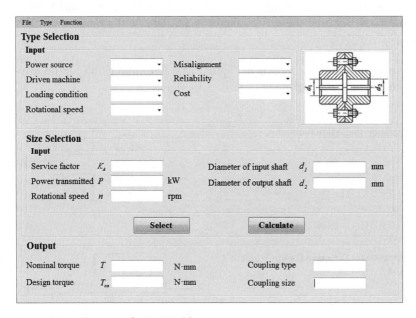

Figure P13.6 Illustration for CAD Problem 2.

14

Springs

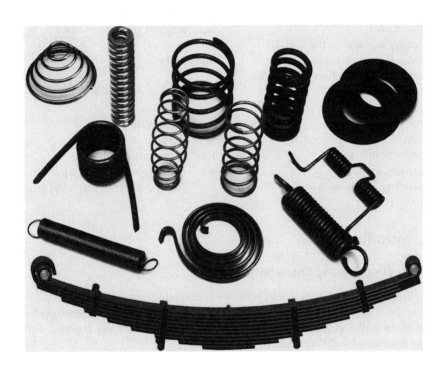

Nomenclature

b	slenderness ratio	g	gravity acceleration, m s^{-2}
C	spring index	H_f	free height/length of spring, mm
D	mean diameter, mm	H_h	length of hook, mm
D_i	inside diameter, mm	H_i	installed height/length, mm
D_o	outside diameter, mm	H_o	operating height/length, mm
d	spring wire diameter, mm	H_s	solid height/length, mm
E	modulus of elasticity, MPa	h	working distance of a spring, mm
F	external axial force, N	I	moment of inertia of spring wire, mm^4
f_n	fundamental natural frequency, Hz	J	polar moment of inertia, mm^4
G	shear modulus, MPa	K_w	Wahl factor
		K_{wT}	Wahl factor for torsion spring

Analysis and Design of Machine Elements, First Edition. Wei Jiang.
© 2019 John Wiley & Sons Singapore Pte. Ltd. Published 2019 by John Wiley & Sons Singapore Pte. Ltd.
Companion website: www.wiley.com/go/Jiang/analysis_of_machine_elements

k	spring rate, N mm^{-1}, N m^{-1}	σ_b	ultimate tensile strength, MPa
k_φ	spring rate of torsion spring, $\text{N mm}/(°)$	$[\sigma_b]$	allowable bending stress for torsion spring, MPa
L	length of wire, mm	σ_e	elastic limit strength, MPa
M	bending moment, N mm	σ_s	yield strength, MPa
m	mass of spring, kg	τ	shear stress, MPa
n	number of active coils	τ_i	initial torsional stress, MPa
n_{total}	total number of coils	τ_s	transverse shear stress, MPa
p	spring pitch, mm	τ_T	torsional shear stress, MPa
S	shear force, N	$[\tau]$	allowable shear stress of spring wire, MPa
S_{ca}	fatigue strength safety factor		
T	torsional moment, N mm	φ	angular deflection, rad or °
U	strain energy, J	ω_n	angular frequency, rad s^{-1}
U_0	energy loss due to friction, J		
W	section modulus, mm^3	**Subscripts**	
w	work for a deflection under a load, J		
α	pitch angle, °	i	load or deflection at initial installation
δ	coil clearance, mm		
δ_1	coil clearance at the maximum load, mm	m	mean value
		max	maximum value
λ	deflection, mm	min	minimum value
ρ	density, kg m^{-3}	o	load or deflection at operation
σ	bending stress, MPa	s	load or deflection at solid length

14.1 Introduction

14.1.1 Applications and Characteristics

Springs are resilient elements that experience appreciable deflection without permanent deformation. They usually have predefined linear or nonlinear relationships between applied loads and associated deflections. Therefore, springs are used to introduce controllable flexibility by deflection under applied loads.

Springs can be configured to exert desired force or torque to control the motion of machine, like valve springs in internal combustion engines. They can also be used to absorb vibration and impact energy, such as buffer springs in suspension systems for cars and trucks; or to store energy for subsequent release, like spiral springs used in watches. Finally, when the relationship between the force and deflection is linear, springs can be used for force measurement, such as helical springs in scales.

14.1.2 Types of Spring and Structures

Because of variations in spring configurations, applied loads and employed materials, the classification of springs may have some overlap [1–4]. According to the configuration, springs can be classified into helical springs, Belleville springs, spiral springs, leaf springs and ring springs. While depending on the loads they carry, springs include extension springs, compression springs, torsion springs and bending springs. Springs can be made from wire stock or from flat stock. Helical compression, extension and torsion springs are wire springs; while flat springs are made from flat strip materials, like leaf springs. Springs are mass produced and are commercially available from spring manufacturers.

14.1.2.1 Helical Coil Springs

Helical coil springs, including cylindrical, conical, hourglass and barrel helical springs, as shown in Figure 14.1, are probably the most widely used types. A cylindrical helical coil spring, or cylindrical helical spring, is made of a round wire, wrapped into a cylindrical form with a constant (Figure 14.1a) or variable (Figure 14.1b) pitch between adjacent coils [2]. The load-deflection relationship of cylindrical helical springs with constant pitch is linear, while the latter, together with conical, hourglass and barrel helical springs are nonlinear, and they are designed to solve special problems.

When a cylindrical helical spring is subjected to a compressive force, a tensile force or a torsional moment, the spring is termed as cylindrical helical compression spring, cylindrical helical extension spring and cylindrical helical torsion spring, respectively. The basic structure of these cylindrical helical springs is exactly the same except for the end configurations.

Cylindrical helical compression springs

Figure 14.2 shows cylindrical helical compression springs with various end configurations. Figure 14.2a has closed and ground ends and usually for medium- to large-size springs. Such end configuration provides flat surfaces on which to seat the spring. The support end surfaces are vertical to the spring axis, or the direction of compressive loads. Figure 14.2b has closed ends by deforming the ends to a zero-degree pitch angle without grinding. The closed end coils greatly enhanced the stability of springs. Figure 14.2c has plain ends as if a long spring has been cut into sections without

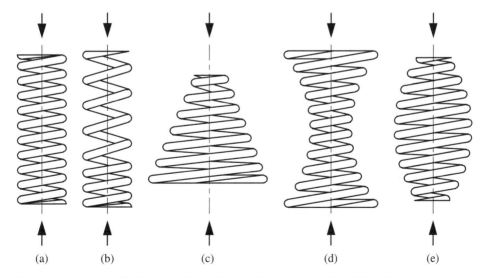

(a)   (b)   (c)   (d)   (e)

Figure 14.1 Various helical compression springs. (a) Constant pitch, (b) variable pitch, (c) conical, (d) hourglass and (e) barrel. Source: Adapted from Collins 2002, figure 14.1, p 516. Reproduced with permission of John Wiley & Sons, Inc.

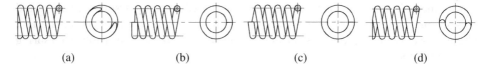

(a)   (b)   (c)   (d)

Figure 14.2 Cylindrical helical compression springs with various end configurations. (a) Closed and ground end, (b) closed end, (c) plain end and (d) plain end and ground.

changing the spacing of end coils. The plain end springs are unstable, while plain end and ground springs in Figure 14.2d have improved stability because of partial grinding of the end coils. Cylindrical helical compression springs for important applications should be both closed and ground to ensure a better load transfer [1].

Cylindrical helical extension springs

A cylindrical helical extension spring is designed to carry tensile force and to store energy. Extension springs appear similar to compression springs, yet they are usually closely wound, with all coils in tight contact in a free state [3]. When a sufficient external tensile load is applied at end loops or hooks, the coils separate.

Figure 14.3 shows several end configurations of extension springs to facilitate attaching springs to mating machine elements. Twist hook and loop in Figure 14.3a,b are widely used due to ease of fabrication, low cost and high strength. However, high bending stress will generate near the hook or loop. Figure 14.3c has an extended hook.

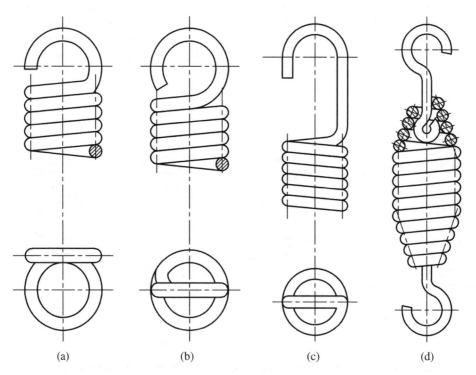

(a)   (b)   (c)   (d)

Figure 14.3 Helical extension springs with various end configurations. (a) Side hook (b) Twist loop, (c) extended hook and (d) coned end with swivel hook.

Figure 14.3d has a swivel hook, which tends to reduce stress concentration in hooks, as the hook can turn freely. The cost of a spring is affected by end configurations.

Cylindrical helical torsion springs

A cylindrical helical torsion spring is used to exert rotational moment, or torque. Torsion springs are wound in the same way as for compression and extension springs, except that the ends are configured for transmitting torque conveniently [5]. As a torque is applied, the spring deflects about its axis by rotation. Figure 14.4 presents typical ends of cylindrical helical torsion springs. Torsion springs are used where a torque needs to be applied, or rotational energy needs to be stored, such as in clothespins to provide gripping force, in spring hinges to close doors, or in animal traps and so on. Due to limited space, Figures 14.2–14.4 only present typical end configurations of cylindrical helical springs. More end configurations can be found in design handbooks [6, 7].

14.1.2.2 Belleville Springs

A Belleville spring is essentially a circular disc in a conical shape with a central hole, as shown in Figure 14.5a. They are capable of obtaining a high spring rate within limited axial space. Belleville springs are used either in series or in parallel. Stacking in series in Figure 14.5b provides a larger deflection for a given load, while stacking in parallel as shown in Figure 14.5c provides a higher load for a specified deflection [1]. Different combinations of Belleville springs can be designed to obtain the desired load-deflection relationship, that is, a characteristic curve, similar to Figure 14.5d. They have wide applications when space is limited and high loads with small deflections are required.

14.1.2.3 Spiral Springs

Spiral springs are wound from flat strips of metal with each turn wrapped tightly on its inner neighbour in the form of a spiral, as shown in Figure 14.6a. The strip extends when the free end is loaded with a torque. The load-deflection relationship of a spiral spring is shown in Figure 14.6b. A spiral spring works in confined spaces, capable of storing far more energy than a compression spring. They are used in clocks, watches and other storage devices.

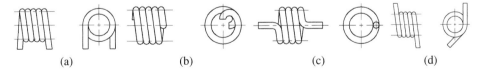

(a)	(b)	(c)	(d)

Figure 14.4 Cylindrical helical torsion springs with various end configurations.

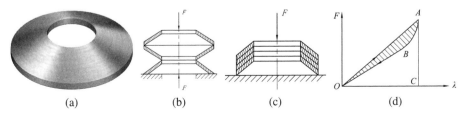

(a)	(b)	(c)	(d)

Figure 14.5 Belleville spring and its characteristic curve.

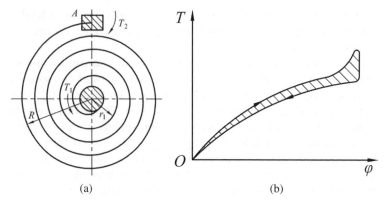

(a)   (b)

Figure 14.6 Spiral spring and its characteristic curve.

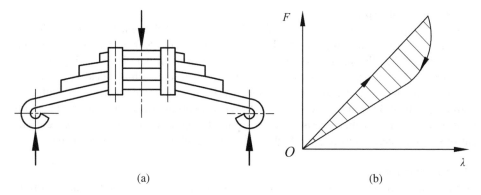

(a)   (b)

Figure 14.7 Leaf spring and its characteristic curve.

14.1.2.4 Leaf Springs

Leaf springs are essentially made of flat strips and are stressed in bending (Figure 14.7a). Large forces can be exerted within a small space by leaf springs. By stacking leaves with different geometry and dimensions, a desired characteristic curve similar to Figure 14.7b can be obtained. Leaf springs are commonly used as a part of automotive suspension systems as well in many other industrial products.

14.2 Working Condition Analysis

14.2.1 Geometry and Terminology

Most helical springs are wound from wire of solid round cross section. The main variables and terminology used to describe the geometry and to analyse the performance characteristics of helical compression springs are illustrated in Figure 14.8 and are explained as follows:

1) Wire diameter, d. Cross sectional diameter of wire. The specification of required wire diameter is one of the most important outcomes of spring design. It is a

Figure 14.8 Geometry and terminology of a helical compression spring.

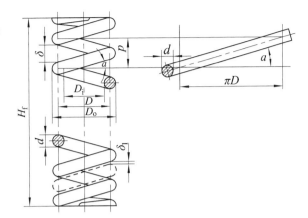

standardized value and decided by strength calculation. The standard values of wire diameter listed here are selected from GB/T 1358-2009 [8].

...0.2	0.4	0.6	0.8	1	1.2	1.6	2	2.5	3	3.5	4
4.5	5	6	8	10	12	15	16	20	30	40	50...

2) Mean diameter, D. The average of inside and outside coil diameters, $D = Cd$. It is involved in stress and deflection calculation. The standard values of mean diameter listed here are selected from GB/T 1358-2009 [8].

...5	5.5	6	6.5	7	7.5	8	8.5	9	10	12	14	16	18	20	22
25	28	30	32	38	42	45	48	50	52	55	58	60	65	70	75...

The calculated values of wire diameter d, mean diameter D, need to be rounded to standard values decided by various standards.

3) Inside diameter, D_i. Coil diameter measured on the inside of wire perpendicular to the spring axis, $D_i = D - d$.
4) Outside diameter, D_o. Coil diameter measured on the outside of wire perpendicular to the spring axis, $D_o = D + d$.
5) Spring index, C. The ratio of mean diameter D to wire diameter d, that is, $C = D/d$. The stresses and deflections in springs are dependent on C. It is a measure of relative severity of coil curvature.
6) Spring rate, k. The ratio of load variation to deflection variation. It reflects the stiffness of spring.
7) Number of active coils, n. The number of turns contributes to the axial deflection. It is determined by deflection calculation.
8) Total number of coils, n_{total}. The total number of turns from one end of a wire to the other. It depends on the number of active coils and end coil configurations, as listed in Table 14.1.
9) Pitch, p. The axial distance from a point on one coil to the corresponding point on the adjacent coil. For a compression spring, $p = (0.28–0.5) D$, and for an extension spring, $p = d$ [7].
10) Deflection, λ. Reduction or extension in axial length of a spring under an external load.

Table 14.1 Effect of end configurations on the number of coils and spring length [1, 7].

Term	Compression springs				Extension springs
	Closed and ground ends	Closed ends not ground	Plain ends	Plain ends and ground	
Total coils, n_{totoal}	$n+2$	$n+2$	n	$n+1$	n
Free length, H_f	$H_f = pn + (1.5 \sim 2)d$	$H_f = pn + (3 \sim 3.5)d$	$H_f = pn + d$	$H_f = p(n+1)$	$H_f = (n+1)d + (1 \sim 2)D_i$

11) Free height/length, H_f. Spring height in the unloaded condition. The relationships between the free length, pitch, wire diameter and number of active coils are listed in Table 14.1.

12) Solid height/length, H_s. Spring height with all coils touching adjacent ones.

13) Operating height/length, H_o. Spring height during normal operation. For a compression spring, $H_o = H_f - \lambda$ and for an extension spring, $H_o = H_f + \lambda$.

14) Installed height/length, H_i. Spring height after installation.

15) Coil clearance, δ. The space between adjacent coils at spring free length, $\delta = p - d$. A proper coil clearance δ in free state make it possible to generate deflection under compressive loads. To maintain flexibility even after compression, a coil clearance of δ_1, usually recommended as $\delta_1 = 0.1d \geq 0.2$ mm, is still required under the maximum load [4].

16) Pitch angle, α. Unwrap a spring coil onto a surface, the horizontal line is the spring mean circumference and vertical line is the pitch p. The pitch angle is defined as $\alpha = \arctan(p/\pi D)$. For a compression spring, pitch angle is usually $\alpha = 5 - 9°$. Springs can be right- or left-handed. Use a right-handed spring whenever possible.

17) Wire length, L. The length of spring wire is $L = \pi D n_{total}/\cos\alpha$ for compression springs and $L = \pi D n$ plus the length of hooks for extension springs.

18) Slenderness ratio, b. The ratio of spring free length H_f to mean diameter D. Usually, slenderness ratio is $b = 1 - 5.3$.

The effect of end configurations on the total number of coils and spring length can be found in design handbooks [6, 7] and are briefly listed in Table 14.1.

14.2.2 Spring Characteristic Curves

Unlike most machine elements, springs exhibit large and visible deflections under applied loads. The relationships between the applied load and corresponding deflection are presented as spring characteristic curves, as shown in Figure 14.9.

The slope of a spring characteristic curve is spring rate. It is the ratio of change in load to the corresponding change in deflection. For an extension or compression spring, the applied load refers to tension or compression and deflection is extension or contraction. The spring rate is defined as

$$k = \frac{dF}{d\lambda}$$

(14.1)

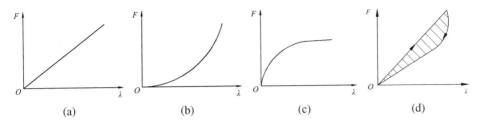

Figure 14.9 Typical spring characteristic curves.

While for a torsion spring, the applied load is torque and deflection is angular deflection and the spring rate is

$$k_\varphi = \frac{dT}{d\varphi} \tag{14.2}$$

Spring rate reflects the stiffness of spring. Under the same loads, a stiff spring has small deflection, while a flexible spring has a large deflection. The variation of spring stiffness is revealed clearly by characteristic curves. It is an important information for the design, manufacturing and testing of spring.

Depending on the structure of spring, a spring may have constant, progress, degressive or combined stiffness. The characteristic curve of a constant stiffness spring is a line, as shown in Figure 14.9a. The deflection is directly proportional to the applied load, following Hooke's law. It is a fundamental relationship in spring design. Cylindrical helical compression, extension and torsion springs have constant stiffness characteristic curves.

Nonlinear characteristic relations are desirable for some special applications. Figure 14.9b shows a progressive characteristic curve, exhibiting an exponential increase in resistance as deflection increases [9]. This kind of characteristic is useful in damping harmful vibrations. Conical springs possess this characteristic and are often used in vehicle suspension systems to control the maximum deflection. Figure 14.9c present a degressive spring characteristic, that is, the load maintained is nearly constant as deflection increases. Figure 14.9d shows combined stiffness characteristic curves, that is, the loading and unloading characteristic curves do not coincide. They are represented by various types of springs, more specifically illustrated from Figures 14.5–14.7. There is partial energy release during unloading.

14.2.3 Storage and Dissipation of Energy

When a spring deflects under an external load, the work associated with the load is stored as strain energy. The input work is equal to the area under the given characteristic curve. The strain energy for a compression or extension spring is defined as

$$U = \int_0^\lambda F(\lambda)d\lambda \tag{14.3}$$

While for a torsion spring, the strain energy is defined as

$$U = \int_0^\varphi T(\varphi)d\varphi \tag{14.4}$$

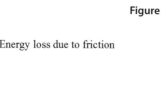

Figure 14.10 Energy storage and dissipation.

When friction is absent, the stored energy is fully recovered as the spring returns to its original length; however, when present, the area surrounded by loading and unloading characteristic curve represents energy loss U_0, as illustrated in Figure 14.10. The lost energy is for overcoming friction and may generate undesirable heat, however, it can be desirable whenever impact or vibration must be reduced. The larger the ratio of U_0 to U, the stronger the capability of shock and vibration absorption. Ring springs, Belleville springs and multilayer leaf springs are functioned according to this principle and are used as buffing springs widely, especially in heavy machinery.

14.2.4 Potential Failure Modes

Springs are elastic elements which can produce large deflection under applied loads. They are expected to exert desired force or to deliver stored energy essential for the machine operation over long periods of time. Common potential failure modes that prohibit proper functioning of springs include yield, fatigue, buckling, resonance, corrosion, creep, fretting fatigue and so on [5].

Springs are usually subjected to static or fluctuating loads during operation. Excessive static loads may cause the induced stress to exceed the yield strength of spring material, resulting permanent dimensional change. Fluctuating loads may cause stress variation and eventually fatigue failure. Extreme axial load on a slender compression spring may cause buckling. Resonance or surging may occur if cyclic operating frequencies are close to the resonant frequency of spring.

Operating conditions may also affect failure modes. Corrosive environments may reduce the strength and surface hardness of spring, leading to accelerated corrosion failure. Elevated temperatures may cause thermal relaxation and creep, producing unacceptable dimensional change or reduction of load carrying capability. As always, it is, therefore, extremely important to identify potential failure modes and propose proper measures to prevent failures at the design stage.

14.3 Load Carrying Capacities

14.3.1 Analysis of Helical Compression Springs

14.3.1.1 Load-Deflection Relationship

Figure 14.11 shows the load-deflection relationship at different loading conditions. When a cylindrical helical compression spring carries no load, the free length is H_f. At installation, an initial preload F_i is applied and the spring reaches installed length

Figure 14.11 Characteristic curve of a helical compression spring.

(a)

(b)

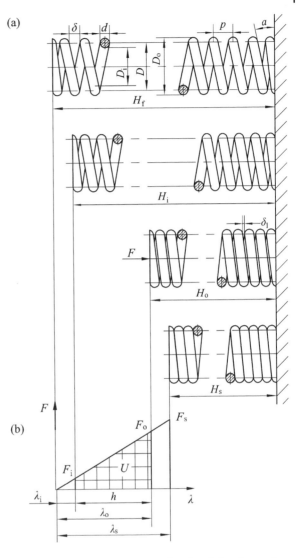

H_i, and the deflection is λ_i. When the spring carries an operating load F_o, the spring operating length is H_o, and the deflection is λ_o. The force varies between F_o and F_i, for a reciprocating spring, the working distance is $h = \lambda_o - \lambda_i$. When the load increases to the maximum load F_s, the correspond spring length is solid length H_s and the deflection is λ_s. This is the shortest possible length that the spring can have and the internal stress reaches the elastic limit of spring wire. Springs should avoid to be compressed to the solid length during operation. For a constant pitch compression spring, this load and deflection satisfies

$$\frac{F_i}{\lambda_i} = \frac{F_o}{\lambda_o} = \ldots\ldots = const \tag{14.5}$$

The preload F_i is usually selected as $F_i = (0.1\text{–}0.5)\,F_o$. The maximum operating F_o is usually decided by the working condition, satisfying $F_o \leq 0.8\,F_s$ [10].

14.3.1.2   Force Analysis

The force analysis of cylindrical helical spring is similar whether it carries a tension or compression load. Figure 14.12a shows a helical compression spring loaded by an axial force F. Cut the spring and remove lower portion of spring. The remaining partial spring is in equilibrium under the action of external axial force F, the transverse sheer force S and the torsional moment T. Summing the axial forces, thus

$$S = F \tag{14.6}$$

Summing the moments with respect to the wire centre yields

$$T = \frac{FD}{2} \tag{14.7}$$

Therefore, the entire length of active wire in the helix is subjected to a torque.

14.3.1.3   Strength Analysis

1) Stress analysis

Whether an axial load produces extension or compression, the wire is twisted. The primary stress developed in the cross section of wire is torsional shear stress, calculated by (Figure 14.12b)

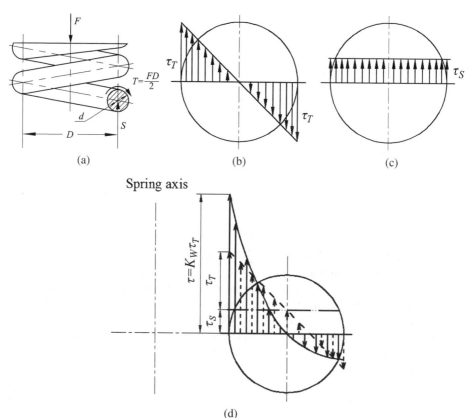

(a)   (b)   (c)

(d)

Figure 14.12 Force and stress analysis of a helical compression spring.

$$\tau_T = \frac{FD/2}{\pi d^3/16} \tag{14.8}$$

The torsional shear stress is inversely proportional to the cube of wire diameter. This implies the great effect of the variation of wire diameter on the spring performance. The transverse shear stress induced by shear force S, which is equivalent to F, is (Figure 14.12c)

$$\tau_S = \frac{F}{\pi d^2/4} \tag{14.9}$$

The torsional shear stress τ_T and transverse shear stress τ_S oppose at the outer coil radius but coincide at the inner coil radius. The maximum combined shear stress in the coil wire, which occurs at the inner surface of wire, may be computed by super-position of the torsional shear stress and transverse shear stress as

$$\tau = \tau_T + \tau_S = \frac{FD/2}{\pi d^3/16} + \frac{F}{\pi d^2/4} = \frac{8FD}{\pi d^3}\left(1 + \frac{1}{2C}\right) \approx \frac{8FC}{\pi d^2} \tag{14.10}$$

where $C = D/d$ is spring index, usually within the range of 4–12. It is a measure of coil curvature.

The maximum combined shear stress in the coil wire expressed in Eq. (14.10) is for springs under static or low cycle loading and is indicated by the dashed line in Figure 14.12d. However, for a spring subjected to high cycle of loading, the actual stress at wire cross section is slightly higher on the inside of coil due to the influence of pitch angle and coil curvature, as shown by the solid line in Figure 14.12d. The Wahl factor, K_w, is introduced to account for the curvature of wire. For a circular cross section spring wire, the Wahl factor is expressed as:

$$K_w \approx \frac{4C - 1}{4C - 4} + \frac{0.615}{C} \tag{14.11}$$

Since spring index C is within 4–12, Wahl factor K_w is within the range of 1.1–1.4. A midrange value of 1.2 is often selected as an initial estimate for K_w in spring design calculations.

2) Static strength analysis

Considering the Wahl factor and strength conditions at the inner radius of coil wire cross section, from Eq. (14.10), the shear stress and shear strength can then be expressed as

$$\tau = K_w\tau_T = K_w\frac{8FC}{\pi d^2} \leq [\tau] \tag{14.12}$$

The wire diameter is designed by

$$d \geq \sqrt{\frac{8K_wCF}{\pi[\tau]}} = 1.6\sqrt{\frac{K_wCF}{[\tau]}} \tag{14.13}$$

where the allowable shear stress $[\tau]$ can be found in Table 14.3.

3) Fatigue strength analysis

Springs are almost always subject to fatigue loading. If the number of load cycles is less than 1000, or the load is constant during operation, static strength analysis

Table 14.2 Spring materials and their properties [1, 2, 7, 11, 12].

Material	Material type	GB specification	ASTM/AISI specification	Shear modulus G (GPa)	Elastic modulus E (GPa)	Service temperature (°C)	Feature and applications
High carbon steels	Hard-drawn wire	25 ~ 80	AISI 1066 ASTM A227-47	80	206	−40 ~ +120	General purpose spring steel
	Music wire	65Mn	AISI 1085 ASTM A228-51				Good strength and fatigue performance, suitable for small springs
	Oil tempered	60, 65, 65Mn	AISI 1065 ASTM A229-41			−40 ~ +150	Suitable for large springs
Alloy steels	Chromium-vanadium	50CrVA	AISI 6150 ASTM A231-41	77	203	−40 ~ +210	Good static, fatigue and impact strength, good high-temperature performance
	Chromium-silicon	55CrSi	AISI 9254 ASTM A401			−40 ~ +250	Very high strength, good fatigue and shock resistance
Stainless steels	Stainless steel	12Cr18Ni9	ASTM A313(302)	70	193	−200 ~ +300	Good corrosion and creep resistance, high-temperature performance
		07Cr17Ni7Al	ASTM A313(631)	75	200		Good high-temperature performance
	Spring brass	Qsi3-1	ASTM B134	40	93	−40 ~ +120	Good corrosion resistance, good electrical conductivity
Copper alloys	Phosphor bronze	QSn6.5-0.1	ASTM B159	40	103	−250 ~ +120	
	Beryllium copper	QBe2	ASTM B197	42	120	−200 ~ 120	
Nickel alloys	Monel		Monel 400	66	179	320	corrosion-resistant, good high- and low-temperature properties, nearly nonmagnetic
	Inconel		Inconel 600	76	214	230	

is sufficient. For important springs working under variable stresses with millions of cycles during lifetime, it is necessary to calculate fatigue strength.

When the load varies from preload F_i to operating load F_o, according to Eq. (14.12) the maximum stress is

$$\tau_{max} = \frac{8K_w D}{\pi d^3} F_o \tag{14.14}$$

And the minimum stress is

$$\tau_{min} = \frac{8K_w D}{\pi d^3} F_i \tag{14.15}$$

For the variation of stress with the feature of constant mean stress, that is, $\tau_m = Const.$ or constant minimum stress, that is, $\tau_{min} = Const.$, the fatigue strength and static strength can be calculated by referring discussions in Chapter 2.

When $\tau_m = Const.$, from Eq. (2.29), we have the fatigue strength safety factor as

$$S_{ca} = \frac{\tau_{-1} + (K_\tau - \psi_\tau)\tau_m}{K_\tau(\tau_m + \tau_a)} \geq [S] \tag{14.16}$$

While for $\tau_{min} = Const$, from Eq. (2.30), the fatigue strength safety factor is

$$S_{ca} = \frac{2\tau_{-1} + (K_\tau - \psi_\tau)\tau_{min}}{(K_\tau + \psi_\tau)(2\tau_a + \tau_{min})} \geq [S] \tag{14.17}$$

The static strength safety factor for both cases is

$$S_{ca} = \frac{\tau_s}{\tau_m + \tau_a} \geq [S] \tag{14.18}$$

The meaning of variables from Eqs. (14.16)–(14.18) can be referred to in Chapter 2. The allowable safety factor is usually selected within the range of 1.2–2.5.

14.3.1.4 Rigidity Analysis

1) Deflection

Assume an initially unloaded spring carries an external axial force F and generates a linear deflection λ. The work w done by force F is expressed as

$$w = \frac{1}{2}F\lambda \tag{14.19}$$

The primary load on the wire of helical compression spring is torsion. Under the influence of induced torque T, the wire will twist, generating a twist angle of φ. The strain energy is

$$U = \frac{1}{2}T\varphi \tag{14.20}$$

The angle of twist φ, from the *Mechanics of Materials* [13], is

$$\varphi = \frac{TL}{GJ}$$

where the length of wire is approximate $L = \pi D n$ when pitch angle is less than $10°$, and the polar moment of inertia of round wire is

$$J = \frac{\pi d^4}{32}$$

These two forms of energy are equal. Therefore

$$\frac{1}{2}F\lambda = \frac{1}{2}T\varphi$$

By substitution and rearrangement, we have

$$F\lambda = T\varphi = \left(\frac{1}{2}FD\right)^2 \frac{L}{GJ}$$

Therefore, the linear deflection is

$$\lambda = \frac{8nFD^3}{Gd^4} = \frac{8nFC^3}{Gd} \tag{14.21}$$

where n is the number of active coils, G is shear modulus of spring material and is usually 8×10^4 N mm^{-2} for steel, and 4×10^4 N mm^{-2} for bronze [10]. Again, wire diameter is noticed to have strong effect on spring performance.

2) Spring rate

Incorporating Eq. (14.21), the spring rate can be derived as

$$k = \frac{F}{\lambda} = \frac{Gd}{8nC^3} = \frac{Gd^4}{8nD^3} \tag{14.22}$$

Therefore, spring rate is inversely proportional to the cube of spring index C. Small spring index increases spring rate k and lead to a stiff spring, while large spring index C reduces spring rate and produces a flexible spring. Hence, proper selection of spring index C can control the spring flexibility. In addition, the effects of G, d, n on the spring rate k also need to be considered.

3) Number of active coils

For a compression spring or an extension spring without preload, the number of active coils of cylindrical helical spring can be obtained from Eq. (14.21) as

$$n = \frac{G\lambda d^4}{8FD^3} = \frac{G\lambda d}{8FC^3} = \frac{Gd}{8C^3 k} \tag{14.23}$$

For extension springs with preload, the number of active coils of a cylindrical helical spring is

$$n = \frac{Gd}{8(F - F_0)C^3}\lambda \tag{14.24}$$

If $n < 15$, n is selected as a multiple of 0.5. If $n > 15$, n is selected as an integer [4].

14.3.1.5 Buckling Analysis

Elastic instability, or buckling, occurs when a long, slender cylindrical helical spring is loaded by a compressive force and produces large axial deflections. Buckling is affected by spring free length, mean diameter and spring end constraint [2, 5].

The ratio of spring free length H_f to the mean diameter D is defined as slenderness ratio b. Buckling could be prevented by selecting proper slenderness ratios. Spring end

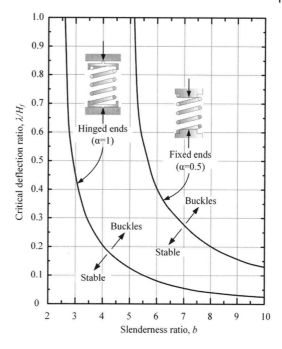

Figure 14.13 Buckling limits for helical compression springs under different end configurations. Source: Collins 2002, figure 14.11, p 528. Reproduced with permission of John Wiley & Sons, Inc.

configurations also affect the stability of helical compression springs. Plain ends, as in Figure 14.2c, are unstable, while closed and ground ends, as in Figure 14.2a, could greatly enhance stability. Figure 14.13 plots the relations between buckling limits and slenderness ratios at two end configurations, that is, both ends hinged (i.e. ends are permitted to rotate) and both ends fixed (i.e. closed and ground ends on flat, parallel surfaces). Buckling occurs above and to the right of presented curves.

Practically, a spring can be installed either inside or outside a cylindrical guide mandrel to prevent buckling. For either method, a minimum diametrical clearance of approximately one-tenth of coil diameter is recommended to avoid friction and rubbing between the spring and the guide [5].

14.3.1.6 Critical Frequency Analysis

Resonant frequency response or surging occurs if operating frequency approaches the natural frequency of cylindrical helical spring. When a spring works in an application requiring a rapid reciprocating motion, such as an engine valve spring, it is important to ensure that the frequency of applied force is not close to the fundamental natural frequency of spring, otherwise resonance may occur resulting in damaging stresses [1]. Spring surge also reduces the motion control capability of spring.

The fundamental natural frequency f_n of cylindrical helical spring with both ends fixed can be calculated from the basic equation [14],

$$f_n = \frac{\omega_n}{2\pi} = \frac{1}{2}\sqrt{\frac{k}{m}} \tag{14.25}$$

The mass of active coils of helical spring is

$$m = \frac{\pi d^2}{4}\pi D n \rho \tag{14.26}$$

where ρ is density and for steel ρ is $7700\,\mathrm{kg\,m^{-3}}$, for beryllium bronze ρ is $8100\,\mathrm{kg\,m^{-3}}$.

14.3.2 Analysis of Helical Extension Springs

Extension springs may have initial tension or without initial tension. Most extension springs are made with adjacent coils contact tightly, imposing an initial preload on the spring. Only when the applied load overcomes the initial tension, that is, $F > F_i$, do spring coils begin to separate and the spring deflects in a linear relationship at a uniform rate, as shown in Figure 14.14. The characteristic curve of an extension spring without initial tension is exactly the same as that of a compression spring.

The initial tension in an extension spring is created during winding process by twisting wire. The value of initial tension depends on the material, wire diameter, spring index and manufacturing process, which is typically 10–25% of the maximum design force [2]. It can be estimated from Eq. (14.12), as

$$F_i = \frac{\pi d^3 \tau_i}{8K_w D} \tag{14.27}$$

where τ_i is initial torsional stress, and can be selected by [3]

$$\tau_i = (0.4 \sim 0.8)\frac{\sigma_b}{C} \tag{14.28}$$

Except for the initial tension consideration, the stresses for an extension spring can be computed by the same formula as that for a compression spring. The critical stresses in an extension spring often occur in the end loops or hooks. Therefore, bending and torsion in the hook must be included in the analysis while designing a spring with a hook end. Interested readers can refer to references [1, 2] for detailed stress analysis.

The deflections for an extension spring can be computed by the same formula as that for a compression spring. Referring to Eq. (14.21), the maximum deflection for a compression spring or extension spring without preloading is

$$\lambda_{\max} = \frac{8F_{\max} C^3 n}{Gd} \tag{14.29}$$

while for an extension spring with preloading, the maximum deflection is

$$\lambda_{\max} = \frac{8(F_{\max} - F_i)C^3 n}{Gd} \tag{14.30}$$

Contrary to a compression spring, all coils in an extension spring are active.

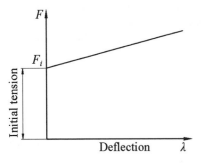

Figure 14.14 Characteristic curve of a helical extension spring with initial tension.

14.3.3 Analysis of Helical Torsion Springs

14.3.3.1 Load-Deflection Relationship

Like extension and compression springs, torsion springs also work properly within the elastic range. Figure 14.15a shows geometrical variables and linear load-deflection relationship. Similar to a compressive spring, T_s is the torque limit where the stress in the spring wire reaches its elastic limit; T_o is the operating torque; T_i is the initial installation torque, usually select $T_i = (0.1 \sim 0.5) \, T_o$. φ_s, φ_o, φ_i, are the corresponding angular deflection.

14.3.3.2 Force Analysis

Figure 14.15b shows a helical torsion spring carrying a torque T about the spring axis. Imagine the spring is cut by a normal plane B-B to the spring wire and remove one portion of spring. The remaining partial spring is in equilibrium under the action of external torque T. Summing the moments, and considering the small pitch angle α, thus

$$M \approx T \tag{14.31}$$

where M is the bending moment acting on the circular cross section of spring wire.

14.3.3.3 Strength Analysis

The primary stress in the coils of helical torsion spring is bending stress rather than torsional shear stress encountered in helical compression and extension springs. The

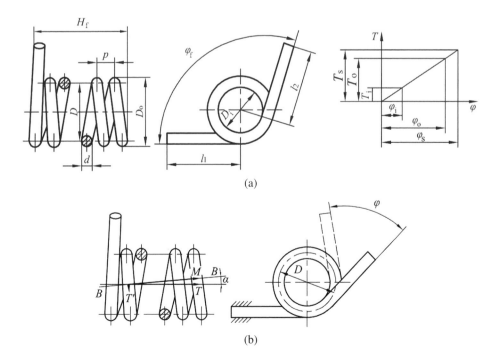

Figure 14.15 Characteristic curve and force analysis of a helical torsion spring.

bending stress on the inside of wire can be obtained from [13] and modified to account for the curved wire expressed in the form

$$\sigma = \frac{K_{wT}M}{W} \approx \frac{K_{wT}T}{0.1d^3} \leq [\sigma_b] \tag{14.32}$$

where W is the section modulus of round spring wires,

$$W = \frac{\pi d^3}{32} \approx 0.1d^3 \tag{14.33}$$

K_{wT} is Wahl factor for torsion spring. For a round wire [2],

$$K_{wT} = \frac{4C^2 - C - 1}{4C(C - 1)} \tag{14.34}$$

The wire diameter for a torsional spring is designed by

$$d \geq \sqrt[3]{\frac{K_{wT}T}{0.1[\sigma_b]}} \tag{14.35}$$

14.3.3.4 Rigidity Analysis

The deflection of helical torsion spring can be expressed in radians or in degrees. Angular deflection φ in radians under the applied moment, or torque, can be calculated approximately by [13]

$$\varphi = \frac{TL}{EI} = \frac{T\pi Dn}{EI} \tag{14.36}$$

The angular deflection φ in degrees is then

$$\varphi = \frac{180TDn}{EI} \tag{14.37}$$

where I is moment of inertia of spring wire section, mm⁴. For a round wire spring,

$$I = \frac{\pi d^4}{64}$$

The spring rate of torsion spring is expressed in units of torque per degree, expressed as

$$k_\varphi = \frac{T}{\varphi} = \frac{EI}{180Dn} \tag{14.38}$$

And the number of active coils is

$$n = \frac{EI\varphi}{180DT} \tag{14.39}$$

14.4 Design of Springs

14.4.1 Introduction

The spring design is inherently an iterative process, involving the selection of spring type, materials and determination of spring dimensions, that is, wire diameter d, mean diameter D, the number of active coils n, free length H_f, end configurations and other variables. The aim of spring design is to specify the spring geometry to obtain the desired load-deflection response and the required load in defined spatial confines with expected life under operating conditions. Additional considerations include natural frequency, shock absorption, corrosion resistance and so on. A well-designed spring functions properly and is manufactured easily.

The manufacturing of coil springs mainly includes four steps. First, straighten coils of spring wire and wind on a mandrel to form springs. Spring ends are then processed. For compression springs, two ends are ground to ensure end supporting surfaces are vertical to the spring axis, while for extension and torsion springs hooks are made for connection or loading. Followed by proper heat treatment to reduce residual stresses induced by the winding process, impact or fatigue tests are required to check spring quality [4].

After springs are formed, a shot peening or presetting operation [1] may be called for in important applications to improve fatigue strength by inducing favourable compressive residual stresses. Coating, plating or painting are used to enhance corrosion resistance. During manufacturing process, scratches on spring surfaces should be avoided.

14.4.2 Materials and Allowable Stresses

While selecting spring materials, material properties, application conditions, manufacturing process, as well as costs and availability are factors that need to be considered. For reliable functioning of springs, candidate spring materials should have high strength (including ultimate, yield and fatigue strength), high resilience and good resistance to corrosion and creep.

Application conditions include load magnitude and cyclic features, operating temperatures and environments. Hard-drawn or oil tempered carbon steel spring wire is adequate for statically loaded springs, while an alloy steel with restricted surface quality is preferred for cyclically loaded springs [9]. When it is required to prevent corrosion and magnetism, nonferrous materials are preferred. For springs operate at elevated temperatures, nickel alloys are the best choice.

Spring materials may be formed into bar, wire or strip by hot-working or cold-working processes. Normally, cold-working is used for spring wires with diameters less than 8–10 mm, while hot-working is used for greater wire diameters. Square or rectangular wire may also be used.

A great variety of spring materials are available, including high carbon steels, alloy steels and stainless steels, as well as nonferrous materials such as brass, phosphor bronze, beryllium copper and nickel alloys.

Table 14.2 lists material properties and applications for commercially available spring wires with good surface finishes.

The allowable stress of spring materials depends on the type of load, materials, processing methods and wire size. Both torsional and tensile strengths of spring material are inversely proportional to wire diameter. Due to limited data, tensile elastic limit strength in Figure 14.16 can be roughly used to start a spring design.

Since elastic limit σ_e is between 60 and 90% of tensile strength σ_b [1], the tensile strengths of various spring materials can be roughly derived from elastic limits from Figure 14.16. The allowable stresses expressed by tensile strength σ_b are listed in Table 14.3. More detailed and precise data can be found in design handbooks or manufacturer's catalogues [7].

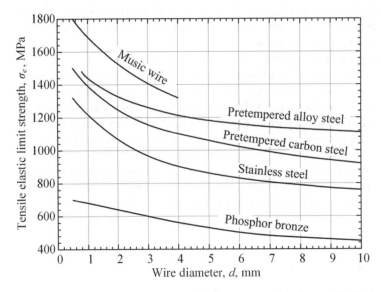

Figure 14.16 Tensile elastic limit strength of spring materials. Source: Hindhede et al. 1983, figure 8-3, p179. Reproduced with permission of John Wiley & Sons, Inc.

Table 14.3 Allowable stress of spring materials [1, 2, 7].

Stress types		Materials			
		Oil tempered	High carbon steels	Stainless steels	Copper alloys
Allowable shear stress for compression springs $[\tau]$ (MPa)	Static loads $(N < 10^3)$	$0.55\sigma_b$	$0.50\sigma_b$	$0.45\sigma_b$	$0.40\sigma_b$
	Finite fatigue life $(10^3 \leq N \leq 10^6)$	$0.45\sigma_b$	$0.42\sigma_b$	$0.36\sigma_b$	$0.35\sigma_b$
	Infinite fatigue life $(N > 10^6)$	$0.38\sigma_b$	$0.36\sigma_b$	$0.32\sigma_b$	$0.32\sigma_b$
Allowable shear stress for extension springs $[\tau]$ (MPa)	Static loads $(N < 10^3)$	Multiply 80% of the above corresponding data			
	Finite fatigue life $(10^3 \leq N \leq 10^6)$				
	Infinite fatigue life $(N > 10^6)$				
Allowable bending stress for torsion springs $[\sigma_b]$ (MPa)	Static loads $(N < 10^3)$	$0.80\sigma_b$	$0.78\sigma_b$	$0.75\sigma_b$	$0.75\sigma_b$
	Finite fatigue life $(10^3 \leq N \leq 10^6)$	$0.64\sigma_b$	$0.62\sigma_b$	$0.60\sigma_b$	$0.60\sigma_b$
	Infinite fatigue life $(N > 10^6)$	$0.55\sigma_b$	$0.54\sigma_b$	$0.50\sigma_b$	$0.50\sigma_b$

14.4.3 Design Criteria

To ensure proper function of spring, spring design must satisfy both strength and load-deflection requirements. The design criteria involve two basic equations, the strength equation Eq. (14.12) and the deflection equation Eq. (14.21). The static strength calculation determines mean diameter D and spring wire diameter d, while the rigidity calculation decides the number of active coils n. When a spring is subjected to fluctuating loads, Eqs. (14.16)–(14.18) are used against fatigue failure.

Besides, for a long and slender cylindrical helical compression spring, a proper slenderness ratio should be selected by Figure 14.13 to prevent buckling. Springs works in an application requiring rapid reciprocating motion are recommended to have a fundamental natural frequency at least 15 times the operating frequency to avoid resonances [5]. Otherwise the spring needs to be redesigned by increasing the spring rate k or reducing mass m.

14.4.4 Design Procedures and Guidelines

Spring design, like the design of other elements, is inherently an iterative process where design criteria need to be satisfied and desired load-deflection response to be obtained. The design procedure and guidelines are provided next, and design cases are presented in the next section.

1) Select spring materials and heat treatments according to operating conditions and decide the allowable stresses $[\tau]$ and $[\sigma_b]$.
2) Select spring index C and compute Wahl factor K_w.
   The recommended range of spring index C is between 4 and 12, and preferably around 5–8. Springs with an index C less than 4 are difficult to form; while those with C exceeding 12 tend to cause tangling [5].
   Assume a trial mean diameter D according to the installation space, and from Eq. (14.22)
   $$k = \frac{F}{\lambda} = \frac{Gd}{8C^3 n} = \frac{Gd^4}{8D^3 n}$$
   We know that spring rate k is proportional to the reverse of C^3. With the increase of spring index C, spring rate k decreases dramatically. Spring index C will affect the stresses and deflection in springs. And a large C value helps eliminate the spring buckling tendency [2]. Spring rate k is in proportion to the reverse of n, that is, with the increase of n, spring rate k will decrease. Also, for a constant deflection λ, with the increase of spring rate k, the load F also increases.
3) Compute the wire diameter d
   To satisfy the stress requirement, the wire diameter d can be determined by Eq. (14.13) and selected from a series of standard diameters.
4) Compute the number of active coils n and total number of coils n_{total}
   The number of active coils is the number of turns over which the wire twists under load, and therefore contributes to spring deflection. For preloaded extension springs, it is calculated by
   $$n = \frac{Gd}{8(F_{max} - F_0)C^3} \lambda_{max} \geq 3$$

While for ordinary extension springs or compression springs,

$$n = \frac{Gd}{8F_{max}C^3}\lambda_{max} \geq 3$$

Normally, the recommended number of active coils should be within the range of 3–15. The total number of coils should account for end coil configurations, which can be found in Table 14.1. This influences overall spring flexibility.

5) Check rigidity, load, deflection, static and fatigue strength, stability of compression spring and potential surging if required
6) Decide the dimension of springs
   For a compression spring, closed and ground ends are usually a good choice; while for an extension spring, a full end loop over the centre is preferred. A clearance between coils δ_1 should be allowed to avoid coil-to-coil contact when a spring is at its maximum operating deflection. After calculation, it is required to check whether spring variables satisfy installation requirements.
7) Produce working drawing for the spring

14.4.5 Design Cases

Example Problem 14.1
Design a cylindrical helical compression spring with a mean diameter of 18 mm. The spring deflects 5 mm when the applied load varies from $F_{min} = 150\,\text{N}$ to $F_{max} = 250\,\text{N}$. Loading is essentially static. The spring end is closed but not ground and both ends are fixed.

Solution

Steps	Computation	Results	Units
1. Select material, the diameter of spring wire and allowable stress	Select oil tempered carbon steel for spring wire. Select the diameter of spring wire as $d = 3$ mm. From Figure 14.16, select $\sigma_e = 1150\,\text{MPa}$, $\sigma_b = \dfrac{\sigma_e}{0.6 \sim 0.9} = (1277 \sim 1916)\,\text{MPa}$ Select $\sigma_b = 1600\,\text{MPa}$ and from Table 14.3, we have $[\tau] = 0.55 \times \sigma_b = 880\,\text{MPa}$	$[\tau] = 880$	MPa
2. Select mean diameter for the coil	The mean diameter of spring coil is $D = 18$ mm.		
3. Compute spring index C	$C = \dfrac{D}{d} = \dfrac{18}{3} = 6$	$C = 6$	
4. Compute the Wahl factor K_w	From Eq. (14.11), we have $K_w \approx \dfrac{4C-1}{4C-4} + \dfrac{0.615}{C} = \dfrac{4\times 6-1}{4\times 6-4} + \dfrac{0.615}{6} = 1.2525$	$K_w = 1.2525$	

(*continued*)

Steps	Computation	Results	Units
5. Compute spring wire diameter	From Eq. (14.13) $d \geq \sqrt{\dfrac{8K_w CF}{\pi[\tau]}} = 1.6\sqrt{\dfrac{K_w CF}{[\tau]}} = 1.6\sqrt{\dfrac{1.2525 \times 6 \times 250}{880}}$ $= 2.33\ mm$ Select $d = 3$ mm.	$d = 3$ $D = 18$	mm mm
6. Compute the number of active coils	From Table 14.2, select $G = 80\,000$ MPa. From Eq. (14.21) $\lambda = \dfrac{8nFC^3}{Gd}$ We have $\lambda = \lambda_{max} - \lambda_{min} = \dfrac{8n(F_{max} - F_{min})C^3}{Gd}$ $n = \dfrac{Gd}{8(F_{max} - F_{min})C^3}\lambda$ $= \dfrac{80000 \times 3}{8 \times (250 - 150) \times 6^3} \times 5 = 6.95$ Select $n = 7$	$n = 7$	
7. Check the rigidity, load and deflection	The required spring rate is $k = \dfrac{\Delta F}{\Delta \lambda} = \dfrac{F_{max} - F_{min}}{\lambda_{max} - \lambda_{min}} = \dfrac{250 - 150}{5}$ $= 20 N/mm$ The spring rate of designed spring is calculated from Eq. (14.22) as $k = \dfrac{Gd}{8nC^3} = \dfrac{80000 \times 3}{8 \times 7 \times 6^3} = 19.84\ \text{N/mm}$ Close to the required spring rate. The deflection at F_{min} $\lambda_{min} = \dfrac{F_{min}}{k} = \dfrac{150}{19.84} = 7.56\ \text{mm}$ The deflection at F_{max} is $\lambda_{max} = 5 + 7.56 = 12.56\ \text{mm}$ The maximum load is $F_{max} = k\lambda_{max} = 19.84 \times 12.56 = 249.2 \text{N}$ Close to the maximum design load.		

8. Decide the dimensions of the spring	(1) Outside diameter of spring
	     $D_o = D + d = 18 + 3 = 21$ mm
	 (2) Inside diameter of spring
	     $D_i = D - d = 18 - 3 = 15$ mm
	 (3) Total number of coils
	     The number of end coils is 2. Therefore,
	     $n_{total} = n + 2 = 7 + 2 = 9$
	 (4) The pitch
	     $p = (0.28 \sim 0.5)\,D = 5.04 \sim 9$ mm
	 (5) Free length
	     $H_f = pn + 3d = (5.04 \sim 9) \times 7 + 3 \times 3$
	     $= 44.28 \sim 72.0$ mm
		$D_o = 21$
		
		 $D_i = 15$
		
		 $n_{total} = 9$
		
		 $p = 5.04 \sim 9$
		
		 $H_f = 44.28$
		 ~ 72.0
			mm
			
			 mm
			
			 mm
			
			 mm

(continued)

Steps	Computation	Results	Units
9. Check stability	Slenderness ratio $$b = \frac{H_f}{D} = \frac{72}{18} = 4$$ $$\frac{\lambda}{H_f} = \frac{5}{44.28} = 0.11 < 0.2$$ From Figure 14.13, the spring will not buckle.		
10. Produce drawings	Omitted		

Example Problem 14.2

Design a cylindrical helical extension spring. The spring carries a static axial load. The deflection is $\lambda_1 = 9.0$ mm at the force of $F_1 = 200$ N, and $\lambda_2 = 18.0$ mm as the force increases to $F_2 = 320$ N. The mean diameter of spring is $D = 18$ mm and outside diameter $D_o \leq 22$ mm.

Solution

Steps	Computation	Results	Units
1. Decide the diameter of the spring wire	According to $D_o - D = 22 - 18 = 4$ mm, the diameter of spring wire is initially selected as $d = 3$ mm.		
2. Select material and allowable stress	Select oil tempered carbon steel for spring wire. From Figure 14.16, select $\sigma_e = 1150$ MPa, $$\sigma_b = \frac{\sigma_e}{0.6 \sim 0.9} = (1277 \sim 1916) \ MPa$$ Select $\sigma_b = 1600$ MPa and from Table 14.3, we have $$[\tau] = 0.55 \times 0.8 \times \sigma_b = 704 \ MPa$$	$[\tau] = 704$	MPa
3. Compute spring index C and the Wahl factor K_w	Select $C = D/d = 18/3 = 6$, from Eq. (14.11), we have $$K_w \approx \frac{4C-1}{4C-4} + \frac{0.615}{C} = \frac{4\times 6 - 1}{4 \times 6 - 4} + \frac{0.615}{6}$$ $$= 1.2525$$	$K_w = 1.2525$	
4. Determine the diameter of the spring wire	From Eq. (14.13) $$d \geq \sqrt{\frac{8K_w CF}{\pi[\tau]}} = 1.6\sqrt{\frac{K_w CF}{[\tau]}} = 1.6\sqrt{\frac{1.2525 \times 6 \times 320}{704}}$$ $$= 2.95 \ mm$$ Select the diameter $d = 3$ mm.	$d = 3$ $D = 18$	mm mm
5. Compute the number of active coils	From Eq. (14.1), the spring rate is $$k = \frac{dF}{d\lambda} = \frac{F_2 - F_1}{\lambda_2 - \lambda_1} = \frac{320 - 200}{18-9}$$ $$= 13.33 \ N/mm$$	$n = 11$	

(continued)

Steps	Computation	Results	Units
	From Table 14.2, select $G = 80\,000$ MPa. From Eq. (14.23) $$n = \frac{Gd}{8C^3 k} = \frac{Gd^4}{8D^3 k} = \frac{80000 \times 3^4}{8 \times 18^3 \times 13.33} = 10.42$$ Select $n = 11$, and the stiffness of the designed spring is: $$k = \frac{Gd^4}{8D^3 n} = \frac{80000 \times 3^4}{8 \times 18^3 \times 11} = 12.63 \text{ N/mm}$$ Close to the required spring rate.	$k = 12.63$	N mm^{-1}
6. Check	(1) Initial tension $$F_i = F_1 - k\lambda_1 = 200 - 12.63 \times 9$$ $$= 86.36 \text{ N}$$ Calculate initial stress by Eq. (14.12) $$\tau_i = K_w \frac{8F_i C}{\pi d^2} = 1.2525 \times \frac{8 \times 86.36 \times 6}{\pi \times 3^2}$$ $$= 183.73 \text{ MPa}$$ From Eq. (14.28), the initial torsional stress can be selected by $$\tau_i = (0.4 \sim 0.8)\frac{\sigma_b}{C} = (0.4 \sim 0.8)\frac{1600}{6}$$ $$= 107 \sim 213 \text{ MPa}$$ Therefore, the initial stress is within the required range.	$F_i = 86.36$ $\tau_i = 183.73$	N MPa
	(2) The load limit is $$F_s = \frac{\pi d^2 [\tau]}{8CK_w} = \frac{\pi \times 3^2 \times 704}{8 \times 6 \times 1.2525} = 331 \text{ N}$$ The applied load is less than the load limit.	$F_s = 331$	N
	(3) Deflection at load limit $$\lambda_s = \frac{F_s - F_i}{k} = \frac{331 - 86.36}{12.63} = 19.34 \text{ mm}$$		
7. Decide the dimensions of the spring	(1) Outside diameter $$D_o = D + d = 18 + 3 = 21 < 22 \text{ mm}$$	$D_o = 21$	mm
	(2) Inside diameter of spring $$D_i = D - d = 18 - 3 = 15 \text{ mm}$$	$D_i = 15$	mm
	(3) The pitch $$p \approx d = 3 \text{ mm}$$	$p = 3$	mm
	(4) Free length $$H_f = (n + 1)d + 2D_i = (11 + 1) \times 3 + 2 \times 15 = 66 \text{ mm}$$	$H_f = 66$	mm
	(5) Installed height/length H_i $$H_i = H_f + \lambda_1 = 66 + 9 = 75 \text{ mm}$$	$H_i = 75$	mm
	(6) Operating height/length H_o $$H_o = H_f + \lambda_2 = 66 + 18 = 84 \text{ mm}$$	$H_o = 84.0$	mm
	(7) Height/length H_s at load limit $$H_s = H_f + \lambda s = 66 + 19.34 = 85.34 \text{ mm}$$	$H_s = 85.34$	mm
	(8) Pitch angle α $$\alpha = \arctan \frac{p}{\pi D} = \arctan \frac{3}{18\pi} = 3.04°$$	$\alpha = 3.04$	°
	(9) Wire length L $$L = \pi D n + 2\pi D = \pi(18 \times 11 + 2 \times 18) = 734.76 mm$$	$L = 734.76$	mm
8. Produce drawings	Omitted		

Example Problem 14.3

Design a torsion spring with an initial torque of $T_1 = 1000\,\text{N mm}$ and maximum torque of $T_2 = 3000\,\text{N mm}$. The minimum angular deflection is $\varphi_1 = 25°$ and the maximum angular deflection is $\varphi_2 = 75°$. The free angular deflection is $\varphi_f = 120°$. The length of spring arm is 40 mm. The spring is to be installed in a steadily operated machine.

Solution

Steps	Computation	Results	Units
1. Select materials and determine the allowable bending stress	Select oil tempered carbon steel for spring wire. Initially select the diameter of spring wire as $d = 3\,\text{mm}$. From Figure 14.16, select $\sigma_e = 1150\,\text{MPa}$, $$\sigma_b = \frac{\sigma_e}{0.6 \sim 0.9} = (1277 \sim 1916)MPa$$ Select $\sigma_b = 1600\,\text{MPa}$ and from Table 14.3, we have $$[\sigma_b] = 0.8 \times \sigma_b = 1280 MPa$$	$[\sigma_b] = 1280$	MPa
2. Select spring index C and calculate the curvature factor K_{wT}	Initially select $C = 8$ $$K_{wT} = \frac{4C^2 - C - 1}{4C(C-1)} = \frac{4 \times 8^2 - 8 - 1}{4 \times 8 \times (8-1)} = 1.1$$	$C = 8$ $K_{wT} = 1.1$	
3. Calculate the spring wire diameter	Therefore, from Eq. (14.35), the spring's wire diameter is $$d \geq \sqrt[3]{\frac{K_{wT}T}{0.1[\sigma_b]}} = \sqrt[3]{\frac{1.1 \times 3000}{0.1 \times 1280}} = 2.95$$ Select spring wire diameter as $d = 3.5\,\text{mm}$.	$d = 3.5$	mm
4. Calculate spring diameters	$D = Cd = 8 \times 3.5 = 28\,\text{mm}$ $D_o = D + d = 28 + 3.5 = 31.5\,\text{mm}$ $D_i = D - d = 28 - 3.5 = 24.5\,\text{mm}$	$D = 28$ $D_o = 31.5$ $D_i = 24.5$	mm mm mm
4. Calculate the number of active coils	From Table 14.2, select $E = 206\,000\,\text{MPa}$. Since the moment of inertia of spring wire is $$I = \frac{\pi d^4}{64} = \frac{\pi \times 3.5^4}{64} = 7.36\,\text{mm}^4$$ From Eq. (14.39), the number of active coils is $$n = \frac{EI\varphi}{180TD} = \frac{206000 \times 7.36 \times 50}{180 \times (3000 - 1000) \times 28}$$ $= 7.52$ Select $n = 7.5$	$n = 7.5$	
5. Calculate spring rate	From Eq. (14.2) $$k_\phi = \frac{dT}{d\varphi} = \frac{3000 - 1000}{75 - 25} = 40 N \cdot mm/°$$ From Eq. (14.38) $$k_\phi = \frac{EI}{180Dn} = \frac{206000 \times 7.36}{180 \times 28 \times 7.5}$$ $= 40.11 N \cdot mm/(°)$ The spring rate of the designed spring is close to the requirement.	$k_\varphi = 40.11$	N mm/(°)

(*continued*)

Steps	Computation	Results	Units
6. Calculate spring dimensions	(1) Spring pitch p Select $\delta_0 = 0.5$ mm, therefore $p = d + \delta_0 = 3.5 + 0.5 = 4.0$ mm	$p = 4.0$	mm
	(2) Pitch angle α $\alpha = \arctan \dfrac{p}{\pi D} = \arctan \dfrac{4}{\pi \times 28} = 2.6°$	$\alpha = 2.6°$	
	(3) Free height H_f The length of spring arm $H_h = 40$ mm. The free length is $H_f = n(d + \delta_0) + d = 7.5 \times (3.5 + 0.5)$ $+ 3.5 = 33.5$ mm	$H_h = 40$ $H_f = 33.5$	mm mm
	(4) The length of spring wire L $L = \pi D n + L_h = \pi \times 28 \times 7.5 + 40 \times 2$ $= 739.4$ mm	$L = 739.4$	mm
7. Produce drawings	Omitted		

References

1 Budynas, R.G. and Nisbett, J.K. (2011). *Shigley's Mechanical Engineering Design*, 9e. USA: McGraw-Hill.

2 Mott, R.L. (2003). *Machine Elements in Mechanical Design*, 4e. Prentice Hall.

3 Juvinall, R.C. and Marshek, K.M. (2011). *Fundamentals of Machine Component Design*, 5e. New York: Wiley.

4 Pu, L.G. and Ji, M.G. (2006). *Mechanical Design*, 8e. Beijing: Higher Education Press.

5 Collins, J.A. (2002). *Mechanical Design of Machine Elements and Machines: A Failure Prevention Perspective*, 1e. New York: Wiley.

6 Oberg, E. (2012). *Machinery's Handbook*, 29e. New York: Industrial Press.

7 Wen, B.C. (2015). *Machine Design Handbook*, 5e, vol. 3. Beijing: China Machine Press.

8 GB/T 1358-2009 Cylindrical helical spring seriate sizes. Standardization Administration of the People's Republic of China, Beijing, 2009.

9 Hindhede, U., Zimmerman, J.R., Hopkins, R.B. et al. (1983). *Machine Design Fundamentals: A Practical Approach*. New York: Wiley.

10 Qiu, X.H. (1997). *Mechanical Design*, 4e. Beijing: Higher Education Press.

11 GB/T 23935-2009 Design of cylindrical helical springs. Standard Press of China. Standardization Administration of the People's Republic of China, Beijing, 2009.

12 Shigley, J.E., Mischke, C.R., and Brown, T.H. Jr., (2004). *Standard Handbook of Machine Design*, 3e. McGraw-Hill Professional.

13 Gere, J.M. and Timoshenko, S.P. (1996). *Mechanics of Materials*, 4e. CL Engineering.

14 Géradin, M. and Rixen, D.J. (2015). *Mechanical Vibrations: Theory and Application to Structural Dynamics*, 3e. New York: Wiley.

Problems

Review Questions

1   What is the characteristic curve of a spring? Give examples of springs corresponding to different types of characteristic curves.

2   What factors affect the strength, rigidity and stability of a spring? What measures can be taken to improve the strength, rigidity and stability of a spring?

3   What are the requirements for spring wire materials? List commonly used spring materials.

4   How should one decide on the allowable stress of spring wires?

5   Two cylindrical helical springs have a number of active coils of n_A and n_B, and $n_A = 2n_B$. Other specifications are the same.
   (a) If the maximum stress in spring A under load F is τ, what is the maximum stress in spring B under the same load?
   (b) If the deflection of spring A under the load F is λ, what is the deflection of spring B under the same load?
   (c) If the spring rate of spring A is k_A, what is the spring rate of spring B?

Objective Questions

1   If spring index C is selected as too small, the spring will _____.
   (a) have low rigidity and easily vibrate
   (b) be difficult to wind
   (c) easily cause instability
   (d) have large dimensions

2   The spring wire diameter of a cylindrical helical spring is calculated by _____ analysis and the number of active coils is calculated by _____ analysis.
   (a) strength
   (b) stability
   (c) rigidity
   (d) structure

3   While calculate stress at the cross section of a spring wire, Wahl factor is introduced to consider _____.
   (a) inner stress generated during spring wind
   (b) possible defects on the spring wire surface
   (c) stress concentration on the wire close to the spring axis
   (d) the effect of rising angle and coil curvature on stress

4   Shot peening on spring wires is for the purpose of improving _____.
   (a) static strength
   (b) fatigue strength

(c) rigidity

(d) high-temperature properties

5 When a cylindrical helical spring has its number of active coil doubled, the deflection will be _____ times the original deflection if the load and other spring specifications are unchanged.

(a) 1/2

(b) 2

(c) 4

(d) 8

Calculation Questions

1 A helical compression spring has a mean diameter of $D = 40$ mm, and spring wire diameter of $d = 5$ mm. The number of active coils is $n = 8$. The spring wire is made of steel wire with shear module of $G = 80\,000$ GPa. If the maximum load is $F = 600$ N:

a) What are the basic torsional stress τ and the combined stresses τ_1 and τ_2 for static and cyclic loading, respectively?

b) What is the spring rate?

c) What is the deflection under the load of 600 N?

2 A helical compression spring will be used in an application where a normal operating load is to be 100 N. The spring is made of music wire with a shear module of $G = 80\,000$ MPa and allowable stress of $[\tau] = 450$ MPa. The spring has an outside coil diameter of 33 mm, wire diameter of $d = 3$ mm and free length of 48 mm. The ends are closed and ground and there are total of eight turns. Compute the following:

(1) The mean diameter, inside diameter, spring index and Wahl factor.

(2) The stress and spring deflection at the operating load of 100 N.

(3) The operating length and spring rate.

(4) Check the spring for buckling.

3 A cylindrical helical torsion spring is used as a hinge on a door, as illustrated in Figure P14.1. A force of $F = 5$ N is applied to open the closed door. And a maximum force of $F_{max} = 20$ N is used to rotate the door to $180°$. The spring is made from alloy steel with the allowable bending stress of $[\sigma_b] = 900$ MPa and elastic module $E = 203\,000$ MPa. Determine the diameter of spring wire and mean diameter, and the number of active coils n.

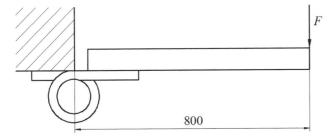

F

800

Figure P14.1 Illustration for calculation Question 3.

Design Problems

1 Design a compression spring with closed and ground ends using hard-drawn wire. The deflection is to be 60 mm when the force is 80 N and to close solid when the force is 100 N. The spring must fit inside a 40 mm diameter hole. Loading is essentially static.

2 A value spring operates under a variable load, with one end fixed and one end hinged. The spring is needed to resist a dynamic load that varies from 200 to 1000 N while the end deflection varies from 5 to 25 mm. The outer diameter of spring is less than 40 mm and the life is 10^5. Design the helical compression spring.

3 Design a helical extension spring. The spring has a length of 100 mm at the load of 150 N and a length of 125 mm at the load of 400 N. The outside diameter is 30 mm and allow 0.5 coil extra for hook flexibility.

4 Design a torsion spring with an initial torque of $T_1 = 2000\,\text{N\,mm}$ and a maximum torque of $T_2 = 8000\,\text{N\,mm}$. The minimum angular deflection is $\varphi_1 = 10°$ and the maximum angular deflection is $\varphi_2 = 40°$. The spring is to be installed in a steadily operating machine.

CAD Problems

1 Write a flow chart for spring design process to complete the Example Problem 14.1.

2 Develop a program to implement a user interface similar to Figure P14.2 and complete the Example Problem 14.1.

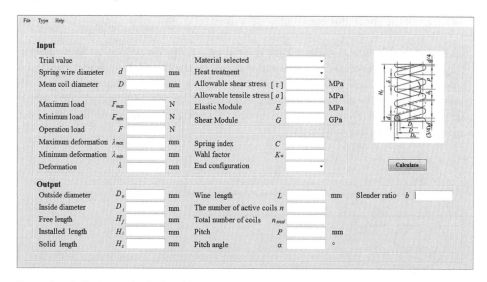

Figure P14.2 Illustration for CAD Problem 2.

Index

Analysis and Design of Machine Elements, First Edition. Wei Jiang.
© 2019 John Wiley & Sons Singapore Pte. Ltd. Published 2019 by John Wiley & Sons Singapore Pte. Ltd.
Companion website: www.wiley.com/go/Jiang/analysis_of_machine_elements